Cerebral Amyloid Angiopathy in Alzheimer's Disease and Related Disorders

Cerebral Amyloid Angiopathy in Alzheimer's Disease and Related Disorders

Edited by

Marcel M. Verbeek

Departments of Neurology and Pathology,
University Medical Center St. Radboud,
Nijmegen, The Netherlands

Robert M.W. de Waal

Department of Pathology,
University Medical Center St. Radboud,
Nijmegen, The Netherlands

and

Harry V. Vinters

UCLA Medical Center,
Los Angeles, CA, U.S.A.

SPRINGER-SCIENCE+BUSINESS MEDIA, B.V.

A C.I.P. Catalogue record for this book is available from the Library of Congress.

ISBN 978-90-481-5480-7 ISBN 978-94-017-1007-7 (eBook)
DOI 10.1007/978-94-017-1007-7

Printed on acid-free paper

Contents

SECTION II: GENETICS OF CAA

SECTION III: CELLULAR AND MOLECULAR PATHOLOGY OF CAA

Contributors

Egbert Bakker
Department of Neurology and Clinical Genetics, Leiden University Medical Center,
Leiden, The Netherlands

Ulrich Bickel
Institute of Pharmacology and Toxicology, Philipps University, Marburg, Germany

Marjolijn Bornebroek
Department of Neurology and Clinical Genetics, Leiden University Medical Center, Leiden, The Netherlands

Monique M.B. Breteler
Department of Epidemiology & Biostatistics, Erasmus Medical Center Rotterdam

Greg M. Cole
Department of Medicine and Neurology, UCLA and Sepulveda VAMC, North Hills, CA, USA

Mark R. Emmerling
Department of Neuroscience and Therapeutics, Parke-Davis Pharmaceutical Research, Division of Warner-Lambert Company, Ann Arbor, MI, USA

Blas Frangione
New York University School of Medicine, New York, NY, USA

Bernardino Ghetti
Indiana University School of Medicine, Indianapolis, IN, USA

Jorge Ghiso
New York University School of Medicine, New York, NY, USA

Warren J. Goux
Department of Chemistry, University of Texas at Dallas, Richardson, TX, USA

Steven M. Greenberg
Department of Neurology, Massachusetts General Hospital and Harvard Medical School, Boston, MA, USA

Joost Haan
Department of Neurology, Leiden University Medical Center, Leiden, The Netherlands

Raj N. Kalaria
Institute for Health of the Elderly, Newcastle General Hospital, Westgate Road, and Department of Psychiatry, University of Newcastle, Newcastle upon Tyne, United Kingdom

Yu-Min Kuo
Haldeman Laboratory for Alzheimer Disease Research, Sun Health Research Institute, Sun City, AZ, USA

Marion L.C. Maat-Schieman
Department of Neurology, Leiden University Medical Center, Leiden, The Netherlands

Mark O. McCarron
University of Glasgow Department of Neuropathology, Institute of Neurological Sciences, Southern General Hospital, Glasgow, Scotland, UK

Michael Mullan
Roskamp Institute, University of South Florida, Tampa, FL, USA

Remco Natté
Department of Neurology, Leiden University Medical Center, Leiden, The Netherlands

James A.R. Nicoll
University of Glasgow Department of Neuropathology, Institute of Neurological Sciences, Southern General Hospital, Glasgow, Scotland, UK

Ísleifur Ólafsson
Department of Clinical Biohemistry, Reykjavík Hospital and deCODE Genetics Inc., Reykjavík, Iceland

Daniel Paris
Roskamp Institute, University of South Florida, Tampa, FL, USA

P. Piccardo
Indiana University School of Medicine, Indianapolis, IN, USA,

Reinhard Prior
Department of Neurology, University of Duesseldorf, Duesseldorf, Germany

Alexander A. Roher
Haldeman Laboratory for Alzheimer Disease Research, Sun Health Research Institute, Sun City, AZ, USA

Alex E. Roher
Haldeman Laboratory for Alzheimer Disease Research, Sun Health Research Institute, Sun City, AZ, USA

Raymund A.C. Roos
Department of Neurology, Leiden University Medical Center,Leiden, The Netherlands

Dennis J. Selkoe
Harvard Medical School and Center for Neurologic Diseases, Brigham & Women's Hospital, Boston, MA, USA

Leifur Thorsteinsson
Department of Clinical Biohemistry, Reykjavík Hospital and deCODE Genetics Inc., Reykjavík, Iceland

Terrence Town
Roskamp Institute, University of South Florida, Tampa, FL, USA

Britta Urmoneit
Department of Neurology, University of Duesseldorf, Duesseldorf, Germany

Sjoerd G. Van Duinen
Department of Pathology, Leiden University Medical Center Leiden, The Netherlands

William E. Van Nostrand
Departments of Medicine and Pathology, Health Sciences Center, State University of New York, Stony Brook, NY, USA

Marcel M. Verbeek
Departments of Pathology and Neurology, University Medical Center St. Radboud, Nijmegen, The Netherlands

R. Vidal
New York University School of Medicine, New York, NY, USA

Harry V. Vinters
Department of Pathology & Laboratory Medicine, Section of Neuropathology, Brain Research Institute & Neuropsychiatric Institute, UCLA Medical Center, Los Angeles, CA, USA

Jean-Paul G. Vonsattel
Department of Pathology and Neuroscience Center, Massachusetts General Hospital and Harvard Medical School, Charlestown, MA, USA

Robert M.W. de Waal
Department of Pathology, University Medical Center St. Radboud, Nijmegen, The Netherlands

Larry C. Walker
Neuropathology Laboratory, Neuroscience Therapeutics, Parke-Davis Pharmaceutical Research Division, Warner-Lambert, Ann Arbor, MI, USA

Haruyasu Yamaguchi
Gunma University School of Health Sciences, Maebashi 371-8514, Japan

Fusheng Yang
Dept. Medicine and Neurology, UCLA and Sepulveda VAMC, North Hills, CA, USA

Berislav V. Zlokovic
Department of Neurological Surgery, USC School of Medicine, Los Angeles, CA, USA

Dedication

Dedicated to the late dr. George Glenner, dr. Henryk Wisniewski and dr. Thaddeus Mandybur, pioneers in research on amyloid in the nervous system, including its blood vessels.

Preface

Since at least the time of Scholz, students of Alzheimer's disease have been grappling with the role of microvascular pathology in the cause and mechanism of this complex and devastating disorder. A number of historical milestones in the elucidation of Alzheimer's disease have derived directly from scientists' desire to understand the whys and wherefores of the striking cerebrovascular amyloidosis that accompanies the disorder. Perhaps the most notable example of this connection is the original discovery by George Glenner of the amyloid beta protein. Glenner was understandably fascinated by the parallels between Alzheimer's disease and other human amyloidoses, and he purposely focused on the high prevalence of meningovascular amyloid deposition in approaching AD, both conceptually and experimentally. While neurobiologists were generally more interested in the neurofibrillary tangles and senile plaques that filled the cerebral cortex, Glenner correctly surmised that understanding the biochemical nature of meningovascular amyloid would provide a simpler and more tractable route to understanding Alzheimer's disease. Applying the methods that had been well-developed by pathologists and biochemists studying other human amyloidoses, Glenner and Wong successfully purified and partially sequenced the subunit of the meningovascular amyloid deposit of both Alzheimer's disease and Down's syndrome. What Glenner instinctively knew quickly became apparent to others: that deciphering the microvascular amyloid would be an excellent first step in unraveling the pathogenesis of Alzheimer's. We now know that his instincts were prescient.

It is both timely and compelling that Marcel Verbeek, Harry Vinters and Rob de Waal and their many talented co-authors have now decided to weave together the many threads of knowledge about cerebrovascular amyloidosis in Alzheimer's disease and related disorders. As I have examined the brains of Alzheimer victims under the microscope over the years, I have been impressed by the striking relationship of beta-amyloidosis to the cerebral microvasculature. In a series of brains of patients dying with Alzheimer's disease, Cathy Joachim and I observed the invariant occurrence of at least some and often many meningeal and intracortical microvessels bearing amyloid β-protein deposits. Not infrequently, the immediately surrounding brain tissue showed a neuritic and gliotic response that closely resembled that found in senile plaques. Numerous investigators have pointed to the potential importance of vascular basement membrane constituents as a substrate for β-amyloidosis in the brain. While it has been exceedingly difficult to confirm unequivocally a seminal pathogenetic role for microvessel constituents, there can be no doubt that a full understanding of Alzheimer's disease will require a detailed knowledge of the contribution of altered vascular form and function. The comprehensive review of current information about many aspects of this subject provided by this volume will no doubt serve as a highly useful resource to investigators seeking to make further progress.

The attention devoted to the genetics and clinicopathology of hereditary cerebral hemorrhage with amyloidosis (HCHWA) of the Dutch type in this volume seems particularly appropriate. Those of us working in the field of Alzheimer pathogenesis sometimes forget that the first specific genetic abnormality that could be credibly linked to the Alzheimer phenotype was the missense mutation in the beta-amyloid precursor protein causing the "Dutch disease." This discovery made it highly likely that APP would turn out to be the site of missense mutations in conventional Alzheimer's disease, a prediction that was soon shown to be true. HCHWA-D epitomizes the intimate relationship between vascular and parenchymal β-amyloidosis in the human brain that has fascinated neuropathologists for so long.

The current volume cogently places the Dutch disorder into the broader context of "sporadic" cerebral amyloid angiopathy, a syndrome which we now know occurs more frequently and with more disastrous consequences than was once believed. Thus, the old and new information brought together in one place by dr. Verbeek, Vinters and de Waal and their colleagues is meaningful not just for those concerned about Alzheimer's disease but for basic and applied biologists interested in understanding how the cerebral microvasculature works. I commend the editors and authors for taking on this important task and believe that you, the reader, will find their efforts to be both highly informative and useful to your own work.

Dennis J. Selkoe,

Professor of Neurology and Neuroscience,

Harvard Medical School,

Director of Center for Neurologic Diseases,

Brigham & Women's Hospital,

Boston, MA, USA

SECTION I

CLINICAL ASPECTS OF CAA AND CAA-RELATED HEMORRHAGE

Chapter 1

CLINICAL ASPECTS AND DIAGNOSTIC CRITERIA OF SPORADIC CAA-RELATED HEMORRHAGE

STEVEN M. GREENBERG
Department of Neurology, Massachusetts General Hospital and Harvard Medical School, Boston, MA, USA

Cerebral amyloid angiopathy (CAA) is a major cause of hemorrhagic stroke in the elderly, including hemorrhages that occur with anticoagulant treatment. Though fundamentally a neuropathologic diagnosis, CAA can be identified with good reliability during life by the presence on gradient-echo MRI of multiple, strictly lobar hemorrhages without other definite cause. The relatively good recovery from a first CAA-related hemorrhage is counterbalanced by the tendency for these hemorrhages to recur at approximately 10% per year. Higher rates of recurrence are predicted by a history of previous hemorrhage and the presence of the apolipoprotein E ε2 and ε4 alleles. Improving tools for the diagnosis and staging of CAA, together with an emerging understanding of β-amyloid deposition and toxicity, form a strong foundation for future trials aimed at prevention of hemorrhage recurrence.

1. INTRODUCTION

Given the tremendous contributions made by pathologists and pathological investigation towards our understanding of cerebral amyloid angiopathy (CAA), one might easily wonder whether the clinician's only role in investigating this disorder is as discussant at clinical-pathologic exercises [1-3]. There are in fact several compelling reasons for studying CAA from a clinical perspective. First and most obvious is the practical goal of translating our growing knowledge of the molecular pathology of CAA into techniques for diagnosing and treating this important (and heretofore untreatable) disorder. Another rationale for a clinically based approach is to

M.M. Verbeek et al. (eds.),
***Cerebral Amyloid Angiopathy in Alzheimer's Disease and Related Disorders*, 3–19.**

gain a window on the disease's dynamic features (e.g. recurrence), that are "frozen" or obscured at postmortem examination.

2. CLINICAL MANIFESTATIONS OF CAA

2.1 HEMORRHAGIC STROKE

Spontaneous cerebral hemorrhage is the most clinically dreaded consequence of CAA and the most commonly encountered presentation in any series of symptomatic cases. Much of what we have learned about clinically manifest CAA pertains specifically to this presentation.

2.1.1 Epidemiology and Risk Factors

While the true incidence of CAA-related hemorrhage can only be estimated, various calculations all point to CAA as an important contributor to stroke in the elderly. Primary hemorrhagic stroke occurs in approximately 100 per 100,000 elderly Americans and Europeans per year [4-6] or roughly 10% of strokes in this age group. Estimates for the proportion of these hemorrhages that are related to CAA can be obtained from either autopsy or clinical series, though each approach has its flaws. *Autopsy* series, which probably underestimate the true proportion because of the tendency for fewer lobar hemorrhages to come to autopsy [7], have found 12-15% of hemorrhages in the elderly to be due to CAA [8,9]. The *clinical* approach is to count all lobar hemorrhages in the elderly as being due to CAA, which overestimates the disease's frequency by including some non-CAA entities (e.g. hypertensive hemorrhages that extend to the cortex, vascular malformations). The latter method has yielded estimates of 38-45% (based on series describing patient age and hemorrhage location) [6,10].

Age represents the strongest risk factor for sporadic CAA-related hemorrhage, as predicted by the steep age-dependence of the underlying pathology [11-14]. Among 26 CAA patients identified in autopsy series of consecutive hemorrhages [8,9,15], all were over age 60 and 23 (88%) over age 70. Similarly, among 95 patients diagnosed with CAA during life by our research group, 93 (98%) were over age 60 at first hemorrhage, 68 (72%) over 70, and 32 (34%) over age 80. *Gender* does not appear to play an important role in CAA-related hemorrhage [16]. The status of *hypertension* (HTN) as a risk factor for CAA-related hemorrhage is less clear, in particular the question of whether HTN in conjunction with CAA confers greater risk for lobar hemorrhage than CAA alone. Among the 107 pathologic cases of

CAA compiled by Vinters, a history of HTN was noted in 32% [16], not substantially more frequent than the general elderly population [17]. Similarly, a pathological study of 25 brains with CAA and Alzheimer's disease found no association of HTN with the presence of hemorrhages [18]. Clinical series of elderly patients with lobar hemorrhage have yielded somewhat higher estimates for HTN of 40-70% [6,10,19-21], a figure which may be inflated by inclusion of some true hypertensive hemorrhages. HTN is nonetheless generally [6,19-21] (though not invariably [10]) found to be less frequent in geriatric patients with lobar hemorrhage compared with those who have hemorrhages of the basal ganglia, thalamus, cerebellum or pons. These studies, as well as data regarding recurrent lobar hemorrhage and HTN (see below), argue for a relatively minor role of HTN in CAA-related hemorrhage.

Among potential *genetic risk factors* for CAA, the apolipoprotein E (*APOE*) gene has emerged as the strongest predictor of the sporadic form of the disease. As described by McCarron and Nicoll in this volume, clinical and pathologic series have suggested both the *APOE* ε2 and ε4 alleles as possible risk factors for the presence [21-24] and earlier onset [21,24,25] of CAA-related hemorrhage. The altered *APOE* allele frequencies appear to be specific for CAA as opposed to HTN hemorrhages [21]. Smaller studies have been performed to search for whether mutations associated with familial CAA are also present in patients with sporadic hemorrhages (Table 1). With a single exception [26], the reported mutations in the amyloid precursor protein (APP) [27,28] and cystatin C [29] have been found to be absent [30-33].

Table 1. Familial CAA mutations in sporadic cerebral hemorrhage

Series [ref]	Subjects	CAA-associated mutations APP [27,28]	(#/total) Cystatin C [29]
Graffagnino, 1994 [30]	Sporadic hemorrhage (mean age 61; 16/48 lobar)	0/48	0/48
Graffagnino, 1995 [26]	Sporadic CAA	0/1	1/1
Anders, 1997 [31]	Sporadic CAA with giant cell reaction	0/5	0/5
Itoh, 1997 [32]	Sporadic CAA	0/11	0/11
Nagai, 1998 [33]	CAA (1 familial, 4 sporadic)	NA	0/5

NA: not available

A final risk factor to consider is the iatrogenic *coagulopathy* resulting from anticoagulant (e.g. warfarin) or thrombolytic (e.g. tissue-type plasminogen activator) treatment. Though "spontaneous" cerebral hemorrhage in the presence of coagulopathy is often classified as a separate entity [6,10], there is growing evidence that coagulopathy may in many cases unmask underlying CAA. Rosand and colleagues studied 29 patients age >60 with lobar hemorrhage while on warfarin and found an elevated frequency of *APOE* ε2 relative to warfarin patients without hemorrhage [34]. Pathologic findings in 6 of 8 cases demonstrated CAA. Similarly, severe CAA has been noted in several instances of cerebral hemorrhage following thrombolysis for acute myocardial infarction [35-38], including 2 of 5 consecutive postmortem examinations in the TIMI II trial [39]. The contribution of CAA to both anticoagulant- and thrombolytic-related hemorrhage may help explain certain features of these hemorrhages, including their age dependence and increased likelihood of being multiple and lobar [39-43].

Several factors point to likely future increases in both the relative and absolute clinical impact of CAA-related hemorrhage. These include the rapid aging of industrialized populations, growing indications for anticoagulation and thrombolysis in the elderly [44,45], and the current paucity of preventative or acute treatment strategies for CAA.

2.1.2 Clinical presentation and course

CAA-related hemorrhages favor the same cortical or corticosubcortical ("lobar") regions preferentially affected by the CAA deposits themselves. The acute lobar hemorrhages related to CAA appear to behave like other types of lobar hemorrhage, with symptoms dictated by factors such as hematoma size and location [7,46,47]. A slight predilection for frontal and parietal locations has been noted (accounting for 61% of 171 hematomas reviewed by Vinters [16] and 56% of 103 hematomas in patients diagnosed with definite or probable CAA at our institution).

Lobar hemorrhages often extend to the subarachnoid space, less frequently rupturing into ventricles [9]. Like hemorrhages related to HTN, lobar hemorrhages frequently grow in size during the first hours after onset [48]. A distinctive feature is their tendency to cause seizures along with the other complications of an acute mass lesion [49].

Non-lobar locations for CAA-related hemorrhages are uncommon. Cerebellum contains variable amounts of vascular amyloid and has been reported as an occasional site of hemorrhage [9,50]. Somewhat surprisingly, primary subarachnoid hemorrhage related to CAA is rare [51] despite extensive pathologic involvement of leptomeningeal vessels. CAA and

CAA-related hemorrhage tend to spare regions favored by hypertensive vasculopathy such as pons, thalamus, and putamen.

Studies of clinical *outcome* in lobar hemorrhage have pointed to several favorable features - their superficial location and tendency to spare the ventricular system - and several unfavorable prognostic factors - older age of the patients and larger hematoma volumes [20,21,52]. Overall mortality in lobar hemorrhage is in the range of 10-30% [6,20,46,52,53], with best prognosis for patients with smaller hematomas (<~50 ml) and higher level of consciousness on admission. Acute treatment of CAA-related hemorrhage follows the same principles as other types of intracerebral hemorrhage, typically focusing on control of intracranial pressure. Despite theoretical concerns about increased vessel fragility to surgical trauma, surgical resection of a CAA-related hematoma appears to carry little or no additional risk compared with other types of intracerebral hemorrhage (Table 2).

Table 2. Surgery in CAA-related hemorrhage.

Series [ref]	Patients*	Symptomatic rebleeding at surgical site
Greene, 1990 [96]	11	0
Matkovic, 1991 [97]	8	1
Minakawa, 1995 [98]	10	0
Izumihara, 1999 [99]	37	2
Total reported	66	3 (4.5 %)

*All patients had hematoma evacuation with exception of 2 biopsies reported in ref [96].

Offsetting the relatively favorable outcome associated with initial CAA-related hemorrhage is the high risk of *recurrence*. Lobar hemorrhages in the elderly recur at a rate of close to 10% per year (Table 3; [54,55]), approximately 4-5-fold more frequently than after hypertensive hemorrhage [54,56] and 100-fold increased relative to the general elderly population. Most (>80%) recurrences that follow a lobar hemorrhage are also lobar (Table 3), typically at a site separate from the initial lesion. Recurrent hemorrhage appears to carry greater mortality and disability than the initial stroke [54,57-59].

Among potential risk factors, a history of HTN appears to have little effect on recurrence of lobar hemorrhage [54,55,58]. *APOE* genotype, however, was significantly associated with recurrence in a series of consecutive lobar hemorrhage patients [55]. Patients in this series who carried the *APOE* ε2 or ε4 alleles (n=38) had a 2 year cumulative recurrence rate of 28%, compared with only 10% among those with the common ε3/ε3 genotype (n=32). If confirmed, this observation would raise the possibility that *APOE* genotype marks those CAA patients with more biologically aggressive disease and increased risk of progression.

2.2 OTHER CAA-RELATED SYNDROMES

CAA has been noted in association with several neurologic syndromes other than hemorrhagic stroke, listed and well discussed in several previous reviews [16,60]. Though fully described only in the last decade, *transient neurologic symptoms* may be the most common neurologic syndrome related to CAA other than hemorrhagic stroke. Transient symptoms were noted prior to or at presentation in 21 of 95 (22%) consecutive patients at our institution with a diagnosis of CAA and adequate historical information. The symptoms consist of brief (minutes), often stereotyped spells of focal weakness, numbness, paresthesias, or language disturbances; sensory or motor symptoms are often described as spreading across contiguous body parts over seconds to minutes [3,61,62]. Although CAA is associated pathologically with ischemic as well as hemorrhagic damage [63,64], we have ascribed the transient symptoms to focal seizure or spreading depression caused by small cortical hemorrhages. Supporting this interpretation are the demonstration of small hemorrhages in cortical regions corresponding to the spells, the smooth rather than stepwise spread of the sensory and motor symptoms (atypical for true transient ischemic attacks), and their response to anticonvulsant medication [62]. Because of the risk of catastrophic hemorrhage if CAA patients are inadvertently treated with anticoagulants, differentiating these spells from transient ischemic attacks is of considerable practical importance.

An uncommon, but potentially treatable clinical presentation of CAA is *vasculitis of the central nervous system*. Patients with this disorder, like those with central nervous system vasculitis unrelated to CAA, can demonstrate mental status changes, headache, multifocal neurologic deficits, multiple hemorrhagic and nonhemorrhagic lesions (occasionally with a mass-like appearance [65-67]), and inflammatory cells in the cerebrospinal fluid [68]. Several reported cases responded to immunosuppressive therapy [67-69]. Vascular β-amyloid appears to be the trigger for giant cell reaction and vasculitis in these cases, suggested by the observation of internalized β-amyloid within the giant cells [31,70,71]. True CAA-related vasculitis should be distinguished, however, from those pathologic cases with "incidental" giant cell reaction to β-amyloid without substantial inflammation or clinical signs of vasculitis [72].

Table 3. Hemorrhage recurrence following lobar hemorrhage.

Series [ref]	Recurrences following lobar hemorrhages Total	Lobar (%)	2 year recurrence (recur/total [%])
Passero, 1995 [54]	16	3 (81)	8/42 (19)
Counsell, 1995 [57]	4	4 (100)	
Neau, 1997 [58]	17	14 (82)	
Gonzalez-Duarte, 1998 [59]	6	4 (67)	
O'Donnell, 1999 [55]	16	16 (100)	12/71 (17)
Total reported	59	51 (86)	20/113 (18)

*Because of patient drop-out, these figures slightly under-estimate the cumulative 2 year recurrence rate.

The relationship between CAA and *dementia* is a complex one and probably encompasses a range of different pathogenetic mechanisms. At one end of the spectrum are those relatively infrequent instances in which dementia appears to be a direct result of CAA itself. A number of patients in this category have presented with *subacute cognitive decline* (over weeks to months) in the setting of a variable combination of seizures, bilateral motor signs, and severe white matter destruction [2,62,73-75]. The observed white matter pathologic change has been ascribed to ischemia from diffuse narrowing of the amyloid-laden cortical penetrating vessels, leading to destructive changes similar to those in subcortical arteriosclerotic leukoencephalopathy (Binswanger's disease) [76,77]. Indeed, CAA has been listed as a cause of Binswanger's disease [78], though the multiplicity of pathologic changes in CAA cases (including small hemorrhages, small cortical ischemic infarctions, and Alzheimer's-type changes along with the white matter destruction [2,62,73-75]) make it difficult to link the clinical syndrome to any single lesion.

Cognitive decline with leukoencephalopathy appears to be most prominent in the Dutch-type familial CAA, where it may precede onset of hemorrhagic strokes (see ref. [79] and Bornebroek *et al*, this volume). The full-blown syndrome of subacute cognitive decline seems to be uncommon in sporadic CAA, however, appearing in only 1 or 2 of the 95 consecutive patients diagnosed with CAA at our institution.

The more common cause of dementia in sporadic CAA is coexistent *Alzheimer's disease* (AD). AD and CAA have long been observed to overlap at a higher frequency than chance, a likely result of closely related biologies and shared genetic risk factors such as *APOE* ε4 [22,80,81]. CAA-related hemorrhages were identified in a relatively high proportion (6 of 117, 5.1%) of AD brains in a recent postmortem series [82]. The converse relationship is also present, with pathologic changes of AD appearing in approximately 40-

60% of patients with CAA-related hemorrhage [9,16,24]. The frequency of clinically manifest dementia (noted prior to first CAA-related hemorrhage) appears somewhat below this range, on the order of 25-40% [16,21].

The extent to which the various CAA-related pathologies (including vasculopathy [71,83,84], hemorrhage, and ischemia) not only *coexist* with AD but actually *contribute* to its clinical presentation is a question of considerable importance to the field. This issue is discussed by Kalaria and others elsewhere in this volume.

Table 4. Boston criteria for diagnosis of CAA-related hemorrhage (modified from [3]).

1. Definite CAA

Full postmortem examination demonstrating:
a. Lobar, cortical, or corticosubcortical hemorrhage
b. Severe CAA with vasculopathy [71,83,84]
c. Absence of other diagnostic lesion

2. Probable CAA with Supporting Pathology

Clinical data and *pathologic tissue* (evacuated hematoma or cortical biopsy) demonstrating:
a. Lobar, cortical, or corticosubcortical hemorrhage
b. Some degree of CAA in specimen
c. Absence of other diagnostic lesion

3. Probable CAA

Clinical data and *magnetic resonance imaging* demonstrating:
a. Multiple hemorrhages restricted to the lobar, cortical, or corticosubcortical region
b. Age over 60
c. Absence of other cause of hemorrhage*

4. Possible CAA

Clinical data and *magnetic resonance imaging* demonstrating:
a. Single lobar, cortical, or corticosubcortical hemorrhage
b. Age over 60
c. Absence of other cause of hemorrhage*

*Potential causes of intracerebral hemorrhage: excessive warfarin (INR>3.0); antecedent head trauma or ischemic stroke; CNS tumor, vascular malformation

3. PREMORTEM DIAGNOSIS OF CAA

Studies of the clinical behavior and response to treatment of CAA depend on the investigator's ability to diagnose the disease with accuracy during life. An important goal has thus been to develop and validate diagnostic criteria that could serve as a basis for clinical investigation and decision-making. A successful example of this approach is the clinical diagnosis of "probable" AD [85], which despite less than perfect specificity [86,87], has proved extremely useful for clinicians and clinical investigators.

Diagnostic criteria for CAA-related hemorrhage were proposed by the Boston CAA Group (Table 4). By these guidelines, the diagnosis of "definite" CAA is reserved for cases with full postmortem examination and demonstration of advanced CAA without other causative pathology. "Probable" CAA conversely is a diagnosis that can be made during life in patients with either 1) radiographic evidence of two or more hemorrhages entirely restricted to the characteristic locations for CAA, or 2) pathologic evidence of CAA in specimens obtained during hematoma evacuation or biopsy ("probable CAA with supporting pathology").

Several lines of data have emerged since these criteria were proposed, that generally support their validity [88]. The criteria for probable CAA with supporting pathology were explored in a study of biopsy-sized tissue samples taken randomly from postmortem brains with varying extent of CAA [14]. This study demonstrated at least some vascular amyloid (detected by Congo red) in each of the specimens from brains with CAA-related hemorrhage. The finding of mild vascular amyloid in a specimen was not entirely specific for CAA-related hemorrhage, however, with calculated values ranging from 95% specificity when the amyloid was detected in a patient under age 65 to 77% specificity for individuals over 85. More severe CAA in a specimen was associated with higher specificity.

In the more common situation in which pathologic tissue is not available, two lines of evidence support use of the proposed radiographic criteria for probable CAA. One comes from direct clinical-pathologic correlation. To date, we have encountered 11 patients with clinical and radiographic evidence for probable CAA (two or more strictly lobar hemorrhages) who also underwent pathologic examination via autopsy [4], hematoma evacuation [4], or cortical biopsy [3]. Vessels from 10 of the 11 patients (91%) demonstrated CAA (including severe CAA in each of the 4 full autopsy cases). The one specimen without congophilic staining was a biopsy from a patient with multiple hemorrhages and increased white blood cells in the cerebrospinal fluid; no diagnostic lesion was identified in this patient. A second, more indirect approach has been to compare *APOE* genotypes among clinically and pathologically diagnosed patients, relying on aggregate differences in *APOE* between CAA patients and the general elderly population. The similar elevations in frequency of the *APOE* ε2 and ε4 alleles between clinical and pathologic cases [88] again supports use of the proposed criteria. If replicated in other series, these data would suggest that the Boston criteria can serve as the basis for further clinical investigations in CAA.

The most sensitive radiologic technique for identifying the iron deposits left by small intracerebral hemorrhages is gradient-echo MRI [89]. The chronic hemorrhages demonstrated by this technique in a substantial proportion of patients with an acute hemorrhage [90,91] are often undetectable by other radiologic methods such as CT or conventional MRI (Figure 1). It is thus the technique of choice for identifying the presence or absence of multiple, strictly lobar lesions required for diagnosis of probable CAA.

Clinical diagnosis of CAA may ultimately be supplemented or replaced by a more sensitive and specific diagnostic test. Advances in directly visualizing vascular amyloid deposits are discussed in this volume by Bickel. Another approach has been to measure biological markers of CAA such as loss of soluble APP [92] and cystatin C [93] in cerebrospinal fluid. The sensitivity and specificity of these markers for sporadic CAA-related hemorrhage remain to be established.

4. TREATMENT OF CAA

Other than the infrequent instances of CAA-related vasculitis, treatment of CAA is largely limited to avoidance of hemorrhage-promoting agents such as anticoagulants or antiplatelet drugs. Warfarin increases the frequency (~7-10 fold) and severity (~60% mortality) of cerebral hemorrhage [94], while aspirin appears to raise the risk of hemorrhage to a lesser extent (~2 fold) [95]. Given the high rate of recurrence in CAA (even without coagulopathy), use of these medicines should be weighed against the considerable risk of catastrophic hemorrhage.

It is obviously the hope of those studying CAA that there will be specific and effective treatments to discuss when future reviews are written. There are grounds for optimism in this regard. Advances in our understanding of the biological properties of β-amyloid, together with the emerging clinical tools for diagnosing and marking progression in CAA, form a promising foundation for future clinical drug trials [88].

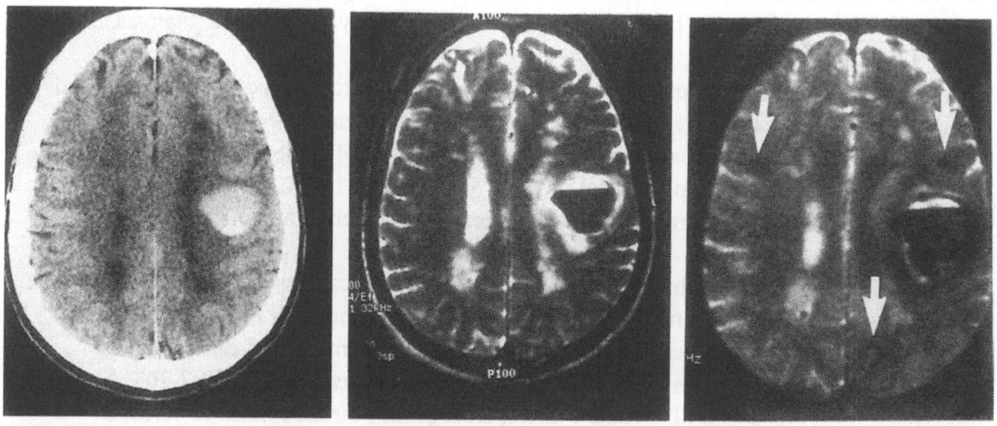

Figure 1. Gradient-echo MRI in the diagnosis of CAA. CT scan (left), T2-weighted MRI (center), and gradient-echo MRI (right) were obtained within days of lobar hemorrhage in a 69 year old man. The gradient-echo sequence shows both the acute lesion seen on all studies as well as three lesions indicative of prior hemorrhage (arrows). The appearance is consistent with probable CAA (see Table 4). Taken from ref [90].

REFERENCES

1. Kase, C.S., Vonsattel, J.P. and Richardson, E.P. (1988) Case Records of the Massachusetts General Hospital, case 10-1988, *N. Engl. J. Med.* **318**, 623-631.
2. DeWitt, L.D. and Louis, D.N. (1991) Case records of the Massachusetts General Hospital, case 27-1991, *N. Engl. J. Med.* **325**, 42-54.
3. Greenberg, S.M. and Edgar, M.A. (1996) Case records of the Massachusetts General Hospital, case 22-1996, *N. Engl. J. Med.* **335**, 189-196.
4. Brott, T., Thalinger, K. and Hertzberg, V. (1986) Hypertension as a risk factor for spontaneous intracerebral hemorrhage, *Stroke* **17**, 1078-1083.
5. Broderick, J.P., Brott, T., Tomsick, T., Huster, G. and Miller, R. (1992) The risk of subarachnoid and intracerebral hemorrhages in blacks as compared with whites, *N. Engl. J. Med.* **326**, 733-736.
6. Schutz, H., Bodeker, R.H., Damian, M., Krack, P. and Dorndorf, W. (1990) Age-related spontaneous intracerebral hematoma in a German community, *Stroke* **21**, 1412-1418.

7. Kase, C.S. (1994) Lobar hemorrhage, in C.S. Kase and L.R. Caplan (eds.), *Intracerebral Hemorrhage*, Butterworth-Heinemann, Boston, pp. 363-382.
8. Lee, S.S. and Stemmermann, G.N. (1978) Congophilic angiopathy and cerebral hemorrhage, *Arch. Pathol. Lab. Med.* **102**, 317-321.
9. Itoh, Y., Yamada, M., Hayakawa, M., Otomo, E. and Miyatake, T. (1993) Cerebral amyloid angiopathy: a significant cause of cerebellar as well as lobar cerebral hemorrhage in the elderly, *J. Neurol. Sci.* **116**, 135-141.
10. Broderick, J., Brott, T., Tomsick, T. and Leach, A. (1993) Lobar hemorrhage in the elderly. The undiminishing importance of hypertension, *Stroke* **24**, 49-51.
11. Tomonaga, M. (1981) Cerebral amyloid angiopathy in the elderly, *J. Am. Geriatric Soc.* **29**, 151-157.
12. Vinters, H.V. and Gilbert, J.J. (1983) Cerebral amyloid angiopathy: incidence and complications in the aging brain. II. The distribution of amyloid vascular changes, *Stroke* **14**, 924-928.
13. Masuda, J., Tanaka, K., Ueda, K. and Omae, T. (1988) Autopsy study of incidence and distribution of cerebral amyloid angiopathy in Hisayama, Japan, *Stroke* **19**, 205-210.
14. Greenberg, S.M. and Vonsattel, J.-P.G. (1997) Diagnosis of cerebral amyloid angiopathy. Sensitivity and specificity of cortical biopsy., *Stroke* **28**, 1418-1422.
15. Jellinger, K. (1977) Cerebrovascular amyloidosis with cerebral hemorrhage, *J. Neurol.* **214**, 195-206.
16. Vinters, H.V. (1987) Cerebral amyloid angiopathy. A critical review., *Stroke* **18**, 311-324.
17. Sytkowski, P.A., D'Agostino, R.B., Belanger, A.J. and Kannel, W.B. (1996) Secular trends in long-term sustained hypertension, long-term treatment, and cardiovascular mortality. The Framingham Heart Study 1950 to 1990, *Circulation* **93**, 697-703.
18. Ferreiro, J.A., Ansbacher, L.E. and Vinters, H.V. (1989) Stroke related to cerebral amyloid angiopathy: the significance of systemic vascular disease, *J. Neurol.* **236**, 267-272.
19. Bahemuka, M. (1987) Primary intracerebral hemorrhage and heart weight: a clinicopathologic case-control review of 218 patients, *Stroke* **18**, 531-536.
20. Massaro, A.R., Sacco, R.L., Mohr, J.P., Foulkes, M.A., Tatemichi, T.K., Price, T.R., Hier, D.B. and Wolf, P.A. (1991) Clinical discriminators of lobar and deep hemorrhages: the Stroke Data Bank, *Neurology* **41**, 1881-1885.
21. Greenberg, S.M., Briggs, M.E., Hyman, B.T., Kokoris, G.J., Takis, C., Kanter, D.S., Kase, C.S. and Pessin, M.S. (1996) Apolipoprotein E ε4 is associated with the presence and earlier onset of hemorrhage in cerebral amyloid angiopathy., *Stroke* **27**, 1333-1337.
22. Greenberg, S.M., Rebeck, G.W., Vonsattel, J.P.V., Gomez-Isla, T. and Hyman, B.T. (1995) Apolipoprotein E ε4 and cerebral hemorrhage associated with amyloid angiopathy, *Ann. Neurol.* **38**, 254-259.
23. Premkumar, D.R., Cohen, D.L., Hedera, P., Friedland, R.P. and Kalaria, R.N. (1996) Apolipoprotein E-epsilon4 alleles in cerebral amyloid angiopathy and cerebrovascular pathology associated with Alzheimer's disease, *Am. J. Pathol.* **148**, 2083-2095.
24. Nicoll, J.A., Burnett, C., Love, S., Graham, D.I., Dewar, D., Ironside, J.W., Stewart, J. and Vinters, H.V. (1997) High frequency of apolipoprotein E epsilon 2 allele in hemorrhage due to cerebral amyloid angiopathy, *Ann. Neurol.* **41**, 716-721.
25. Greenberg, S.M., Vonsattel, J.P., Segal, A.Z., Chiu, R.I., Clatworthy, A.E., Liao, A., Hyman, B.T. and Rebeck, G.W. (1998) Association of apolipoprotein E epsilon2 and vasculopathy in cerebral amyloid angiopathy, *Neurology* **50**, 961-965.

26. Graffagnino, C., Herbstreith, M.H., Schmechel, D.E., Levy, E., Roses, A.D. and Alberts, M.J. (1995) Cystatin C mutation in an elderly man with sporadic amyloid angiopathy and intracerebral hemorrhage, *Stroke* **26**, 2190-2193.
27. Levy, E., Carman, M.D., Fernandez Madrid, I.J., Power, M.D., Lieberburg, I., van Duinen, S.G., Bots, G.T., Luyendijk, W. and Frangione, B. (1990) Mutation of the Alzheimer's disease amyloid gene in hereditary cerebral hemorrhage, Dutch type, *Science* **248**, 1124-1126.
28. Hendriks, L., van Duijn, C.M., Cras, P., Cruts, M., Van Hul, W., van Harskamp, F., Warren, A., McInnis, M.G., Antonarakis, S.E., Martin, J.J., Hofman, A. and Van Broeckhoven, C. (1992) Presenile dementia and cerebral hemorrhage linked to a mutation at codon 692 of the beta-amyloid precursor protein gene, *Nat. Genet.* **1**, 218-221.
29. Palsdottir, A., Abrahamson, M., Thorsteinsson, L., Arnason, A., Olafsson, I., Grubb, A. and Jensson, O. (1988) Mutation in cystatin C gene causes hereditary brain hemorrhage, *Lancet* **2**, 603-604.
30. Graffagnino, C., Herbstreith, M.H., Roses, A.D. and Alberts, M.J. (1994) A molecular genetic study of intracerebral hemorrhage, *Arch. Neurol.* **51**, 981-984.
31. Anders, K.H., Wang, Z.Z., Kornfeld, M., Gray, F., Soontornniyomkij, V., Reed, L.A., Hart, M.N., Menchine, M., Secor, D.L. and Vinters, H.V. (1997) Giant cell arteritis in association with cerebral amyloid angiopathy: immunohistochemical and molecular studies, *Hum. Pathol.* **28**, 1237-1246.
32. Itoh, Y. and Yamada, M. (1997) Cerebral amyloid angiopathy in the elderly: the clinicopathological features, pathogenesis, and risk factors, *Journal of Medical and Dental Sciences* **44**, 11-19.
33. Nagai, A., Kobayashi, S., Shimode, K., Imaoka, K., Umegae, N., Fujihara, S. and Nakamura, M. (1998) No mutations in cystatin C gene in cerebral amyloid angiopathy with cystatin C deposition, *Mol. Chem. Neuropathol.* **33**, 63-78.
34. Rosand, J., Hylek, E.M., O'Donnell, H.C. and Greenberg, S.M. (1999) Cerebral amyloid angiopathy and warfarin-associated hemorrhage. A genetic and neuropathological study., *Neurology (abs)* **52**, A503.
35. Ramsay, D.A., Penswick, J.L. and Robertson, D.M. (1990) Fatal streptokinase-induced intracerebral hemorrhage in cerebral amyloid angiopathy, *Can. J. Neurol. Sci.* **17**, 336-341.
36. Pendlebury, W.W., Iole, E.D., Tracy, R.P. and Dill, B.A. (1991) Intracerebral hemorrhage related to cerebral amyloid angiopathy and t-PA treatment, *Ann. Neurol.* **29**, 210-213.
37. Leblanc, R., Haddad, G. and Robitaille, Y. (1992) Cerebral hemorrhage from amyloid angiopathy and coronary thrombolysis, *Neurosurgery* **31**, 586-590.
38. Wijdicks, E.F. and Jack, C.R.J. (1993) Intracerebral hemorrhage after fibrinolytic therapy for acute myocardial infarction, *Stroke* **24**, 554-557.
39. Sloan, M.A., Price, T.R., Petito, C.K., Randall, A.M.Y., Solomon, R.E., Terrin, M.L., Gore, J., Collen, D., Kleinman, N., Feit, F., Babb, J., Herman, M., Roberts, W.C., Sopko, G., Bovill, E., Forman, S. and Knatterud, G.L. (1995) Clinical features and pathogenesis of intracerebral hemorrhage after rt-PA and heparin therapy for acute myocardial infarction: The Thrombolysis in Myocardial Infarction (TIMI) II Pilot and Randomized Clinical Trial combined experience., *Neurology* **45**, 649-658.
40. Kase, C.S., Robinson, R.K., Stein, R.W., DeWitt, L.D., Hier, D.B., Harp, D.L., Williams, J.P., Caplan, L.R. and Mohr, J.P. (1985) Anticoagulant-related intracerebral hemorrhage, *Neurology* **35**, 943-948.
41. Gore, J.M., Sloan, M., Price, T.R., Randall, A.M.Y., Bovill, E., Collen, D., Forman, S., Knatterud, G.L., Sopko, G. and Terrin, M.L. (1991) Intracerebral hemorrhage, cerebral

infarction, and subdural hematoma after acute myocardial infarction and thrombolytic therapy in the Thrombolysis in Myocardial Infarction study: Thrombolysis in Myocardial Infarction, Phase II, Pilot and Clinical Trial, *Circulation* **83**, 448-459.
42. Kase, C.S., Pessin, M.S., Zivin, J.A., del Zoppo, G.J., Furlan, A.J., Buckley, J.W., Snipes, R.G. and Little John, J.K. (1992) Intracranial hemorrhage after coronary thrombolysis with tissue plasminogen activator, *Am. J. Med.* **92**, 384-390.
43. Hylek, E.M. and Singer, D.E. (1994) Risk factors for intracranial hemorrhage in outpatients taking warfarin, *Ann. Intern. Med.* **120**, 897-902.
44. McMechan, S.R. and Adgey, A.A. (1998) Age related outcome in acute myocardial infarction. Elderly people benefit from thrombolysis and should be included in trials (editorial), *BMJ* **317**, 1334-1335.
45. Mendelson, G. and Aronow, W.S. (1998) Underutilization of warfarin in older persons with chronic nonvalvular atrial fibrillation at high risk for developing stroke, *J. Am. Geriatr. Soc.* **46**, 1423-1424.
46. Ropper, A.H. and Davis, K.R. (1980) Lobar cerebral hemorrhages: acute clinical syndromes in 26 cases, *Ann. Neurol.* **8**, 141-147.
47. Weisberg, L.A., Stazio, A., Shamsnia, M. and Elliott, D. (1990) Nontraumatic parenchymal brain hemorrhages, *Medicine* **69**, 277-295.
48. Brott, T., Broderick, J., Kothari, R., Barsan, W., Tomsick, T., Sauerbeck, L., Spilker, J., Duldner, J. and Khoury, J. (1997) Early hemorrhage growth in patients with intracerebral hemorrhage, *Stroke* **28**, 1-5.
49. Weisberg, L.A., Shamsnia, M. and Elliott, D. (1991) Seizures caused by nontraumatic parenchymal brain hemorrhages, *Neurology* **41**, 1197-1199.
50. Cuny, E., Loiseau, H., Rivel, J., Vital, C. and Castel, J.P. (1996) Amyloid angiopathy-related cerebellar hemorrhage., *Surg. Neurol.* **46**, 235-239.
51. Yamada, M., Itoh, Y., Otomo, E., Hayakawa, M. and Miyatake, T. (1993) Subarachnoid hemorrhage in the elderly: a necropsy study of the association with cerebral amyloid angiopathy, *J. Neurol. Neurosurg. Psychiatry* **56**, 543-547.
52. Kase, C.S., Williams, J.P., Wyatt, D.A. and Mohr, J.P. (1982) Lobar intracerebral hematomas: clinical and CT analysis of 22 cases, *Neurology* **32**, 1146-1150.
53. Helweg-Larsen, S., Sommer, W., Strange, P., Lester, J. and Boysen, G. (1984) Prognosis for patients treated conservatively for spontaneous intracerebral hematomas, *Stroke* **15**, 1045-1048.
54. Passero, S., Burgalassi, L., D'Andrea, P. and Battistini, N. (1995) Recurrence of bleeding in patients with primary intracerebral hemorrhage, *Stroke* **26**, 1189-1192.
55. O'Donnell, H.C., Rosand, J., Knudsen, K.A., Furie, K.L., Segal, A.Z., Chiu, R.I., Ikeda, D. and Greenberg, S.M. (2000) Apolipoprotein E genotype and risk of recurrent lobar hemorrhage. A prospective study, *N. Engl. J. Med.* in press.
56. Arakawa, S., Saku, Y., Ibayashi, S., Nagao, T. and Fujishima, M. (1998) Blood pressure control and recurrence of hypertensive brain hemorrhage, *Stroke* **29**, 1806-1809.
57. Counsell, C., Boonyakarnkul, S., Dennis, M., P., S., Bamford, J., Burn, J. and Warlow, C. (1995) Primary intracerebral hemorrhage in the Oxfordshire Community Stroke Project. 2. Prognosis., *Cerebrovasc. Dis.* **5**, 26-34.
58. Neau, J.P., Ingrand, P., Couderq, C., Rosier, M.P., Bailbe, M., Dumas, P., Vandermarcq, P. and Gil, R. (1997) Recurrent intracerebral hemorrhage, *Neurology* **49**, 106-113.
59. Gonzalez-Duarte, A., Cantu, C., Ruiz-Sandoval, J.L. and Barinagarrementeria, F. (1998) Recurrent primary cerebral hemorrhage: frequency, mechanisms, and prognosis, *Stroke* **29**, 1802-1805.

60. Kase, C.S. (1994) Cerebral amyloid angiopathy, in C.S. Kase and L.R. Caplan (eds.), *Intracerebral Hemorrhage*, Butterworth-Heinemann, Boston, pp. 179-200.
61. Smith, D.B., Hitchcock, M. and Philpott, P.J. (1985) Cerebral amyloid angiopathy presenting as transient ischemic attacks. Case report, *J. Neurosurg.* **63**, 963-964.
62. Greenberg, S.M., Vonsattel, J.P., Stakes, J.W., Gruber, M. and Finklestein, S.P. (1993) The clinical spectrum of cerebral amyloid angiopathy: presentations without lobar hemorrhage, *Neurology* **43**, 2073-2079.
63. Okazaki, H., Reagan, T.J. and Campbell, R.J. (1979) Clinicopathologic studies of primary cerebral amyloid angiopathy, *Mayo Clin. Proc.* **54**, 22-31.
64. Olichney, J.M., Hansen, L.A., Hofstetter, C.R., Grundman, M., Katzman, R. and Thal, L.J. (1995) Cerebral infarction in Alzheimer's disease is associated with severe amyloid angiopathy and hypertension, *Arch. Neurol.* **52**, 702-708.
65. Briceno, C.E., Resch, L. and Bernstein, M. (1987) Cerebral amyloid angiopathy presenting as a mass lesion, *Stroke* **18**, 234-239.
66. Le Coz, P., Mikol, J., Ferrand, J., Woimant, F., Masters, C., Beyreuther, K., Haguenau, M., Cophignon, J. and Pepin, B. (1991) Granulomatous angiitis and cerebral amyloid angiopathy presenting as a mass lesion, *Neuropathology & Applied Neurobiology* **17**, 149-155.
67. Mandybur, T.I. and Balko, G. (1992) Cerebral amyloid angiopathy with granulomatous angiitis ameliorated by steroid-cytoxan treatment, *Clin. Neuropharm.* **15**, 241-247.
68. Fountain, N.B. and Eberhard, D.A. (1996) Primary angiitis of the central nervous system associated with cerebral amyloid angiopathy: report of two cases and review of the literature., *Neurology* **46**, 190-197.
69. Ginsberg, L., Geddes, J. and Valentine, A. (1988) Amyloid angiopathy and granulomatous angiitis of the central nervous system: a case responding to corticosteroid treatment, *J. Neurol.* **235**, 438-440.
70. Gray, F., Vinters, H.V., Le Noan, H., Salama, J., Delaporte, P. and Poirier, J. (1990) Cerebral amyloid angiopathy and granulomatous angiitis: immunohistochemical study using antibodies to the Alzheimer A4 peptide, *Hum. Pathol.* **21**, 1290-1293.
71. Vonsattel, J.P., Myers, R.H., Hedley-Whyte, E.T., Ropper, A.H., Bird, E.D. and Richardson, E.P. (1991) Cerebral amyloid angiopathy without and with cerebral hemorrhages: a comparative histological study, *Ann. Neurol.* **30**, 637-649.
72. Powers, J.M., Stein, B.M. and Torres, R.A. (1990) Sporadic cerebral amyloid angiopathy with giant cell reaction, *Acta Neuropathol.* **81**, 95-98.
73. Hendricks, H.T., Franke, C.L. and Theunissen, P.H. (1990) Cerebral amyloid angiopathy: diagnosis by MRI and brain biopsy, *Neurology* **40**, 1308-1310.
74. Yoshimura, M., Yamanouchi, H., Kuzuhara, S., Mori, H., Sugiura, S., Mizutani, T., Shimada, H., Tomonaga, M. and Toyokura, Y. (1992) Dementia in cerebral amyloid angiopathy: a clinicopathological study, *J. Neurol.* **239**, 441-450.
75. Silbert, P.L., Bartleson, J.D., Miller, G.M., Parisi, J.E., Goldman, M.S. and Meyer, F.B. (1995) Cortical petechial hemorrhage, leukoencephalopathy, and subacute dementia associated with seizures due to cerebral amyloid angiopathy, *Mayo Clin. Proc.* **70**, 477-480.
76. Gray, F., Dubas, F., Roullet, E. and Escourolle, R. (1985) Leukoencephalopathy in diffuse hemorrhagic cerebral amyloid angiopathy, *Ann. Neurol.* **18**, 54-59.
77. Bogucki, A., Papierz, W., Szymanska, R. and Staniaszczyk, R. (1988) Cerebral amyloid angiopathy with attenuation of the white matter on CT scans: subcortical arteriosclerotic encephalopathy (Binswanger) in a normotensive patient, *J. Neurol.* **235**, 435-437.
78. Caplan, L.R. (1995) Binswanger's disease--revisited., *Neurology* **45**, 626-633.

79. Bornebroek, M., Haan, J., van Buchem, M.A., Lanser, J.B., de Vries, V.D., Weerd, M.A., Zoeteweij, M. and Roos, R.A. (1996) White matter lesions and cognitive deterioration in presymptomatic carriers of the amyloid precursor protein gene codon 693 mutation, *Arch. Neurol.* **53**, 43-48.
80. Schmechel, D.E., Saunders, A.M., Strittmatter, W.J., Crain, B.J., Hulette, C.M., Joo, S.H., Pericak-Vance, M.A., Goldgaber, D. and Roses, A.D. (1993) Increased amyloid beta-peptide deposition in cerebral cortex as a consequence of apolipoprotein E genotype in late-onset Alzheimer disease, *Proc. Natl. Acad. Sci. USA* **90**, 9649-9653.
81. Olichney, J.M., Hansen, L.A., Galasko, D., Saitoh, T., Hofstetter, C.R., Katzman, R. and Thal, L.J. (1996) The apolipoprotein E epsilon 4 allele is associated with increased neuritic plaques and cerebral amyloid angiopathy in Alzheimer's disease and Lewy body variant, *Neurology* **47**, 190-196.
82. Ellis, R.J., Olichney, J.M., Thal, L.J., Mirra, S.S., Morris, J.C., Beekly, D. and Heyman, A. (1996) Cerebral amyloid angiopathy in the brains of patients with Alzheimer's disease: the CERAD experience, Part XV, *Neurology* **46**, 1592-1596.
83. Mandybur, T.I. (1986) Cerebral amyloid angiopathy: the vascular pathology and complications, *J Neuropathol. Exp. Neurol.* **45**, 79-90.
84. Vinters, H.V., Natte, R., Maat-Schieman, M.L., van Duinen, S.G., Hegeman Kleinn, I., Welling Graafland, C., Haan, J. and Roos, R.A. (1998) Secondary microvascular degeneration in amyloid angiopathy of patients with hereditary cerebral hemorrhage with amyloidosis, Dutch type (HCHWA-D), *Acta Neuropathol.* **95**, 235-244.
85. McKhann, G., Drachman, D., Folstein, M., Katzman, R., Price, D. and Stadlan, E.M. (1984) Clinical diagnosis of Alzheimer's disease: report of the NINCDS-ADRDA Work Group under the auspices of Department of Health and Human Services Task Force on Alzheimer's Disease, *Neurology* **34**, 939-944.
86. Joachim, C.L., Morris, J.H. and Selkoe, D.J. (1988) Clinically diagnosed Alzheimer's disease: autopsy results in 150 cases, *Ann. Neurol.* **24**, 50-56.
87. Gearing, M., Mirra, S.S., Hedreen, J.C., Sumi, S.M., Hansen, L.A. and Heyman, A. (1995) The Consortium to Establish a Registry for Alzheimer's Disease (CERAD). Part X. Neuropathology confirmation of the clinical diagnosis of Alzheimer's disease, *Neurology* **45**, 461-466.
88. Greenberg, S.M. (1998) Cerebral amyloid angiopathy. Prospects for clinical diagnosis and treatment., *Neurology* **51**, 690-694.
89. Atlas, S.W., Mark, A.S., Grossman, R.I. and Gomori, J.M. (1988) Intracranial hemorrhage: gradient-echo MR imaging at 1.5 T. Comparison with spin-echo imaging and clinical applications, *Radiology* **168**, 803-807.
90. Greenberg, S.M., Finklestein, S.P. and Schaefer, P.W. (1996) Petechial hemorrhages accompanying lobar hemorrhages: Detection by gradient-echo MRI, *Neurology* **46**, 1751-1754.
91. Offenbacher, H., Fazekas, F., Schmidt, R., Koch, M., Fazekas, G. and Kapeller, P. (1996) MR of cerebral abnormalities concomitant with primary intracerebral hematomas, *AJNR Am. J. Neuroradiol.* **17**, 573-578.
92. Van Nostrand, W.E., Wagner, S.L., Shankle, W.R., Farrow, J.S., Dick, M., Rozemuller, J.M., Kuiper, M.A., Wolters, E.C., Zimmerman, J., Cotman, C.W. and et, a.l. (1992) Decreased levels of soluble amyloid beta-protein precursor in cerebrospinal fluid of live Alzheimer disease patients, *Proc. Natl. Acad. Sci. USA* **89**, 2551-2555.
93. Shimode, K., Fujihara, S., Nakamura, M., Kobayashi, S. and Tsunematsu, T. (1991) Diagnosis of cerebral amyloid angiopathy by enzyme-linked immunosorbent assay of cystatin C in cerebrospinal fluid, *Stroke* **22**, 860-866.

94. Hart, R.G., Boop, B.S. and Anderson, D.C. (1995) Oral anticoagulants and intracranial hemorrhage, *Stroke* **26**, 1471-1477.
95. He, J., Whelton, P.K., Vu, B. and Klag, M.J. (1998) Aspirin and risk of hemorrhagic stroke: a meta-analysis of randomized controlled trials, *JAMA* **280**, 1930-1935.
96. Greene, G.M., Godersky, J.C., Biller, J., Hart, M.N. and Adams, H.P.J. (1990) Surgical experience with cerebral amyloid angiopathy, *Stroke* **21**, 1545-1549.
97. Matkovic, Z., Davis, S., Gonzales, M., Kalnins, R. and Masters, C.L. (1991) Surgical risk of hemorrhage in cerebral amyloid angiopathy, *Stroke* **22**, 456-461.
98. Minakawa, T., Takeuchi, S., Sasaki, O., Koizumi, T., Honad, Y., Fujii, Y., Ozawa, T., Ogawa, H., Koike, T. and Tanaka, R. (1995) Surgical experience with massive lobar hemorrhage caused by cerebral amyloid angiopathy, *Acta Neurochir.* **132**, 48-52.
99. Izumihara, A., Ishihara, T., Iwamoto, N., Yamashita, K. and Ito, H. (1999) Postoperative outcome of 37 patients with lobar intracerebral hemorrhage related to cerebral amyloid angiopathy, *Stroke* **30**, 29-33.

Chapter 2

DIAGNOSIS OF CAA DURING LIFE

Neuroimaging of CAA

ULRICH BICKEL

Institute of Pharmacology and Toxicology, Philipps University, Marburg, Germany

Intensive efforts are underway to develop diagnostic tests, which enable us to detect cerebral amyloid angiopathy (CAA) at an early preclinical stage. These tests should be noninvasive to allow for broad and repeated application. Specific neuroimaging methods targeted at the hallmark constituent of CAA, the Aβ deposits, are particularly attractive goals. A useful technique has to be significantly more sensitive and specific than the currently available structural and functional imaging modalities. One of the main problems is the limited distribution and accumulation in brain of amyloid specific probes, such as radiolabeled Aβ-peptides or anti-Aβ monoclonal antibodies even after invasive modes of administration, e.g., intracisternal injection. Looking at a systemic approach, the impermeability of the blood-brain barrier for amyloid probes requires the incorporation of an efficient delivery strategy. In pursuit of this, animal studies have demonstrated the feasibility to deliver monoclonal antibodies and peptides to the brain. Physiological transport mechanisms for some peptides (transferrin, insulin and others) are present at the barrier and were utilized to transport radiolabeled antibodies or Aβ peptide into brain. The existence of analogous uptake mechanisms in humans is known. That opens up the future possibility to quickly adapt an imaging method to clinical use once initial validation in a suitable animal model, like aged primates, was performed.

1. INTRODUCTION

As the occurrence of cerebral amyloid angiopathy (CAA) most is often associated with Alzheimer's disease (AD), the rationale behind efforts to develop an *in vivo* imaging technique for CAA is intimately related to the unsolved problem of an early diagnostic test in AD [1]. According to clinico-pathological studies, the clinical diagnosis of *probable* AD (as defined in the NINCDS-ADRDA criteria) can be made with an accuracy of about 85% [2], while the diagnosis of *definite* AD still relies on post mortem

M.M. Verbeek et al. (eds.),
Cerebral Amyloid Angiopathy in Alzheimer's Disease and Related Disorders, 21–41.

neuropathological proof, apart from the rare case where a brain biopsy is performed. Given the high prevalence of AD, estimated at present between 4 and 4.5 million patients in the USA alone [1,3], a frequency of 15% missed diagnosis amounts to a substantial number of individuals. Moreover, there may be as many as 1.8 million demented patients currently going without diagnosis or treatment [4,5], 70% of whom can be expected to suffer from AD alone or in combination with other disease [5].

Because the presence of excessive amounts of Aβ is a histopathological hallmark of AD and most forms of CAA, direct detection and quantification of cerebral amyloid deposits during life is an obvious diagnostic target. A disease specific, sensitive and quantitative diagnostic imaging method appears highly desirable for AD [1] and CAA, where currently no radiographic diagnosis is available. This is true, at least, for the early stages when there is no CAA-related hemorrhage [6]. However, there is precedent in the field of amyloidosis with the diagnostic detection and quantification of systemic amyloid protein deposits which demonstrates the application of a scintigraphic imaging technique [7]. The challenge is to bring together existing modes of neuroimaging with the specificity and sensitivity of neuropathological techniques. The present review will focus on approaches to developing Aβ-directed imaging techniques, after a brief outline of current imaging modalities and biochemical diagnostic developments.

2. BIOCHEMICAL ASSAYS FOR DIAGNOSIS

One line of research, which relies on a diagnostic examination of cerebrospinal fluid (CSF), is aimed at the detection of single β amyloid aggregates in CSF [8]. Others pursue improved diagnosis via the measurement of a combination of surrogate markers such as the concentrations of soluble Aβ peptides and tau protein in CSF [9]. Both amyloid plaques and vascular amyloid reside in the brain extracellular space, and amyloid derived from interstitial fluid is most likely deposited in the course of CAA [10]. It is therefore likely that part of the Aβ is removed via the CSF. Aβ concentrations measured in lumbar CSF samples, though, may not show good quantitative correlation with brain interstitial fluid levels [11].

A direct quantitative measurement of amyloid in brain tissue might correlate better with the disease stage and allow for differential diagnosis from other causes of dementia as proposed recently [12]. However, in order to be broadly applicable in presymptomatic, healthy individuals, the test needs to be non-invasive. Only then, long term serial measurements would make intraindividual monitoring of progression of disease in relation to amyloid load possible and establish whether early diagnosis of AD can be based on this approach. Moreover, the effect of potential therapeutic interventions targeted towards preventing or slowing amyloid deposition, such as future inhibitors of β- or γ-secretases could be evaluated.

3. IMAGING METHODS NOT TARGETING AMYLOID DIRECTLY

Structural neuroimaging methods, such as computed tomography (CT) and magnetic resonance imaging (MRI), are useful in excluding causes of dementia other than AD. They have also shown promising results in preselected groups of patients and controls to discriminate probable AD from normal elderly individuals. Diagnostic criteria are based on the measurement of medial temporal lobe atrophy [13,14] or on the volumetric measurement of the entorhinal cortex and hippocampus [15]. Presymptomatic hippocampal atrophy has also been demonstrated in a prospective, longitudinal MRI study [16]. The value of such measurements for the early diagnosis in an unselected population sample has not yet been established. This is mainly due to the lack of specificity of atrophy measurements.

Functional neuroimaging methods, such as functional MRI, measurement of cerebral blood flow by single photon emission computed tomography (SPECT) or regional cerebral glucose metabolism as measured by positron emission tomography (PET) have not yet demonstrated a sensitivity or specificity that would be superior to clinical diagnosis and structural neuroimaging [17].

4. RELEVANCE OF THE BLOOD-BRAIN BARRIER FOR NEUROIMAGING APPROACHES OF CAA

The blood-brain barrier (BBB) is located at the sites of the endothelial cells in cerebral microvessels. Figure 1 demonstrates microvessels with amyloid deposits at the light and electron microscopic (EM) level. With regard to an *in vivo* detection, one of the most important features of brain amyloid both in parenchymal plaques and cerebrovascular deposits is their localization beyond the BBB. EM studies showed that in early CAA of Alzheimer's disease the Aβ is deposited within the outer basement membrane [21].

The integrity of the blood brain barrier in AD and CAA has been addressed in numerous studies. There is general agreement that at the ultrastructural level the BBB in Alzheimer's disease does show pathologic changes. For example, a morphometric study performed on brain biopsies from AD patients [22] demonstrated diminished mitochondrial density and area fraction within capillary endothelial cells, an increase in pericyte size in AD brains, and alterations of inter-endothelial tight junctions suggesting a leaky BBB. Moreover, atrophic thin vessels, fragmented vessels, glomerular loop formations and twisted or tortuous vessels have been described to occur in AD brains. Thickening of the basement membrane observed in normal aging is exaggerated in AD and CAA and the basement membrane also shows vacuolization. One of the contributing factors to the thickening of the basement membrane is an increased content in collagen IV [23]. The morphological aspects of BBB related changes in cerebral amyloid angiopathy in Alzheimer's disease are reviewed in this volume by Kalaria.

Despite these morphological alterations and the presence of serum proteins, such as amyloid P [24,25], indicating extravasation of serum proteins through the BBB, there is no clear functional evidence for BBB leakage in this condition. The determination of BBB permeability in AD patients using [^{68}Ga]-EDTA as a tracer for PET, yielded no significant difference in the cerebrovascular permeability-surface area product between patients and healthy, age matched controls [26]. A recent study employed meglumin iothalamate as a contrast agent in computed tomography [27]. These authors could also not find significant differences in terms of blood to brain transport, tissue to blood efflux, tissue plasma space and tissue extracellular space between patients with clinically diagnosed probable AD and elderly control subjects.

While the immunohistochemical detection of serum derived components, associated with β-amyloid deposits in paravascular localization within the brain is employed as evidence of BBB damage, its extent remains controversial. Besides the fact that cerebral synthesis of some serum components, such as all members of the complement cascade, has recently been described, it is difficult or impossible to distinguish between potential damage to the BBB during life and post mortem events in autopsy samples [28]. More reliable data can be obtained from biopsy specimens [29]. This study reported that the vast majority of brain microvessels contained in cortical biopsy samples from AD patients showed normal features of the BBB, without leakage of the endogenous vascular marker serum albumin. In 5% to 15 % of microvascular profiles, immunostained albumin was seen within the endothelial cell cytoplasm, in the subendothelial space and the basement membrane, and in the adjacent neuropil after detection with anti-human albumin antiserum and protein A gold complex. These microvessels also showed other signs of damage, such as thickening of the basement membrane, shrinkage, or swelling of the endothelial cells. BBB abnormalities which allowed for slow serum protein deposition on a time scale of many months to years would still not allow the rapid accumulation in sufficient concentration within the amyloid deposits of a tracer. For diagnostic use this should happen within the order of hours to days.

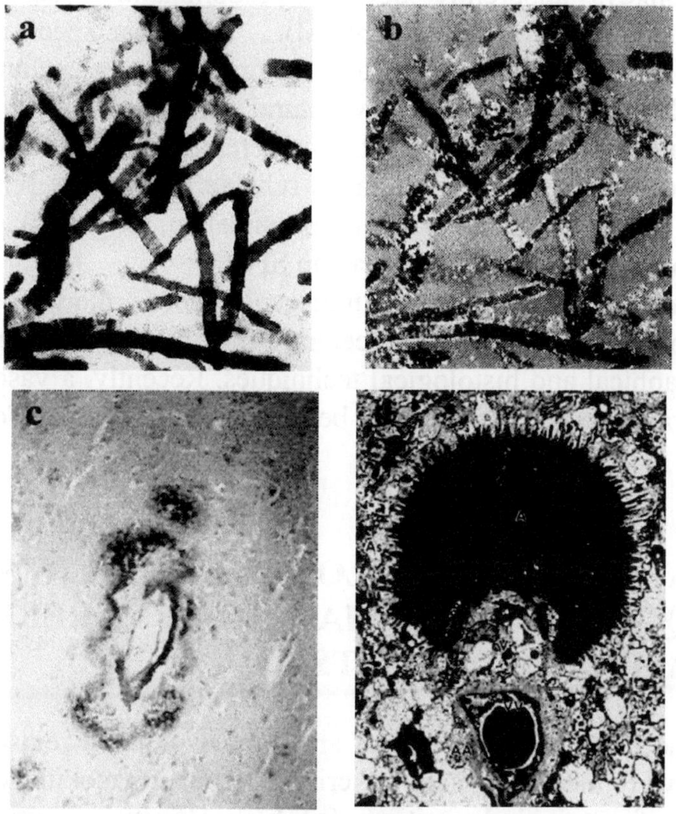

Figure 1. *(a)* Isolated cortical arterioles from a patient with AD are laden with amyloid, as seen under the light microscope after staining with Congo red (x 50). *(b)* shows the same field as in *(a)* under polarized light and reveals the characteristic birefringence phenomenon (originally apple green). Panel *(c)* shows a small artery surrounded by a halo of amyloid which was immunostained with an anti $A\beta_{1\text{-}28}$ antiserum (x 80). The electron photomicrograph *(d)* visualizes star-shaped amyloid fibrils, A, emanating from the wall of an obliterated vessel, V, and minor amyloid infiltration in a second vascular profile, VV. In order to reach the amyloid deposit from the circulation, any tracer must cross the endothelial barrier. Reproduced from [18] *(a,b)*; [19] *(c)*; [20] *(d)*.

5. ANIMAL MODELS OF CAA

Suitable models are indispensable in the preclinical evaluation of diagnostic neuroimaging methods. Among the animal species that have been shown to develop significant amounts of CNS amyloid are aged dogs and non-human primates, including rhesus and squirrel monkeys [30]. Dogs and squirrel monkeys apparently develop predominantly vascular amyloid deposits, which contain more of the $A\beta_{1\text{-}42}$ molecular species than $A\beta_{1\text{-}40}$ [31], while rhesus monkeys seem to develop mostly parenchymal amyloid deposits. Although all models show natural variability both in the amount and localization of amyloid deposits, for cerebral vascular amyloidosis the squirrel monkey appears to be the most reliable model. The relatively large size of the squirrel monkey brain (approximately 25 grams) may also be sufficient to obtain quantitative signals in SPECT.

The availability of a suitable transgenic mouse model could greatly facilitate pilot studies and screening of imaging approaches with autoradiographical and histological techniques. Recently, a vascular type of amyloid deposition in such mice has been described [32] (see Cole *et al*, this volume).

6. AVENUES TO EXPLORE APPROACHES FOR QUANTITATIVE IMAGING OF CEREBRAL AMYLOID DEPOSITS

Radiotracers for the detection of amyloid by external registration (PET, SPECT) have to meet the following criteria: (i) pharmacokinetic distribution to the site of amyloid deposition; (ii) high affinity, specific binding to amyloid; (iii) low non-specific binding to other tissue components, in order to achieve a good signal to noise ratio.

6.1 Small molecule based approach

Attempts to utilize low molecular weight substances have been based on dyes such as Congo red that is routinely used in histological diagnosis of systemic and cerebral amyloid deposits at the light microscopic level. Because Congo red is a highly charged anion unable to pass through the BBB in significant quantity, Chrysamine G, a carboxylic acid analogue of Congo red, was evaluated [33,34]. Chrysamine G is sufficiently lipophilic to cross the BBB, and brain to blood ratios in mice over 10:1 were reported. The binding of a radiolabeled Chrysamine G to homogenates of regions of

AD brains was measured and compared to samples of cerebellum and homogenates of control brain. Values in regions with plaques and tangles were reported to be approximately twofold higher, and tracer binding was displacable with unlabeled Chrysamine G, yielding a submicromolar K_D, similar to the binding of the dye to synthetic β amyloid *in vitro* [34]. Whether the twofold increase seen *in vitro* translates into a sufficiently high signal to noise ratio *in vivo* remains to be tested.

6.2 Anti-amyloid antibody-based approaches

Monoclonal antibodies have been raised successfully against different synthetic peptides derived from the Aβ sequence. These antibodies are invaluable in immunocytochemistry and Western blotting to detect β amyloid with high specificity and sensitivity. As a logical consequence attempts have been made to establish an *in vivo* diagnostic method with these tools. So far the antibodies used in such studies were generated to synthetic peptides from the NH_2-terminal part of the Aβ sequence, which would recognize all Aβ forms rather than between species that vary at their COOH-terminal. At this stage it may be advantageous to aim at assessing total amyloid load, as there are divergent reports regarding the quantitative contribution of the longer ($A\beta_{1\text{-}42/43}$) vs. shorter forms ($A\beta_{1\text{-}39/40}$) in parenchymal and vascular deposits [35,36], respectively. This point is also true with regard to the use of primate models for initial testing of such an imaging approach, because species differences compared to man in the distribution of the Aβ forms have been described. For instance, an antibody specific for Aβ ending at residue 42/43 was found to label vascular amyloid in cynomolgus monkeys at a higher frequency than an antibody recognizing Aβ ending at residue 40 [31]. Similar conditions apparently apply to rhesus and squirrel monkeys [30].

6.2.1 Intrathecal antibody injection

Walker and coworkers [37] chose an intracisternal injection of the monoclonal antibody (MAb) 10D5 in primates in an attempt to label amyloid *in vivo*. The antibody is specific for amino acids 1 to 16 of Aβ peptide [38]. Squirrel monkeys of 16 and 18 years of age and an aged rhesus monkey (34 years old) were used as experimental models. They received either Fab fragments of the monoclonal antibody or a mixture of Fab and whole IgG (in the case of the rhesus monkey). Another aged squirrel monkey (17 years old) received a control injection of nonimmune IgG. All animals were sacrificed 24 h following injection of the antibodies and were then perfused transcardially with buffer solution to clear the vasculature. The brains were perfusion-fixed with paraformaldehyde, removed and processed for immunohistochemistry on frozen sections. In order to visualize Aβ deposits decorated *in vivo* by antibody, the sections were treated with a secondary anti mouse antibody followed by peroxidase-anti-peroxidase and development with diaminobenzidine. Adjacent sections were stained applying the same 10D5 antibody *in vitro* for conventional immunohistochemistry. The *in vitro* labeling verified that all animals had developed significant amounts of vascular and parenchymal amyloid deposits. With the secondary anti-mouse antibody it was also possible to detect a small subset of amyloid deposits in those animals that had received *in vivo* injection of the specific antibody. The deposits labeled with this technique were only seen in a band of tissue extending less than 0.5 mm from the cortical surface into the tissue. Approximately 5% of total deposits detected by the *in vitro* staining could also be seen after *in vivo* labeling in the squirrel monkeys. While the fraction of amyloid deposits labeled after *in vivo* injection in the rhesus monkey was estimated to be higher (15%), it was similarly restricted to a superficial cortical layer. This study demonstrates the principal feasibility of labeling amyloid deposits *in vivo* with a specific Mab. At the same time it is obvious that the limited penetration of the antibody after injection into the cerebrospinal fluid prevents quantitative measurement of amyloid deposits by this approach. Molecules of the size of Fab fragments (25kD) or whole IgG (150 kD) have very low diffusion coefficients into brain tissue [39]. In addition, there is rapid turnover of cerebrospinal fluid (CSF) by secretion, bulk flow and reabsorption into the venous system via the arachnoid villi which results in an exchange of total CSF volume in humans every 4-5 h [40]. Therefore, this invasive approach does not provide a viable basis to establish a clinically useful diagnostic technique.

6.2.2 Intravenous administration of native antibody

Based on the hypothesis that systemically injected anti-amyloid antibodies would have access to sites of amyloid deposition on or adjacent to the endothelium of brain microvessels from the luminal side of the BBB, Majocha *et al.* [41] and Friedland *et al.* [42] evaluated the antibody 10H3 [43]. This antibody was raised against $A\beta_{1\text{-}28}$ and was selected because it displayed good labeling characteristics of plaque and vascular amyloid *in vitro*. In order to achieve a more rapid clearance from the plasma pool, Fab fragments of 10H3 were prepared. These retained sufficient immunoreactivity for Aβ in immunohistochemistry [41]. Radiolabeling with ^{99m}Tc was performed with a diamine dimercaptide chelator, which did not compromise immunoreactivity as tested *in vitro*. The biodistribution and brain accumulation of ^{99m}Tc-labeled 10H3 Fab fragments was evaluated in patients diagnosed with probable AD according to NINCDS-ADRDA criteria following intravenous injection of the tracer at a mean radioactive dose of 33 mCi. The plasma elimination curve and a series of SPECT scans were obtained up to 36 h. The half live of the labeled Fab in plasma was acceptably short (2-3 h). The authors found the presence of label in the cerebral blood pool in all subjects after intravenous injection. However, there was no evidence of measurable uptake into brain parenchyma at later imaging times. Some accumulation of radioactivity in scalp tissue or in the bone marrow of the skull was attributed to a diffuse, nonspecific staining by the 10H3 antibody. The nonspecific binding was also observed in scalp biopsies of non-AD control subjects and was apparently unrelated to the presence of Aβ [44].

The failure to detect cerebral accumulation of the Fab fragment is further evidence that the BBB in AD is not damaged to an extent which would allow the penetration of high molecular weight diagnostic agents by a purely diffusion mediated-mechanism. It also appears unlikely that Aβ peptide is exposed in significant amounts on the luminal side of the BBB.

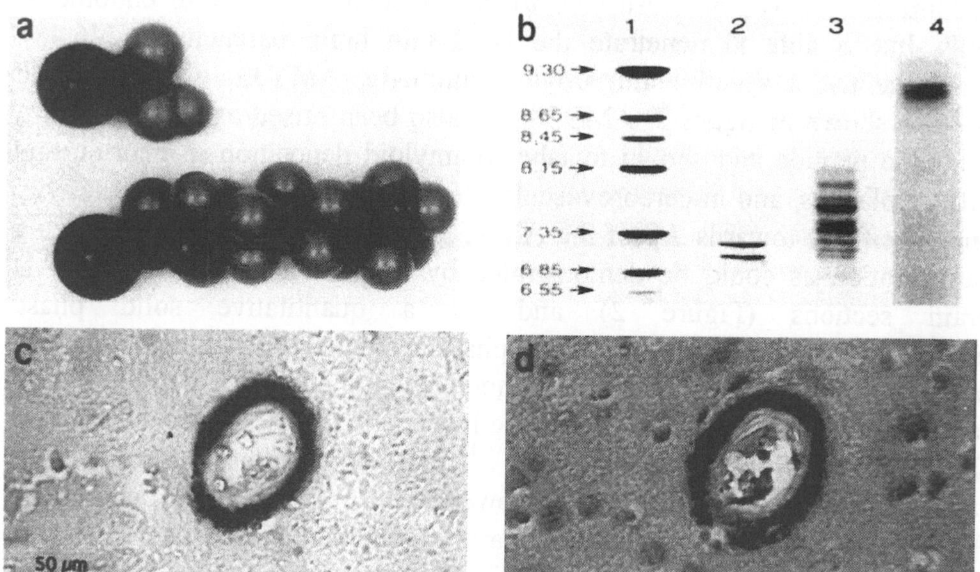

Figure 2. Cationization of an antibody: *(a)* illustrates the chemical principle, whereby a side shain carboxyl group of Asp or Glu residues (upper part) is converted to an extended amino group,*(b)* shows the shift of the isoelectric point (lane 1, standards) from neutral (lane 2, native antibody,) into the basic range; the extent can be titrated by pH, time and concentration (lanes 3 and 4). The native antibody, AMY33, was raised against $A\beta_{1\text{-}28}$ and recognizes vascular amyloid on AD brain sections, shown by the immunocytochemistry in *(c)*. The cationized form of the antibody retained specific immunoreactivity, as shown in *(d)* which was also lightly counterstained with hematoxylin. From Bickel, Lee *et al.* [47].

6.2.3 Intravenous administration of cationized antibody

Natural or chemically derivatized peptides and proteins with an isoelectric point in the basic range undergo absorption mediated endocytosis into cells and have been shown to be subject to absorptive mediated transcytosis at the BBB [45]. That mechanism is mediated by electrostatic interactions between anionic microdomains on the endothelial cell surface and positively charged proteins. Therefore, one possible strategy to overcome restricted BBB permeability of proteins is their cationization. In the case of IgG, transcytosis through the BBB was initially demonstrated using nonspecific polyclonal bovine IgG which was cationized with hexamethylenediamine and carbodiimide activation [46]. The modification results in conversion of superficial carboxyl groups of a protein (Asp or Glu side chains) into primary amino groups and raises the isoelectric point to an extent that can be controlled by adjusting the reaction conditions. After systemic injection of cationized IgG radiolabeled with ^{125}I into rats, it could be shown both by autoradiography of brain sections and by pharmacokinetic analysis of tissue homogenates depleted of the vascular fraction that the cationized IgG was not restricted to the vascular lumen or to endothelial cells, but is able to penetrate the BBB into brain parenchyma [46,48]. Subsequently, a specific anti-amyloid antibody, AMY33, was cationized [47], as shown in figure 2. AMY33 had also been raised against the $A\beta_{1-28}$ synthetic peptide and shown to label β-amyloid deposition in neuritic and diffuse plaques, and in cerebrovascular deposits [49]. The immunoreactivity and specificity towards Aβ of the AMY33 antibody was retained following cationization as could be demonstrated by immunohistochemistry on AD brain sections (Figure 2) and by a quantitative solid phase immunoradiometric assay [47]. The pharmacokinetics and brain uptake of both the native and the cationized monoclonal antibody were quantified after radiolabeling. Conjugation with diethylene triamine penta-acetic acid (DTPA) was used to chelate ^{111}In, an isotope suitable for radioimmunodetection by SPECT. However, the ^{111}In labeling of cationized AMY33 resulted in unfavorable pharmacokinetic characteristics. Plasma concentration time curves and organ distribution in two species (dog, mouse) did not result in the expected increase in tissue uptake. In particular, there was no significant brain uptake [50]. Clearance from plasma and uptake into organs could be improved by an alternative labeling method using radioiodine. On emulsion autoradiography of sections of AD brain it was shown that ^{125}I-cationized AMY33 labels vascular amyloid deposits in leptomeningeal and intraparenchymal vessels specifically. The modified antibody preparation was then evaluated in aged squirrel monkeys for *in vivo* detection of brain amyloid deposits. Although the presence of vascular amyloid in all brains of these animals could unambiguously be verified by later immunohistochemistry, the *in vivo* labeling by injection of iodinated antibody was insufficient to result in a strong specific signal in emulsion autoradiography of brain sections obtained from brains between 8 h and 48 h after intravenous injection of antibody (Bickel and Pardridge, unpublished observations). One possible explanation of this result may the previously described serum inhibition phenomenon seen with some cationized radiolabeled proteins *in vivo* [51]. The serum inhibition resulted in greatly diminished *in vivo* brain uptake despite increased cellular uptake of the cationized proteins *in vitro* into isolated brain microvessels in the absence of serum.

6.3 Strategies based on Amyloid peptide

6.3.1 Serum amyloid P as a tracer

The serum amyloid P component (SAP) is a member of the pentraxin family of serum proteins and is synthesized in the liver. This protein exists as a decamer made up from glycosylated subunits with molecular weights of about 25 kD and has a molecular diameter of approximately 10 nm [52]. SAP is able to bind to all forms of amyloid deposits in tissues irrespective of their primary biochemical origin. It is also one of those serum derived proteins found to be associated ordinarily with vascular and parenchymal amyloid [25]. The labeling of SAP with ^{123}I yielded a tracer which is suitable for the semiquantitative detection by whole body scintigraphy of visceral amyloid deposits in patients suffering from various forms of systemic amyloidosis [7]. In that study, highly specific uptake and retention of labeled SAP in visceral amyloid deposits in liver, spleen and other organs was demonstrated.

The same authors applied this technique to patients suffering from γ-trace amyloidosis (hereditary cerebral hemorrhage with amyloidosis-Islandic type, HCHWA-I) and obtained evidence for systemic amyloidosis in these cases [53]. In the latter study, ^{124}I was used as a radionucleide, which allowed delayed detection due to its longer half-life (4 d vs. 13 h for ^{123}I). The isotope can also be detected by positron emission tomography but comes with the disadvantage of a higher radiation dosage, which precluded the inclusion of healthy controls in the above study. Therefore, the significance of brain radioactivity detected in two AD patients could not be verified.

In a recent study, Lovat *et al* [54] examined 11 patients with probable or definite AD and conventional whole body planar scintigraphy after injection of ^{123}I-SAP. No tracer could be detected in brain 24 h post injection. It was concluded that the extravasation of SAP into brain tissue was too low compared to its relatively slow clearance from plasma (24 h half-life of SAP). The negative result is another piece of evidence for the relative functional integrity of the BBB in AD.

6.3.2 Imaging approaches based on brain transport of Aβ

The demonstration that ^{125}I-labeled $A\beta_{1-40}$ binds specifically to preexisting amyloid in frozen sections of human AD brain [55] and is then protected from protease attack [56] suggested the application of Aβ itself as a tracer for *in vivo* imaging of cerebral amyloid. In a study involving *in vivo* vascular brain perfusion in guinea pigs, a saturable and specific transport of Aβ through the BBB was measured [57]. Capillary depletion was performed for the analysis of binding to the microvascular constituents vs. transport beyond the BBB. A dissociation constant, K_d, of 25 nM for the binding of radioiodinated $A\beta_{1-40}$ to the isolated microvessels was reported. Based on competition with non-labeled peptide, a specific transcellular BBB transport of $A\beta_{1-40}$ was observed with a K_m of 49 nM and a maximal rate V_{max} of 11 fmol/min/g. These results are in contradiction to measurements performed in rats and mice, in which no measurable brain uptake of ^{125}I-$A\beta_{1-40}$ could be found after intravenous injection [58,59]. On the other hand, Mackic *et al.* [60] studied brain uptake of radioiodinated $A\beta_{1-40}$ in adult vs. aged squirrel monkeys. The latter authors found an 18.6-fold higher BBB permeability of intact $A\beta_{1-40}$ compared to the vascular marker inulin. In the aged animals, BBB permeability was increased approximately 2-fold and it was also reported that cortical and leptomeningeal microvessels sequestered higher amounts of the Aβ tracer. It has been proposed that species-specific differences in the Aβ sequence may account for discrepancies among the studies [61], as rodents have an amino acid exchange in the Aβ sequence compared to other species, including primates and guinea pig. Whether the brain delivery of soluble $A\beta_{1-40}$ is sufficient to establish a diagnostic imaging method remains to be investigated. Intracarotid infusion of ^{125}I-$A\beta_{1-40}$ in aged squirrel monkeys was reported to label some cerebral vessels in one of three animals when analyzed by autoradiography of brain sections [62]. Brain delivery *in vivo* may be limited by the relatively high systemic clearance and by the metabolic lability of the peptide in plasma. For example, in the rat an elimination half life of 27 min and a decrease in the TCA precipitability from 92.7% immediately after injection to 41.3% at 30 min was measured [59] and similar data were measured in primates [63].

Another physiological transport mechanism with potential implications for the pathophysiological events responsible for amyloid deposition was suggested by Zlokovic and coworkers [61,64]. In those studies the BBB transport of radiolabeled apolipoproteins (ApoJ, ApoE3 and ApoE4) was compared in the presence or absence of an equimolar amount of bound soluble $A\beta_{1-40}$. $A\beta_{1-40}$ bound to ApoJ displayed a very high permeability surface area product at the BBB when studied in the perfused guinea pig brain preparation. ApoJ was described as a major carrier protein for Aβ peptide in cerebral spinal fluid and plasma [65,66]. Because ApoJ is present at micromolar concentrations in blood, and has an affinity for Aβ in the low nanomolar range, most plasma Aβ *in vivo* is bound to ApoJ [65]. Because the transport of the Aβ-ApoJ complex is supposed to be mediated by gp330/megalin [64], the receptor may be saturated *in vivo* by the endogenous high concentration of ApoJ, as concluded from a study in mice [67]. This saturation parallels the saturation found with transferrin and the transferrin receptor at the BBB. In such cases, the endogenous ligand is unsuitable for delivering significant amounts of tracer into brain following intravenous administration. High uptake rates can only be measured in experimental brain perfusion systems where competition with the endogenous ligand can be eliminated.

6.3.3 Targeted delivery of radiolabeled Aβ peptide

The amount of peptide-based neuropharmaceuticals in brain following systemic administration can be enhanced by vector-mediated drug delivery strategies. Such strategies, which are based on physiological transport mechanisms known to exist at the BBB for a number of peptides and proteins, have been advanced in recent years in preclinical studies. "Chimeric peptides" are synthesized by coupling a drug moiety which lacks intrinsic permeability at the BBB to a vector which undergoes receptor mediated transcytosis through the BBB [39]. The mechanism is analogous to the brain uptake described above for cationized proteins such as cationized monoclonal antibodies. However, the initial endocytosis at the luminal side of the BBB is initiated by the binding of a ligand to its cognate receptor at the plasma membrane. The best-studied examples to date of transcytosis of chimeric peptides at the BBB are the transferrin and the insulin receptor mediated systems.

The transferrin receptor is expressed in high concentrations at the BBB [68] in rodents [68-72], primates [73] and humans [74]. Because the ligand-binding site of the homodimeric receptor protein displays binding affinity for holo-transferrin in the low nanomolar range, it is heavily saturated at physiological transferrin concentrations (25 μmol). For the purpose of drug delivery, receptor antibodies directed at a non-competing binding site on the extracellular domain of the receptor protein can be utilized and have been shown to undergo endocytosis and transcytosis through endothelial cells. In the rat model, brain delivery of antibody after intravenous administration reached a concentration of 0.44 % of the injected dose per gram brain [70]. Coupling of drug moieties to the vector can be accomplished by direct chemical linkage [73] or by biotin avidin technology [75]. A conjugate of the anti-transferrin receptor antibody OX26 and avidin or streptavidin was shown to deliver various biotinylated bioactive peptides through the BBB. In the case of an anologue of the vasoactive intestinal polypeptide, which is involved in the regulation of vascular tone in intracranial blood vessels, the delivery resulted in estimated extracellular brain concentrations in the nanomolar range [76,77]. These concentrations were sufficient to elicit significant pharmacological effects. Therefore, this vector system was evaluated in a modified form for its suitability to deliver a radiopharmaceutical for imaging of amyloid deposits. Saito *et al.* [59] demonstrated that soluble $A\beta_{1-40}$ can be biotinylated and radioiodinated with ^{125}I and then be attached to the OX26-streptavidin vector. In these conjugates $A\beta_{1-40}$ retained its ability to specifically bind to amyloid deposits, as shown on frozen sections of human AD brain. Compared to the radiolabeled free $A\beta$ peptide brain uptake was increased by more than 10-fold to 0.15% of injected dose per gram.

Because OX26 is a monoclonal antibody specific to the rat transferrin receptor, imaging studies in primate AD models cannot be performed with that vector. However, the anti-human insulin receptor α-chain antibody 83-14 has favorable pharmacokinetic characteristics [78]. This antibody crossreacts with the insulin receptor protein in old world monkeys and its uptake rate at the BBB was measured to reach a permability-surface area product of 5.4 ± 0.6 µl/min/g. Using a linkage strategy analogous to the attachment of $A\beta_{1-40}$ to OX26, N-terminally biotinylated $A\beta_{1-40}$ was radiolabeled with ^{125}I and bound to a conjugate of 83-14 and streptavidin (scheme in figure 3a). The anti-insulin receptor antibody mediated uptake of the ^{125}I-biotinylated $A\beta_{1-40}$ into isolated human brain capillaries could be inhibited by an excess of unlabeled antibody [63]. Figure 3b demonstrates that binding to amyloid deposits of the ^{125}I-$A\beta$ peptide is not adversely affected by coupling to the vector. The conjugate was injected into a rhesus monkey and plasma kinetics and brain uptake were studied. As shown in figure 3c, the systemic clearance of the chimeric peptide was accelerated compared to the free ^{125}I-biotinylated $A\beta_{1-40}$. Figure 3d is an autoradiogram obtained from a cryostat brain section of this monkey 3 h after tracer injection. It shows significant delivery of peptide-associated radioactivity into brain. The tissue distribution reflects the higher capillary density in grey versus white matter. The free peptide did not accumulate in brain in a measurable amount.

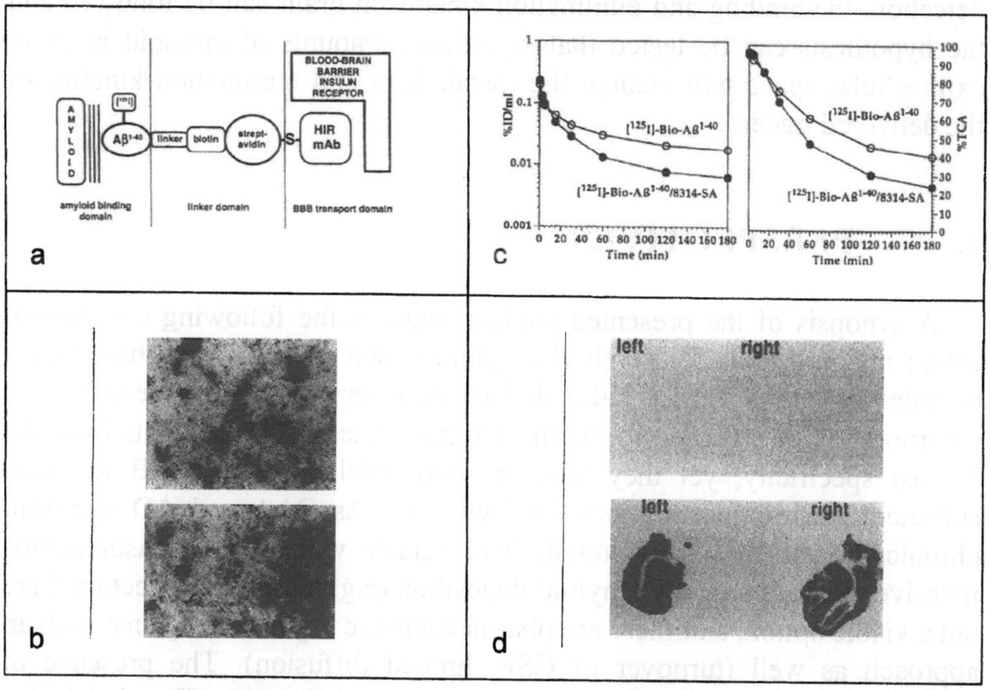

Figure 3. Amyloid imaging approach with vector mediated brain delivery of radiolabeled $A\beta_{1\text{-}40}$: *(a)* The tracer, ^{125}I-labeled Aβ-peptide, is attached through a biotin-streptavidin linker domain, to a monoclonal antibody (HIRmAb = anti-human insulin receptor monoclonal antibody) that provides transcytosis through the BBB. *(b)* Emulsion autoradiographs: Attachment to the vector does not interfere with binding of labeled $A\beta_{1\text{-}40}$ to amyloid deposits on frozen sections of AD brain. Upper panel = free peptide , lower panel = vector bound ^{125}I-Aβ. *(c)* Plasma profile after i.v. bolus injection in rhesus monkeys: (left) More rapid systemic clearance of the chimeric peptide compared to free Aβ, due to vector mediated cellular uptake, is also reflected in more rapid decline in the TCA precipitable fraction (right). *(d)* No radioactivity was seen in brain 3 h after i.v. injection of 300 μCi ^{125}I-$A\beta_{1\text{-}40}$ in a rhesus monkey (upper panel). In contrast, brain uptake is high in another animal after injection of the same dose of Aβ bound to the delivery vector. Quantitative autoradiographies (Phosphorimager) of sections of left and right occipital lobes are shown, respectively. From Wu *et al.* [63].

Other monkeys sacrificed at later time points (24 h, 48 h) revealed a decline in tissue radioactivity concentrations consistent with elimination of the peptide radiopharmaceutical from brain with a half-life of 16 h. These studies were performed in adult rhesus monkeys. Corresponding studies in very old animals who are prone to develop cerebral amyloid have not yet been reported. Based on the *in vitro* evidence that the radiopharmaceutical binds avidly to amyloid deposits and that the chimeric peptide is transcytosed through the BBB into brain extracellular space *in vivo*, it is hypothesized that this vector-mediated delivery of Aβ peptide will label cerebrovascular and parenchymal amyloid. Provided the radiopharmaceutical is labeled with a suitable radionuclide for external detection, the binding and elimination kinetics in brain can be followed and the hypothesis can be tested that significant amounts of amyloid in brain extracellular space will change the metabolism and elimination kinetics of the delivered tracer.

7. PERSPECTIVES

A synopsis of the presented studies suggests the following conclusion: future imaging tools for cerebral amyloid which utilize tracers based on a peptide structure (e.g., radiolabeled Aβ) or antibodies require the incorporation of a successful brain delivery strategy. These agents have the desired specificity, yet they lack the permeability at the BBB to reach sufficient tissue concentrations on their own. As CAA and AD represent chronic diseases requiring longitudinal studies with serial measurements, invasive methods to study amyloid deposition (e.g. intrathecal injections) are not a viable option, and there are pharmacokinetic arguments against such an approach as well (turnover of CSF, limited diffusion). The presence of transport mechanisms at the BBB for peptides and proteins offers the chance to utilize them for delivery purposes. Brain delivery of chimeric peptides exploits this principle and has shown its potential in preclinical studies. In the further development for clinical applications, the choice of a suitable radioisotope and minimization of potential immunogenicity by humanization of monoclonal antibodies used as delivery vectors should be addressed.

REFERENCES

1. Small, G. W. (1998) Differential diagnosis and early detection of dementia, *Am. J. Geriatr. Psychiatry* **6**, S26-33.
2. Geldmacher, D. S. and Whitehouse, P. J., Jr. (1997) Differential diagnosis of Alzheimer's disease, *Neurology* **48**, S2-9.
3. Duncan, B. A. and Siegal, A. P. (1998) Early diagnosis and management of Alzheimer's disease, *J. Clin. Psychiatry* **59**, 15-21.
4. Ross, G. W., Abbott, R. D., Petrovitch, H., Masaki, K. H., Murdaugh, C., Trockman, C., Curb, J. D. and White, L. R. (1997) Frequency and characteristics of silent dementia among elderly Japanese- American men. The Honolulu-Asia Aging Study, *Jama* **277**, 800-805.
5. Doraiswamy, P. M., Steffens, D. C., Pitchumoni, S. and Tabrizi, S. (1998) Early recognition of Alzheimer's disease: what is consensual? What is controversial? What is practical?, *J. Clin. Psychiatry* 59 Suppl 13, 6-18.
6. Greenberg, S. M. (1998) Cerebral amyloid angiopathy: prospects for clinical diagnosis and treatment, *Neurology* **51**, 690-694.

7. Hawkins, P. N., Lavender, J. P. and Pepys, M. B. (1990) Evaluation of systemic amyloidosis by scintigraphy with 123I-labeled serum amyloid P component, *N. Engl. J. Med.* **323**, 508-513.
8. Tjernberg, L. O., Pramanik, A., Bjorling, S., Thyberg, P., Thyberg, J., Nordstedt, C., Berndt, K. D., Terenius, L. and Rigler, R. (1999) Amyloid beta-peptide polymerization studied using fluorescence correlation spectroscopy, *Chem. Biol.* **6**, 53-62.
9. Hulstaert, F., Blennow, K., Ivanoiu, A., Schoonderwaldt, H. C., Riemenschneider, M., De Deyn, P. P., Bancher, C., Cras, P., Wiltfang, J., Mehta, P. D., Iqbal, K., Pottel, H., Vanmechelen, E. and Vanderstichele, H. (1999) Improved discrimination of AD patients using beta-amyloid(1-42) and tau levels in CSF, *Neurology* **52**, 1555-1562.
10. Weller, R. O., Massey, A., Newman, T. A., Hutchings, M., Kuo, Y. M. and Roher, A. E. (1998) Cerebral amyloid angiopathy: amyloid beta accumulates in putative interstitial fluid drainage pathways in Alzheimer's disease, *Am. J. Pathol.* **153**, 725-733.
11. Weller, R. O. (1998) Pathology of cerebrospinal fluid and interstitial fluid of the CNS: significance for Alzheimer disease, prion disorders and multiple sclerosis, *J. Neuropathol. Exp. Neurol.* **57**, 885-894.
12. Kaplan, B., Haroutunian, V., Koudinov, A., Patael, Y., Pras, M. and Gallo, G. (1999) Biochemical assay for amyloid beta deposits to distinguish Alzheimer's disease from other dementias, *Clin. Chim. Acta* **280**, 147-159.
13. Jobst, K. A., Smith, A. D., Szatmari, M., Molyneux, A., M.E., E., King, E., Smith, A., Jaskowski, A., McDonald, B. and Wald, N. (1992) Detection in life of confirmed Alzheimer`s disease using a simple measurement of medial temporal lobe atrophy by computed tomography, *Lancet* **340**, 1179-1182.
14. Jack, C. R., Jr., Petersen, R. C., Xu, Y. C., Waring, S. C., O'Brien, P. C., Tangalos, E. G., Smith, G. E., Ivnik, R. J. and Kokmen, E. (1997) Medial temporal atrophy on MRI in normal aging and very mild Alzheimer's disease, *Neurology* **49**, 786-794.
15. Juottonen, K., Laakso, M. P., Partanen, K. and Soininen, H. (1999) Comparative MR analysis of the entorhinal cortex and hippocampus in diagnosing Alzheimer disease, *Am. J. Neuroradiol.* **20**, 139-144.
16. Fox, N. C., Warrington, E. K., Freeborough, P. A., Hartikainen, P., Kennedy, A. M., Stevens, J. M. and Rossor, M. N. (1996) Presymptomatic hippocampal atrophy in Alzheimer's disease. A longitudinal MRI study, *Brain* **119**, 2001-2007.
17. Scheltens, P. (1999) Early diagnosis of dementia: neuroimaging, *J. Neurol.* **246**, 16-20.
18. Pardridge, W.M., Vinters, H.V., Yang, J., Eisenberg, J., Choi, T.B., Tourtellotte, W.W., Huebner, V. and Shively J.E. (1987) Amyloid angiopathy of Alzheimer's disease: Amino acid composition and partial sequence of a 4,200-Dalton peptide isolated from cortical microvessels. *J.Neurochem.* **49**, 1394-1401.
19. Vinters, H.V., Pardridge, W.M., Secor, D.L. and Ishii, N. (1988) Immunohistochemical study of cerebral amyloid angiopathy. II. Enhancement of immunostaining using formic acid pretreatment of tissue sections. *Am. J. Pathol.* **133**, 150-162.
20. Wisniewski, H. M., Wegiel, J., Wang, K. C. and Lach, B. (1992) Ultrastructural studies of the cells forming amyloid in the cortical vessel wall in Alzheimer´s disease, *Acta Neuropathol.* **84**, 117-127.
21. Yamaguchi, H., Yamazaki, T., Lemere, C. A., Frosch, M. P. and Selkoe, D. J. (1992) Beta amyloid is focally deposited within the outer basement membrane in the amyloid angiopathy of Alzheimer's disease. An immunoelectron microscopic study, *Am. J. Pathol.* **141**, 249-259.
22. Stewart, P. A., Hayakawa, K., Akers, M. A. and Vinters, H. V. (1992) A morphometric study of the blood-brain barrier in Alzheimer's disease, *Lab. Invest.* **67**, 734-742.

23. Kalaria, R. N. and Pax, A. B. (1995) Increased collagen content of cerebral microvessels in Alzheimer's disease, *Brain. Res.* **705**, 349-352.
24. Kalaria, R. N. and Grahovac, I. (1990) Serum amyloid P immunoreactivity in hippocampal tangles, plaques and vessels: implications for leakage across the blood-brain barrier in Alzheimer's disease, *Brain. Res.* **516**, 349-353.
25. Verbeek, M. M., Otte-Holler, I., Veerhuis, R., Ruiter, D. J. and De Waal, R. M. (1998) Distribution of A beta-associated proteins in cerebrovascular amyloid of Alzheimer's disease, *Acta. Neuropathol. (Berl)* **96**, 628-636.
26. Schlageter, N. L., Carson, R. E. and Rapoport, S. I. (1987) Examination of blood-brain barrier permeability in dementia of the Alzheimer type with [68Ga]EDTA and positron emission tomography, *J. Cereb. Blood Flow Metab.* **7**, 1-8.
27. Caserta, M. T., Caccioppo, D., Lapin, G. D., Ragin, A. and Groothuis, D. R. (1998) Blood-brain barrier integrity in Alzheimer's disease patients and elderly control subjects, *J. Neuropsychiatry Clin. Neurosci.* **10**, 78-84.
28. Munoz, D. G., Erkinjuntti, T., Gaytan-Garcia, S. and Hachinski, V. (1997) Serum protein leakage in Alzheimer's disease revisited, *Ann. N.Y. Acad. Sci.* **826**, 173-189.
29. Wisniewski, H. M., Vorbrodt, A. W. and Wegiel, J. (1997) Amyloid angiopathy and blood-brain barrier changes in Alzheimer's disease, *Ann. N.Y. Acad. Sci.* **826**, 161-172.
30. Walker, L. C. (1997) Animal models of cerebral beta-amyloid angiopathy, *Brain Res. Brain Res. Rev.* **25**, 70-84.
31. Nakamura, S., Tamaoka, A., Sawamura, N., Shoji, S., Nakayama, H., Ono, F., Sakakibara, I., Yoshikawa, Y., Mori, H., Goto, N. and *et al.* (1995) Carboxyl end-specific monoclonal antibodies to amyloid beta protein (A beta) subtypes (A beta 40 and A beta 42(43)) differentiate A beta in senile plaques and amyloid angiopathy in brains of aged cynomolgus monkeys, *Neurosci. Lett.* **201**, 151-154.
32. Wyss-Coray, T., Masliah, E., Mallory, M., McConlogue, L., Johnson-Wood, K., Lin, C. and Mucke, L. (1997) Amyloidogenic role of cytokine TGF-beta1 in transgenic mice and in Alzheimer's disease, *Nature* **389**, 603-606.
33. Klunk, W. E., Debnath, M. L. and Pettegrew, J. W. (1994) Development of small molecule probes for the beta-amyloid protein of Alzheimer's disease, *Neurobiol. Aging* **15**, 691-698.
34. Klunk, W. E., Debnath, M. L. and Pettegrew, J. W. (1995) Chrysamine-G binding to Alzheimer and control brain: autopsy study of a new amyloid probe, *Neurobiol. Aging* **16**, 541-548.
35. Roher, A. E., Lowenson, J. D., Clarke, S., Woods, A. S., Cotter, R. J., Gowing, E. and Ball, M. J. (1993) beta-Amyloid-(1-42) is a major component of cerebrovascular amyloid deposits: implications for the pathology of Alzheimer disease, *Proc. Natl. Acad. Sci. U.S.A.* **90**, 10836-10840.
36. Murphy, G. M., Jr., Forno, L. S., Higgins, L., Scardina, J. M., Eng, L. F. and Cordell, B. (1994) Development of a monoclonal antibody specific for the COOH-terminal of beta-amyloid 1-42 and its immunohistochemical reactivity in Alzheimer's disease and related disorders, *Am. J. Pathol.* **144**, 1082-1088.
37. Walker, L. C., Price, D. L., Voytko, M. L. and Schenk, D. B. (1994) Labeling of cerebral amyloid *in vivo* with a monoclonal antibody, *J. Neuropathol. Exp. Neurol.* **53**, 377-383.
38. Seubert, P., Vigo-Pelfrey, C., Esch, F., Lee, M., Dovey, H., Davis, D., Sinha, S., Schlossmacher, M., Whaley, J., Swindlehurst, C. and *et al.* (1992) Isolation and quantification of soluble Alzheimer's beta-peptide from biological fluids, *Nature* **359**, 325-327.
39. Pardridge, W. M. (1991) *Peptide drug delivery to the brain*, Raven Press, New York.

40. Davson, H., Welch, K. and Segal, M. B., Eds. (1987) *Secretion of the cerebrospinal fluid*, The Physiology and Pathophysiology of the Cerebrospinal Fluid, Churchill Livingstone London.
41. Majocha, R. E., Reno, J. M., Friedland, R. P., VanHaight, C., Lyle, L. R. and Marotta, C. A. (1992) Development of a monoclonal antibody specific for beta/A4 amyloid in Alzheimer's disease brain for application to *in vivo* imaging of amyloid angiopathy, *J. Nucl. Med.* **33**, 2184-2189.
42. Friedland, R. P., Majocha, R. E., Reno, J. M., Lyle, L. R. and Marotta, C. A. (1994) Development of an anti-A beta monoclonal antibody for *in vivo* imaging of amyloid angiopathy in Alzheimer's disease, *Mol. Neurobiol.* **9**, 107-113.
43. Majocha, R. E., Benes, F. M., Reifel, J. L., Rodenrys, A. M. and Marotta, C. A. (1988) Laminar-specific distribution and infrastructural detail of amyloid in the Alzheimer disease cortex visualized by computer-enhanced imaging of epitopes recognized by monoclonal antibodies, *Proc. Natl. Acad. Sci. U.S.A.* **85**, 6182-6186.
44. Friedland, R. P., Kalaria, R., Berridge, M., Miraldi, F., Hedera, P., Reno, J., Lyle, L. and Marotta, C. A. (1997) Neuroimaging of vessel amyloid in Alzheimer's disease, *Ann. N.Y. Acad. Sci.* **826**, 242-247.
45. Kumagai, A. K., Eisenberg, J. and Pardridge, W. M. (1987) Absorptive-mediated endocytosis of cationized albumin and a β-endorphin-cationized albumin chimeric peptide by isolated brain capillaries. Model system of blood-brain barrier transport, *J. Biol. Chem.* **262**, 15214-15219.
46. Triguero, D., Buciak, J. B., Yang, J. and Pardridge, W. M. (1989) Blood-brain barrier transport of cationized immunoglobulin G: enhanced delivery compared to native protein, *Proc. Natl. Acad. Sci. U.S.A.* **86**, 4761-4765.
47. Bickel, U., Lee, V. M., Trojanowski, J. Q. and Pardridge, W. M. (1994) Development and *in vitro* characterization of a cationized monoclonal antibody against beta A4 protein: a potential probe for Alzheimer's disease, *Bioconjug. Chem.* **5**, 119-125.
48. Triguero, D., Buciak, J. and Pardridge, W. M. (1990) Capillary depletion method for quantification of blood-brain barrier transport of circulating peptides and plasma proteins, *J. Neurochem.* **54**, 1882-1888.
49. Stern, R. A., Otvos, L., Jr., Trojanowski, J. Q. and Lee, V. M. (1989) Monoclonal antibodies to a synthetic peptide homologous with the first 28 amino acids of Alzheimer's disease beta-protein recognize amyloid and diverse glial and neuronal cell types in the central nervous system, *Am. J. Pathol.* **134**, 973-978.
50. Bickel, U., Lee, V. M. Y. and Pardridge, W. M. (1995) Pharmacokinetic differences between ^{111}In- and ^{125}I-labeled cationized monoclonal antibody against β–amyloid in mouse and dog, *Drug Delivery* **2**, 128-135.
51. Triguero, D., Buciak, J. L. and Pardridge, W. M. (1991) Cationization of immunoglobulin G results in enhanced organ uptake of the protein after intravenous administration in rats and primate, *J. Pharmacol. Exp. Ther.* **258**, 186-192.
52. Ashton, A. W., Boehm, M. K., Gallimore, J. R., Pepys, M. B. and Perkins, S. J. (1997) Pentameric and decameric structures in solution of serum amyloid P component by X-ray and neutron scattering and molecular modelling analyses, *J. Mol. Biol.* **272**, 408-422.
53. Hawkins, P. N., Tyrell, P., Jones, T. and al., e. (1991) Metabolic and scintigraphic studies with radiolabeled serum amyloid P component in amyloidosis: applications to cerebral deposits and Alzheimer disease with positron emission tomography, *Bull. Clin. Neurosci.* **56**, 178-190.

54. Lovat, L. B., O'Brien, A. A., Armstrong, S. F., Madhoo, S., Bulpitt, C. J., Rossor, M. N., Pepys, M. B. and Hawkins, P. N. (1998) Scintigraphy with 123I-serum amyloid P component in Alzheimer disease, *Alzheimer Dis. Assoc. Disord.* **12**, 208-210.
55. Maggio, J. E., Stimson, E. R., Ghilardi, J. R., Allen, C. J., Dahl, C. E., Whitcomb, D. C., Vigna, S. R., Vinters, H. V., Labenski, M. E. and Mantyh, P. W. (1992) Reversible *in vitro* growth of Alzheimer disease beta-amyloid plaques by deposition of labeled amyloid peptide, *Proc. Natl. Acad. Sci. U.S.A.* **89**, 5462-5466.
56. Nordstedt, C., Naslund, J., Tjernberg, L. O., Karlstrom, A. R., Thyberg, J. and Terenius, L. (1994) The Alzheimer A beta peptide develops protease resistance in association with its polymerization into fibrils, *J. Biol. Chem.* **269**, 30773-30776.
57. Zlokovic, B. V., Ghiso, J., Mackic, J. B., McComb, J. G., Weiss, M. H. and Frangione, B. (1993) Blood-brain barrier transport of circulating Alzheimer's amyloid beta, *Biochem. Biophys. Res. Commun.* **197**, 1034-1040.
58. Banks, W. A., Kastin, A. J., Barrera, C. M. and Maness, L. M. (1991) Lack of saturable transport across the blood-brain barrier in either direction for beta-amyloid1-28 (Alzheimer's disease protein), *Brain Res. Bull.* **27**, 819-823.
59. Saito, Y., Buciak, J., Yang, J. and Pardridge, W. M. (1995) Vector-mediated delivery of 125I-labeled beta-amyloid peptide A beta 1- 40 through the blood-brain barrier and binding to Alzheimer disease amyloid of the A beta $_{1-40}$/vector complex, *Proc. Natl. Acad. Sci. U.S.A.* **92**, 10227-10231.
60. Mackic, J. B., Stins, M., McComb, J. G., Calero, M., Ghiso, J., Kim, K. S., Yan, S. D., Stern, D., Schmidt, A. M., Frangione, B. and Zlokovic, B. V. (1998) Human blood-brain barrier receptors for Alzheimer's amyloid-beta 1- 40. Asymmetrical binding, endocytosis, and transcytosis at the apical side of brain microvascular endothelial cell monolayer, *J. Clin. Invest.* **102**, 734-743.
61. Zlokovic, B. V. (1996) Cerebrovascular transport of Alzheimer's amyloid beta and apolipoproteins J and E: possible anti-amyloidogenic role of the blood- brain barrier, *Life Sci.* **59**, 1483-1497.
62. Ghilardi, J. R., Catton, M., Stimson, E. R., Rogers, S., Walker, L. C., Maggio, J. E. and Mantyh, P. W. (1996) Intra-arterial infusion of [125I]A beta $_{1-40}$ labels amyloid deposits in the aged primate brain *in vivo*, *Neuroreport* **7**, 2607-2611.
63. Wu, D., Yang, J. and Pardridge, W. M. (1997) Drug targeting of a peptide radiopharmaceutical through the primate blood-brain barrier *in vivo* with a monoclonal antibody to the human insulin receptor, *J. Clin. Invest.* **100**, 1804-1812.
64. Zlokovic, B. V., Martel, C. L., Matsubara, E., McComb, J. G., Zheng, G., McCluskey, R. T., Frangione, B. and Ghiso, J. (1996) Glycoprotein 330/megalin: probable role in receptor-mediated transport of apolipoprotein J alone and in a complex with Alzheimer disease amyloid beta at the blood-brain and blood-cerebrospinal fluid barriers, *Proc. Natl. Acad. Sci. U.S.A.* **93**, 4229-4234.
65. Ghiso, J., Matsubara, E., Koudinov, A., Choi-Miura, N. H., Tomita, M., Wisniewski, T. and Frangione, B. (1993) The cerebrospinal-fluid soluble form of Alzheimer's amyloid beta is complexed to SP-40,40 (apolipoprotein J), an inhibitor of the complement membrane-attack complex, *Biochem. J.* **293**, 27-30.
66. Koudinov, A., Matsubara, E., Frangione, B. and Ghiso, J. (1994) The soluble form of Alzheimer's amyloid beta protein is complexed to high density lipoprotein 3 and very high density lipoprotein in normal human plasma, *Biochem. Biophys. Res. Commun.* **205**, 1164-1171.

67. Shayo, M., McLay, R. N., Kastin, A. J. and Banks, W. A. (1997) The putative blood-brain barrier transporter for the beta-amyloid binding protein apolipoprotein j is saturated at physiological concentrations, *Life Sci.* **60**, L115-118.
68. Jefferies, W. A., Brandon, M. R., Hunt, S. V., Williams, A. F., Gatter, K. C. and Mason, D. Y. (1984) Transferrin receptor on endothelium of brain capillaries, *Nature* **312**, 162-163.
69. Fishman, J. B., Rubin, J. B., Handrahan, J. V., Connor, J. R. and Fine, R. E. (1987) Receptor-mediated transcytosis of transferrin across the blood-brain barrier, *J. Neurosci. Res.* **18**, 299-304.
70. Pardridge, W. M., Buciak, J. L. and Friden, P. M. (1991) Selective transport of anti-transferrin receptor antibody through the blood-barain barrier *in vivo*, *J. Pharmacol. Exp. Ther.* **259**, 66-70.
71. Moos, T. (1996) Immunohistochemical localization of intraneuronal transferrin receptor immunoreactivity in the adult mouse central nervous system, *J. Comp. Neurol.* **375**, 675-692.
72. Kissel, K., Hamm, S., Schulz, M., Vecchi, A., Garlanda, C. and Engelhardt, B. (1998) Immunohistochemical localization of the murine transferrin receptor (TfR) on blood-tissue barriers using a novel anti-TfR monoclonal antibody, *Histochem. Cell Biol.* **110**, 63-72.
73. Friden, P. M., Walus, L. R., Watson, P., Doctrow, S. R., Kozarich, J. W., Backman, C., Bergman, H., Hoffer, B., Bloom, F. and Granholm, A. C. (1993) Blood-brain barrier penetration and *in vivo* activity of an NGF conjugate, *Science* **259**, 373-377.
74. Pardridge, W. M., Eisenberg, J. and Yang, J. (1987) Human blood-brain barrier transferrin receptor, *Metabolism* **36**, 892-895.
75. Yoshikawa, T. and Pardridge, W. M. (1992) Biotin delivery to brain with a covalent conjugate of avidin and a monoclonal antibody to the transferrin receptor, *J. Pharmacol. Exp. Ther.* **263**, 897-903.
76. Bickel, U., Yoshikawa, T., Landaw, E. M., Faull, K. F. and Pardridge, W. M. (1993) Pharmacologic effects *in vivo* in brain by vector-mediated peptide drug delivery, *Proc. Natl. Acad. Sci. U.S.A.* **90**, 2618-2622.
77. Wu, D. and Pardridge, W. M. (1996) Central nervous system pharmacologic effect in conscious rats after intravenous injection of a biotinylated vasoactive intestinal peptide analog coupled to a blood-brain barrier drug delivery system, *J. Pharmacol. Exp. Ther.* **279**, 77-83.
78. Pardridge, W. M., Kang, Y. S., Buciak, J. L. and Yang, J. (1995) Human insulin receptor monoclonal antibody undergoes high affinity binding to human brain capillaries *in vitro* and rapid transcytosis through the blood-brain barrier *in vivo* in the primate, *Pharm. Res.* **12**, 807-816.

Chapter 3

VASCULAR RISK FACTORS FOR ALZHEIMER'S DISEASE

An epidemiologic perspective

MONIQUE M.B. BRETELER
Department of Epidemiology & Biostatistics, Erasmus Medical Center Rotterdam, Rotterdam, the Netherlands

Vascular disease and Alzheimer's disease are both common disorders, in particularly among elderly subjects. Therefore, it can be expected that the joint occurrence of these two disorders is not a rare phenomenon. In recent years, evidence is increasing that the two may be more closely linked than just by chance. Epidemiological studies have suggested that the risk factors for vascular disease and stroke are associated with cognitive impairment and Alzheimer's disease, and that the presence of cerebrovascular disease intensifies the presence and severity of the clinical symptoms of Alzheimer's disease. In this chapter, current knowledge on the relation between vascular risk factors and Alzheimer's disease is reviewed.

1. INTRODUCTION

Alzheimer's disease is a heterogeneous and multifactorial disorder. Its frequency increases strongly with age, from less than 1% in people aged 65 years to over 25% in those who are aged 85 years or over. Reliable age- and gender-specific estimates of the incidence of Alzheimer's disease exist only for subjects over the age of 60. Gender-specific incidence rates are quite similar till the age of 85 years, after which the incidence rates seem higher for women than for men [1,2]. One in eight men, and almost one in four women, will suffer at least some of their lifetime from Alzheimer's disease. Over the last decade, epidemiologic evidence is accumulating that, in particular in elderly subjects, vascular risk factors and indicators of vascular disease are associated with cognitive impairment and Alzheimer's disease [3,4], and that presence of cerebrovascular disease intensifies the presence and severity of the clincial symptoms of Alzheimer's disease [5]. In this

M.M. Verbeek et al. (eds.),
***Cerebral Amyloid Angiopathy in Alzheimer's Disease and Related Disorders*, 43–58.**

chapter current epidemiological evidence for a relation between vascular risk factors and Alzheimer's disease will be reviewed. The focus will be on 'classical' vascular risk factors, including hypertension, diabetes mellitus, cholesterol, presence of atherosclerosis, atrial fibrillation, and cigarette smoking. In addition, more recently identified vascular risk factors will be reviewed, including *APOE* genotype, serum homocysteine concentration, relative abnormalities in the hemostatic and thrombotic systems, inflammation and alcohol consumption.

2. HYPERTENSION.

Hypertension is one of the most important risk factors for stroke and coronary heart disease and is an important risk factor for vascular dementia. The relation with Alzheimer's disease is less unequivocal. In a longitudinal study reported by Skoog *et al.* both systolic and diastolic blood pressure was increased 10-15 years before the onset of AD [6]. In that study, the risk of developing dementia between the ages of 80 and 85 increased with increasing blood pressure at the age of 70. Interestingly, individuals with hypertension also exhibit increased amounts of senile plaques and neurofibrillary tangles [7]. However, results from cross-sectional studies are inconclusive. Both high [8,9] and low [10] blood pressure have been associated with cognitive impairment, dementia and subtypes of dementia. Others found no association [11,12]. Two longitudinal studies that looked at the relation between blood pressure and cognitive function over a longer time period, reported, consistently with Skoog's findings, that increased systolic blood pressure predicted reduced cognitive function 20 years later [13,14]. But when in these cohorts the blood pressure as measured at the time of cognitive testing at high age was evaluated in a cross-sectional analysis, lower blood pressure were associated with lower cognitive performance or dementia [13-15]. A possible explanation for some of these discrepant findings is that high blood pressure does indeed increase the risk of Alzheimer's disease over a long period, but that in the years directly preceding clinical onset of dementia the blood pressure starts to decline and declines further with further progression of the disease, resulting in a seemingly protective effect of higher blood pressures over short follow-up periods. At first sight, however, this explanation would not fit the results of the Syst-EUR Trial, where antihypertensive treatment in elderly people with isolated systolic hypertension was associated with a lower incidence of dementia over a short, two-year period [16]. However, there is still much debate on whether the results of this trial should be interpreted as supporting the notion that high blood pressure increases the risk of dementia, including

Alzheimer's disease, or rather that calcium antagonists have a protective effect on dementia through a different and hitherto unknown mechanism [17]. Clearly, more prospective studies with a longer follow up are required to settle the issue how blood pressure levels relate to the risk of Alzheimer's disease and how they may change as a result of the disease.

3. DIABETES MELLITUS

Some older cross-sectional case-control studies showed diabetes mellitus to be positively associated with vascular dementia, but inversely with Alzheimer's disease [18-22]. However, these studies were based on selected patients and controls, the presence of diabetes mellitus was assessed from medical records and not actually screened for, and subjects with any indication of vascular disease were rigorously excluded from the patient series. More recent studies, both cross-sectional [23] and longitudinal studies [24-26], show a positive association between diabetes mellitus and Alzheimer's disease. A register-based retrospective cohort study from Rochester, MN, reported an increased risk for dementia and its subtype Alzheimer's disease (RR 1.7; 95% CI 1.3-2.1) [24]. In a prospective, population-based study among 828 elderly Japanese, diabetics had a more than twofold increased risk of AD (RR 2.2; 95% CI 1.0-4.9) [25]. Similarly, among 6370 subjects of the Rotterdam Study who were followed prospectively, a relative risk of Alzheimer's disease of 1.9 (95% CI 1.2-3.1) was found. The risk for Alzheimer's disease was highest among subjects who were treated with insulin (RR 4.3; 95% CI 1.7-10.5) [26]. On the other hand, no association was found in the Honolulu-Asia Aging Study between diabetes 15 or 25 years before the presence of Alzheimer's disease in later life. There was some indication in this study among 3774 Japanese-American men for the risk of vascular dementia to be increased with 50%, however, this was not statistically significant [27].

The mechanism through which diabetes mellitus increases the risk of Alzheimer's disease is not entirely clear. Evidently, diabetes is associated with micro- and macrovascular changes, and it may be that this vascular component adds to the clinical picture in Alzheimer's disease. However, the risk of dementia is increased independently from other vascular risk factors, and there might also be a direct effect of glucose on the cerebral vasculature. It has been suggested that in Alzheimer's disease the hyperglycemia could potentiate the neuronal death produced by other pathological processes taking place, such as amyloid deposition [28]. Another possible explanation points to a crucial role for insulin. Insulin and insulin-like growth factors and their receptors promote neuronal growth in the brain. Studies addressing the pathological mechanisms found evidence for a disturbed insulin signal transduction in AD brains and suggested that insulin-dependent functions may be of pathogenetic relevance in AD [29]. Finally, advanced glycation end products (AGEs) may play a crucial role. The formation of AGEs occurs during normal aging and at accelerated rate in diabetes mellitus. AGEs contribute to the pathogenesis of diabetic complications, but also play an important, although possibly not specific [30], role in the formation of the characteristic protein deposits in AD, including neurofibrillary tangles, beta-amyloid plaques and Hirano bodies [31].

4. CHOLESTEROL

The relation between plasma cholesterol levels and Alzheimer's disease is of interest. The epsilon 4 allele of the apolipoprotein E (ApoE) is associated with Alzheimer's disease (AD) and also with increased plasma cholesterol, low-density lipoprotein levels, atherosclerosis and cardiovascular disease [32,33]. Besides, Sparks *et al.* reported dose-dependent amyloid accumulations in the brains of rabbits that were fed a high-cholesterol diet [34], and more recently it was shown that cholesterol is required for amyloid β (Aβ) formation to occur and that reduction of cellular cholesterol levels inhibited the formation of Aβ [35]. There have been only few epidemiologic reports on the relation between cholesterol and Alzheimer's disease. A very small hospital-based case-control study suggested that Alzheimer patients have higher mean total cholesterol levels than controls [36], and it has been alluded to that the effect of the *APOE* genotype might depend on cholesterol levels [37].

In a population-based sample of 444 men, aged 70-89 years, who were survivors of the Finnish cohorts of the Seven Countries Study, previous high serum cholesterol level (mean level ≥ 6.5 mmol/l) was significantly associated with presence of AD (odds ratio 3.1; 95% CI 1.2-8.5) after controlling for age and *APOE* genotype. In men who subsequently developed AD the serum cholesterol level decreased before the clinical manifestations of AD [38]. Consistent with this latter finding, in the Rotterdam Study no relation was found between serum cholesterol levels and risk of dementia within 2 years (unpublished data). High dietary baseline intake of total fat, saturated fat and cholesterol, that reflects habitual dietary intake, did however increase the risk of dementia and Alzheimer's disease (RR = 2.4, 95% CI 1.1-5.2; RR = 1.9, 95% CI 0.9-4.0; and RR = 1.7, 95% CI 0.9-3.2, respectively). Fish consumption on the other hand, an important source of n-3 polyunsaturated fatty acids, was inversely related to incident dementia (RR = 0.4, 95% CI 0.2-0.91), and in particular to Alzheimer's disease (RR = 0.3, 95% CI 0.1-0.9) [39].

5. EXISTING CARDIOVASCULAR DISEASE

Clinical (cardio)vascular disease (CVD) has hardly been studied in relation to Alzheimer's disease. Because of the diagnostic criteria for Alzheimer's disease, patients with clinical vascular disease are less likely to be diagnosed as Alzheimer patients [40]. However, Aronson *et al* reported that coronary artery disease is common among women developing Alzheimer's disease [41]. In a study among 4,971 subjects those with overt or clinically silent vascular disease performed worse on cognitive tests than did their contemporaries without vascular pathology[3]. In a cross-sectional analysis from the Rotterdam Study, indicators of atherosclerosis of the carotid arteries (wall thickness and plaques as measured by ultra-sonography) and presence of atherosclerosis of the large vessels of the legs (assessed by the ratio of the ankle-to-brachial systolic blood pressure) were associated with Alzheimer's disease. The prevalence of Alzheimer's disease increased with the degree of atherosclerosis. The odds ratio for Alzheimer's disease in those with severe atherosclerosis was 3.0 (95% CI 1.5-6.0) as compared to those without atherosclerosis. A strong interaction between *APOE* ε4 and atherosclerosis was observed. Subjects with at least one *APOE* ε4 allele and severe atherosclerosis had a nearly 20 times increased risk for Alzheimer's disease [4]. The Cardiovascular Health Study followed 5888 subjects for 5-7 years and found that subclinical CVD, including peripheral vascular disease, and atherosclerosis of the common and internal carotid arteries, was associated with the rate of cognitive decline. Those with any *APOE* ε4 allele in combination with atherosclerosis, peripheral vascular disease or diabetes mellitus were at substantially higher risk of cognitive decline that those without the *APOE* ε4 allele or subclinical CVD [42].

6. ATRIAL FIBRILLATION

Cardiac dysrythmias have been suspected since long to aggravate or precipitate dementia [43], yet studies on cognitive performance or risk of dementia in patients with atrial fibrillation are rare. Recently, two studies reported worse cognitive performance in non-demented subjects with atrial fibrillation as compared to subjects without [44,45]. In the Rotterdam Study, atrial fibrillation as assessed in standard 12-lead ECGs, was significantly more frequent among subjects with dementia (age and gender adjusted odds ratio 2.3; 95% CI 1.4-3.7). The relation was slightly stronger for subjects with clinically diagnosed Alzheimer's disease than for subjects with vascular dementia, and could not be explained by a history of stroke. The associations were stronger in women than for men (age adjusted odds ratio for dementia 3.1 (95% CI 1.7-5.5) and 1.3 (95% CI 0.5-3.1), respectively) [46].

7. SMOKING

The relation between smoking and Alzheimer's disease is much disputed. A meta-analysis on case-control studies conducted before 1990 suggested an inverse association between Alzheimer's disease and history of smoking [47]. Studies conducted since then showed no relation [25,48-51], an inverse relation [52,53], or a positive relation [54]. Another meta-analysis yielded again an inverse relation between smoking history and Alzheimer's disease [55]. There are some biologically plausible explanations for a protective effect of smoking on Alzheimer's disease, including a memory and cognition enhancing effect of nicotine, possibly through increasing the density of nicotine receptors or facilitating the release of acetylcholine [56,57]. However, one should consider that at least some of the findings in previous studies of a protective effect of smoking resulted from bias [58]. Most previous studies were cross-sectional, and based on prevalent cases. A recent prospective analysis on elderly incident cases showed that people who smoked cigarettes had a more than twofold increased risk of Alzheimer's disease [59]. Besides bias, an explanation for the discrepant findings may be that some of the earlier studies that showed a protective effect of smoking included much younger cases. It is conceivable that the relation between smoking and Alzheimer's disease is age-dependent, for example because of different genetic susceptibility. Some support for this comes from the observation among early onset patients that the inverse association between smoking and Alzheimer's disease was limited to carriers of the apolipoprotein ε4 allele [60]. Smoking may exert different and opposite effects on the risk of Alzheimer's disease, being generally harmful through for example a vascular mechanism, but also partly beneficial in specific individuals, especially in those who carry an apolipoprotein ε4 allele. This hypothesis may be supported by the observation that in Alzheimer patients carriers of apolipoprotein ε4 have fewer nicotinic receptor binding sites and decreased activity of choline acetyltransferase, as compared to non-carriers of this allele [61]. The finding in the Rotterdam study that the elevated relative risk was particularly present for smokers without the apolipoprotein ε4 allele, is also in line with this hypothesis [59].

8. APOLIPOPROTEIN E

The apolipoprotein E genotype (*APOE*) has been studied in relation to vascular disease because of its central role in lipid metabolism [62]. The ε4 allele is associated with increased serum total cholesterol levels, and with increased risk of atherosclerosis and coronary artery disease [32,33]. The relation of ApoE with cerebrovascular disease is controversial. Much more is known about the relation between *APOE* genotype and Alzheimer's disease. Since the first reports of a link between *APOE* ε4 and Alzheimer's disease in 1993 [63], this association has been confirmed by many authors and in many different populations [64-66]. Several studies suggested possible interaction between *APOE* genotype and environmental risk factors in increasing the risk of Alzheimer's disease [64], including interaction with vascular risk factors [4]. The mechanism by which ApoE exerts its effect in relation to dementia is not clear. Apolipoprotein E is involved in various aspects of neurodegeneration and repair, and possible mechanisms include a role in Aβ deposition in plaques, the development of neurofibrillary tangles, lipid transport, neurotrophic effects, or antioxidant activity [66]. An alternative hypothesis is therefore that not the *APOE**4 polymorphism is causally related to Alzheimer's disease, but rather other genetic defects that are located in or close to the *APOE* gene [67].

9. HOMOCYSTEINE

There is accumulating evidence that an increased blood level of homocysteine is a risk factor for cardiovascular disease [68] as well as cerebrovascular disease and stroke [69,70]. Plasma levels of vitamin B12 and folate are important determinants of plasma homocysteine concentration, and homocysteine concentration is a sensitive marker for vitamin B12 and folate deficiency. It is generally recommended to routinely determine serum vitamin B12 levels as part of the screening of demented patients, on the assumption that vitamin B12 deficiency is a possible cause of reversible dementia. However, this longstanding practice is not evidence-based. Some recent studies showed low vitamin B12 levels in 15 to 25 % of demented and Alzheimer patients [71,72]. Replacement did not result in clinical improvement, although it did have an effect on some neurological abnormalities [71-73]. A few case-control studies suggested a link between higher homocysteine levels and Alzheimer's disease [74] or dementia [75]. Also in a recent case control study among 164 AD patients and 108 controls, serum homocysteine levels were significantly higher and serum folate and vitamin B12 levels were lower in AD patients than in controls [76].

Moreover, some evidence was reported for a relation between higher concentrations of homocysteine and poorer performance on some cognitive tests in non-demented subjects [77,78]. Clearly, the possible link between homocysteine levels and Alzheimer's disease needs further investigation. The mechanisms by which homocysteinemia produces vascular damage are not completely understood, but endothelial injury is probably a central factor [79]. This is of particular interest in the light of suggestions of endothelial involvement in the etiology of Alzheimer's disease [80].

10. THROMBOSIS

Thrombosis plays a central role in the pathogenesis of vascular disease. This raises the question whether hemostatic status is also of importance for the development of dementia. A low anticoagulant response of plasma to activated protein C (APC), or APC resistance, is an abnormality of the coagulation system that increases the risk of venous thrombosis as well as stroke [81,82]. A low APC response is frequently due to the factor V Leiden mutation [83], however, no association between factor V Leiden and stroke has been found [84,85]. In a cross-sectional study based on prevalent cases of dementia of the Rotterdam Study, response to APC was not associated with dementia or its subtype Alzheimer's disease [86]. Carriers of the factor V Leiden mutation, had a more than twofold increased probability of being demented than those without that mutation which was borderline significant. Although the association seemed stronger for vascular dementia, it was present among Alzheimer patients as well [86]. Other abnormalities in the hemostatic and thrombotic systems, including hypercoagulability (as further assessed by thrombin/antithrombin complex (TAT); fibrinogen; prothrombin fragments 1+2) and abnormal endogeneous fibrinolytic activity (as measured through plasminogen activator inhibitor type 1 (PAI-1); tissue type plasminogen activator (t-PA); and D-dimer) were assessed in the same study population. TAT and t-PA were associated with both Alzheimer's disease and vascular dementia, suggesting that predominantly increased thrombin generation was associated with dementia [87].

11. INFLAMMATION

Inflammatory factors play a role in Alzheimer's disease and Aβ induces a local inflammatory reaction that contributes to the progression of the disease [80,88,89]. Epidemiologic studies have reported lower risk of Alzheimer's disease in subjects that used non steroid anti-inflammatory drugs [90,91].

Although these studies can be criticized on methodological grounds, and may be biased, they lend some support to a possible relation between chronic inflammation and Alzheimer's disease. Most basic research has concentrated on pathophysiological processes taking place in the brain. However, peripheral inflammatory processes might also play a role in the etiology of dementia and Alzheimer's disease. First, inflammatory processes are involved in the process of atherosclerosis [92], and there is increasing evidence that atherosclerosis and vascular disease play a role in the etiology of dementia and Alzheimer's disease. Second, it has been suggested that disturbances of the blood brain barrier play a role in the development of Alzheimer's disease, particularly in elderly subjects [93]. Peripheral inflammation factors could then leak to the brain parenchyma and induce the activation of microglial cells. Finally, there may be signalling of the brain by peripheral acute phase proteins through functional receptors on cerebral endothelial cells [94]. Although some studies reported elevated levels of α_1-antichymotrypsine (α_1-ACT), tumor necrosis factor-α (TNF-α) and interleukin 6 (IL-6) in Alzheimer patients, thusfar involvement of peripheral inflammatory processes is not consistently sustained by empirical evidence. However, only few studies have been conducted to date, all small and based on highly selected Alzheimer patients.

Despite several investigations over time, no specific infectious agents have been identified in relation to Alzheimer's disease[95]. It has been suggested, however, that interaction between herpes infections and genetic factors, may play a role in the etiology of Alzheimer's disease [96].

12. ALCOHOL

The relation between alcohol intake and the risk of vascular diseases, including stroke is J-shaped: Moderate alcohol consumption protects from vascular disease, but with increasing alcohol intake the risk gradually increases. Most studies reported no evidence for an altered risk of Alzheimer's disease in people with moderate alcohol intake [47,48,52], but alcohol abuse has been reported to significantly increase the risk of dementia or Alzheimer's disease [97,98]. Recently, the Paquid study, a population-based prospective study in Gironde, France, reported a protective effect from alcohol consumption, which in that study was almost exclusively wine consumption, on the risk of dementia [99]. Preliminary findings from the Rotterdam Study confirm this finding of a protective effect of moderate alcohol consumption on the risk of Alzheimer's disease [100].

13. CONCLUSION

It is relatively novel that etiological research in Alzheimer's disease also focusses on vascular risk factors, and available evidence is still limited. However, evidence is accumulating that vascular risk factors and vascular disease increase the risk of dementia, including Alzheimer's disease. Findings from other research areas corroborate the epidemiological reports of involvement of vascular disease processes in Alzheimer's disease. It is as yet unclear what underlies the associations between vascular risk factors and Alzheimer's disease. It may be that vascular disease and Alzheimer's disease have to some extent a shared etiology, and that the risk factors that they have in common increase the risk of both disorders independently. Other possibilities include that vascular factors are directly etiologically related to Alzheimer's disease or act as contributing factors in causing the disease. Finally, what is considered Alzheimer's disease on clinical ground, may actually be a mixed bag of dementing syndromes due to cerebrovascular pathology, Alzheimer pathology, or a combination of the two, and the association may be with the vascular component of the dementia syndromes rather than with the Alzheimer pathology. What is clear, though, is that what we clinically call Alzheimer's disease is a multifactorial disorder, in particular at older age; and that vascular pathology causes or contributes to the dementia syndrome in at least 50 to 60 % of all demented patients. It is therefore highly important to further elucidate the possible role of vascular pathology in dementia, irrespective of clinical subtypes, and to identify specific mechanisms.

REFERENCES

1. Fratiglioni, L., Viitanen, M., von Strauss, E., Tontodonati, V., Herlitz, A., and Winblad, B. (1997) Very old women at highest risk of dementia and Alzheimer's disease: incidence data from the Kungsholmen Project, Stockholm. *Neurology* **48**, 132-138.
2. Ott, A., Breteler, M.M.B., Van Harskamp, F., Stijnen, T., and Hofman, A. (1998) The incidence and risk of dementia. The Rotterdam Study, *Am. J. Epidemiol.* **197**, 574-580.
3. Breteler, M.M.B., Claus, J.J., Grobbee, D.E., and Hofman, A. (1994) Cardiovascular disease and the distribution of cognitive function in an elderly population. The Rotterdam Study, *BMJ* **308**, 1604-1608.
4. Hofman, A., Ott, A., Breteler, M.M.B., Bots, M.L., Slooter, A.J.C., van Harskamp, F., Van Duijn, C.M., Van Broeckhoven, C., and Grobbee, D.E. (1997) Atherosclerosis, apolopoprotein-E and the prevalence of dementia and Alzheimer's disease in the Rotterdam Study, *Lancet* **349**, 151-154.
5. Snowdon, D.A., Greiner, L.H., Mortimer, J.A., Riley, K.P., Greiner, P.A., and Markesbery, W.R. (1997) Brain infarction and the clinical expression of Alzheimer disease. The Nun Study, *JAMA* **277**, 813-817.

6. Skoog, I., Lernfelt, B., Landahl, S., Palmertz, B., Andreasson, L.A., Nilsson, L., *et al.* (1996) 15-year longitudinal study of blood pressure and dementia, *Lancet* **347**, 1141-1145.
7. Sparks, D.L., Scheff, S.W., Liu, H., Landers, T.M., Coyne, C.M., and Hunsaker III, J.C. (1995) Increased incidence of neurofibrillary tangles (NFT) in non-demented individuals with hypertension, *J. Neurol. Sci.* **131**, 162-169.
8. Cacciatore, F., Abete, P., Ferrara, N., Paolisso, G., Amato, L., Canonico, S., *et al.* (1997) The role of blood pressure in cognitive impairment in an elderly population. Osservatorio Geriatrico Campano Group, *J. Hypertens.* **15**, 135-142.
9. Starr, J.M., Whalley, L.J., Inch, S., and Shering, P.A. (1993) Blood pressure and cognitive function in healthy old people, *J. Am. Geriatr. Soc.* **41**, 753-756.
10. Guo, Z., Viitanen, M., Fratiglioni, L., and Winblad, B. (1996) Low blood pressure and dementia in elderly people: the Kungsholmen project, *BMJ* **312**, 805-808.
11. Scherr, P.A., Hebert, L.E., Smith, L.A., and Evans, D.A. (1991) Relation of blood pressure to cognitive function in the elderly, *Am. J. Epidemiol.* **134**, 1303-1315.
12. Farmer, M.E., White, L.R., Abbott, R.D., Kittner, S.J., Kaplan, E., Wolz, M.M., *et al.* (1987) Blood pressure and cognitive performance. The Framingham Study, *Am. J. Epidemiol.* **126**, 1103-1114.
13. Launer, L.J., Masaki, K., Petrovitch, H., Foley, D., and Havlik, R.J. (1995)The association between midlife blood pressure levels and late-life cognitive function. The Honolulu-Asia Aging Study, *JAMA* **274**, 1846-1851.
14. Kilander, L., Nyman, H., Boberg, M., Hansson, L., and Lithell, H. (1998) Hypertension is related to cognitive impairment: a 20-year follow-up of 999 men, *Hypertension* **31**, 780-786.
15. Skoog, I., Andreasson, L.A., Landahl, S., and Lernfelt, B. A population-based study on blood pressure and brain atrophy in 85-year-olds, *Hypertension* **32**, 404-409.
16. Forette, F., Seux, M.L., Staessen, J.A., Thijs, L., Birkenhager, W.H., Babarskiene, M.R., *et al.* (1998) Prevention of dementia in randomised double-blind placebo-controlled Systolic Hypertension in Europe (Syst-Eur) trial, *Lancet* **352**, 1347-1351.
17. (1999) Prevention of dementia: Syst-EUR trial. Correspondence, *Lancet* **353**, 235-237.
18. Bucht, G., Adolfsson, R., Lithner, R., and Winblad, B. (1983) Changes in blood glucose and insulin secretion in patients with senile dementia of Alzheimer type, *Acta Med. Scand.* **213**, 387-392.
19. Landin, K., Blennow, K., Wallin, A., and Gottfries, C.G. (1993) Low blood pressure and blood glucose levels in Alzheimer's disease. Evidence for a hypometabolic disorder? *J. Intern. Med.* **233**, 357-63.
20. Mortel, K.F., Wood, S., Pavol, M.A., Meyer, J.S., and Rexer, J.L. (1993) Analysis of familial and individual risk factors among patients with ischemic vascular dementia and Alzheimer's disease, *Angiology* **44**, 599-605.
21. Nielson, K.A., Nolan, J.H., Brechtold, N.C., Sandman, C.A., Mulnard, R.A., and Cotman, C.W. (1996) Apolopoprotein-E genotyping of diabetic dementia patients: is diabeters rare in Alzheimer's disease? *J. Am. Geriatr. Soc.* **44**, 897-904.
22. Wolf-Klein, G.P., Siverstone, F.A., Brod, M.S., Levy, A., Foley, C.J., Termotto, V., and Breuer, J. (1988) Are Alzheimer patients healthier? *J. Am. Geriatr. Soc.* **36**, 219-224.
23. Ott, A., Stolk, R.P., Hofman, A., van Harskamp, F., Grobbee, D.E., and Breteler, M.M.B. (1996) Association of diabetes mellitus and dementia: The Rotterdam study, *Diabetologia* **39**, 1392-1397.
24. Leibson, C.L., Rocca, W.A., Hanson, V.A., Cha, R., Kokmen, E., O'Brian, P.C., and Palumbo, P.J. (1997) Risk of dementia among persons with diabetes mellitus: A population-based cohort study, *Am. J. Epidemiol.* **145**, 301-308.

25. Yoshitake, T., Kiyohara, Y., Kato, I., Ohmura, T., Iwamoto, H., Nakayama, K., *et al.* (1995) Incidence and risk factors of vascular dementia and Alzheimer's disease in a defined elderly Japanese population: the Hisayama Study, *Neurology* **45**, 1161-1168.
26. Ott, A., Stolk, R.P., Van Harskamp, F., Pols, H.A.P., Hofman, A., and Breteler, M.M.B. (1999) Diabetes mellitus and the risk of dementia in a elderly population. The Rotterdam Study, *Neurology* (in press).
27. Curb, J.D., Rodrigues, B.L., Abbott, R.D., Petrovitch, H., Ross, G.W., Masaki, K.H., Foley, D., Blanchette, P.L., Harris, T., Chen, R., and White, L.R. (1999) Longitudinal association of vascular and Alzheimer's dementias, diabetes, and glucose tolerance, *Neurology* **52**, 971-975.
28. Messier, C., and Gagnon, M. 1996) Glucose regulation and cognitive functions: relation to Alzheimer's disease and diabetes, *Behav. Brain Res.* **75**, 1-11.
29. Frohlich, L., Blum-Degen, D., Bernstein, H.G., Engelsberger, S., Humrich, J., Laufer, S., Muschner, D., Thalheimer, A., Turk, A., Hoyer, S., Zochling, R., Boissl, K.W., Jellinger, K., and Riedere, P. (1998) Brain insulin and insulin receptors in aging sporadic Alzheimer's disease, *J. Neural. Transm.* **105**, 423-438.
30. Sasaki, N., Fukatsu, R., Tsuzuki, K., Hayashi, Y., Yoshida, T., Fujii, N., Koike, T., Wakayama, I., Yanagihara, R., Garruto, R., Amano, N., and Makita, Z. (1998) Advanced glycation end products in Alzheimer's disease and other neurodegenerative diseases, *Am. J. Pathol.* **153**, 1149-1155.
31. Munch, G., Cunningham, A.M., Riederer, P., and Braak, E. (1998) Advanced glycation endproducts are associated with Hirano bodies in Alzheimer's disease, *Brain. Res.* **796**, 307-310.
32. Van Bockxmeer, F.M., and Mamotte, .C.DS. (1992) Apolipoprotein ε4 homozygosity in young men with coronary heart disease, *Lancet* **340**, 879-880.
33. Davignon, J., Gregg, R.E., and Sing, C.F. (1988) Apolipoprotein E polymorphism and atherosclerosis, *Arteriosclerosis* **8**, 1-21.
34. Sparks, D.L., Scheff, S.W., Hunsaker, J.C. 3rd, Liu, H., Landers, T., and Gross, D.R. (1994) Induction of Alzheimer-like beta-amyloid immunoreactivity in the brains of rabbits with dietary cholesterol, *Exp. Neurol.* **126**, 8-94.
35. Simons, M., Keller, P., De Strooper, B., Beyreuther, K., Dotti, C.G., and Simons, K. (1998) Cholesterol depletion inhibits the generation of beta-amyloid in hippocampal neurons, *Proc. Natl. Acad. Sci.USA* **95**, 6460-6464.
36. Giubilei, F., Antona, R., Antonini, R., Lenzi, G.L., Ricci, G., and Fieschi, C. (1990) Serum lipoprotein pattern variations in dementia and ischemic stroke, *Acta Neurol. Scand.* **81**, 84-86.
37. Jarvik, G.P., Wijsman, E.M., Kukull, W.A., Schellenberg, G.D., Yu, C., andLarson, E.B. (1995) Interactions of apolipoprotein E genotype, total cholesterol level, age, and sex in prediction of Alzheimer's disease: a case-control study, *Neurology* **45**, 1092-1096.
38. Notkola, I.L., Sulkava, R., Pekkanen, J., Erkinjuntti, T., Ehnholm, C,. Kivinen, P., Tuomilehto, J, and Nissinen, A. (1998) Serum total cholesterol, apolipoprotein E epsilon 4 allele, and Alzheimer's disease, *Neuroepidemiology* **17**, 14-20.
39. Kalmijn, S., Launer, L.J., Ott, A., Witteman, J.C.M, Hofman, A., and Breteler, M.M.B. (1997) Dietary fat intake and the risk of incident dementia in the Rotterdam Study, *Ann. Neurol.* **42**, 776-782.
40. Breteler, M.M.B., Claus, J.J., Launer, L.J., van Duijn, C.M., and Hofman, A. (1992) Epidemiology of Alzheimer's disease, *Epidemiologic Reviews* **14**, 59-82.
41. Aronson, M.K., Ooi, W.L., Morgenstern, H., *et al.* (1990) Women, myocardial infarction, and dementia in the very old, *Neurology* **40**, 1102-1106.

42. Haan, M.N., Shemanski, L., Jagust, W.J., Manolio, T.A., and Kuller, L. (1999) The role of APOE epsilon4 in modulating effects of other risk factors for cognitive decline in elderly persons, *JAMA* **282**, 40-46.
43. Anonymous. (1977) Cardiogenic Dementia, *Lancet* **i**, 27-28.
44. O'Connell, J.E., Gray, C.S., French, J.M., and Robertson, I.H. (1998) Atrial fibrillation and cognitive function: case-control study, *J. Neurol. Neurosurg. Psychiatry* **65**, 386-389.
45. Kilander, L., Andren, B., Nyman, H., Lind, L., Boberg, M., and Lithell, H. (1998) Atrial fibrillation is an independent determinant of low cognitive function: a cross-sectional study in elderly men (see comments), *Stroke* **29**, 1816-1820.
46. Ott, A., Breteler, M.M.B, De Bruyne, M.C., Van Harskamp, F., Grobbee, D.E., and Hofman, A. (1997) Atrial Fibrillation and dementia in a Population-Based Study. The Rotterdam Study, *Stroke* **28**, 316-321.
47. Graves, A.B., van Duijn, C.M., Chandra, V., Fratiglioni, L., Heyman ,A., Jorm, A.F., Kokmen, E., Kondo, K., Mortimer, J.A., Rocca, W.A. *et al.* (1991) Alcohol and tobacco consumption as risk factors for Alzheimer s disease: a collaborative re-analysis of case-control studies. EURODEM Risk Factors Research Group, *Int. J. Epidemiol.* **20**, S48-57.
48. Anonymous. (1994) The Canadian Study of Health and Aging: risk factors for Alzheimer's disease in Canada, *Neurology* **44**, 2073-2080.
49. Forster, D.P., Newens, A.J., Kay, D.W., and Edwardson, J.A. (1995) Risk factors in clinically diagnosed presenile dementia of the Alzheimer type: a case-control study in northern England, *J. Epidemiol. Comm. Health* **49**, 253-258.
50. Hebert, L.E., Scherr, P.A., Beckett, L.A., Funkenstein, H.H., Albert, M.S., and Chown, M.J. (1992) Evans DA. Relation of smoking and alcohol consumption to incident Alzheimer's disease, *Am. J. Epidemiol.* **135**, 347-355.
51. Wang, H.X., Fratiglioni, L., Frisoni, G.B., Viitanen, M., and Winblad, B. (1999) Smoking and the occurrence of Alzheimer's disease: cross-sectional and longitudinal data in a population-based study, *Am. J. Epidemiol.* **149**, 640-644.
52. Brenner, D.E., Kukull, W.A., van Belle, G., *et al.* (1993) Relationship between cigarette smoking and Alzheimer's disease in a population-based case-control study, *Neurology* **43**, 293-300.
53. Letenneur, L., Dartigues, J.F., Commenges, D., Burberger-Gateau, P., Tessier, J.F., and Orgogozo, J.M. (1994) Tobacco consumption and cognitive impairment in elderly people. A population-based study, *Ann. Epidemiol.* **4**, 449-454.
54. Prince, M., Cullen, M., and Mann, A. (1994) Risk factors for Alzheimer's disease and dementia: A case-control study based on the MRC elderly hypertension trial, *Neurology* **44**, 97-104.
55. Lee, P.N. (1994) Smoking and Alzheimer's disease: A review of the epidemiological evidence,.*Neuroepidemiology* **13**, 131-144.
56. Newhouse, P.A., Potter, A., Corwin, J., and Lenox, R. (1994) Age-related effects of the nicotinic antagonist mecamylamine on cognition and behavior, *Neuropsychopharmacology* **10**, 930107.
57. Nitta, A., Katono, Y., Itoh, A., Hasegawa, T., and Nabeshima, T. (1994) Nicotine reverses scopolamine-induced impairment of performance in passive avoidance task in rats through its action on the dopaminergic neuronal system, *Pharmacol. Biochem. Behav.* **49**, 807-812.
58. Riggs, J.E. (1993) Smoking and Alzheimer's disease: protective effect or differential survival bias? *Lancet* **342**, 793-794.
59. Ott, A., Slooter, A.J.C., Hofman, A., van Harskamp, F., Witteman, J.C.M., van Broeckhoven, C., van Duijn, C.M., and Breteler, M.M.B. (1998) Smoking and the risk of

dementia and Alzheimer's disease in a population-based cohort study: the Rotterdam Study, *Lancet* **351**, 1840-1843.
60. van Duijn, C.M., Havekes, L.M., van Broeckhoven, C., De Knijff, P., and Hofman, A. (1995) Apolipoprotein E genotype and association between smoking and early onset Alzheimer's disease, *BMJ* **310**, 627-631.
61. Poirier, J., Delisle, M-C., Quirion, R., *et al.* (1995) Apolipoprotein ε4 allele as a predictor of cholinergic deficits and treatment outcome in Alzheimer disease, *Proc. Natl. Acad. Sci.USA* **92**, 12260-12264.
62. Mahley, R.W. (1988) Apolipoprotein E: cholesterol transport protein with expanding role in cell biology, *Science* **240**, 622-630.
63. Strittmatter, W.J., Saunders, A.M., Schmechel, D., Pericak-Vance, M., Enghild, J., Salvesen, G.S., and Roses, A.D. (1993) Apolipoprotein E: high-avidity binding to beta-amyloid and increased frequency of type 4 allele in late-onset familial Alzheimer disease, *Proc. Natl. Acad. Sci.USA* **90**, 1977-1981.
64. van Duijn, C.M. (1996) Epidemiology of the dementias: recent developments and new approaches, *J. Neurol. Neurosurg. Psych.* **60**, 478-488.
65. Farrer, L.A., Cupples, L.A., Haines, J.L., Hyman, B., Kukull, W.A., Mayeux, R., Myers, R.H., Pericak-Vance, M.A., Risch, N., and Van Duijn, C.M. (1997) Effects of age, sex, and ethnicity on the association between apolipoprotein E genotype and Alzheimer disease. A meta-analysis. APOE and Alzheimers Disease Meta Analysis Consortium, *JAMA* **278**, 1349-1356.
66. Slooter, A.J.C., and van Duijn, C.M. (1997) Genetic epidemiology of Alzheimer disease, *Epidemiol. Rev.* **19**, 107-119.
67. De Knijff, P., and van Duijn, C.M. (1998) Role of APOE in dementia: a critical reappraisal, *Haemostasis* **28**, 195-201.
68. Danesh, J., and Lewington, S. (1998) Plasma homocysteine and coronary heart disease: systematic review of published epidemiological studies, *J. Cardiovasc. Risk* **5**, 229-232.
69. Bots, M.L., Launer, J.L., Lindemans, J., Hoes, A.A.W., Hofman, A., Witteman, J.C.M., Koudstaal, P.J., and Grobbee, D.E. (1999) Homocysteine and short term risk of myocardial infarction and stroke in the elderly: The Rotterdam study, *Arch. Intern. Med.* **159**, 38-44.
70. Perry, I.J., Refsum, H., Morris, R.W., Ebrahim, S.B., Ueland, P.M., and Shaper, A.G. (1995) Prospective study of serum total homocysteine concentration and risk of stroke in middle-aged British men, *Lancet* **346**, 1395-1398.
71. Cunha, U.G., Rocha, F.L., Peixoto, J.M., Motta, M.F., and Barbosa, MT. Vitamin B12 deficiency and dementia, *Int. Psychogeriatr.* **7**, 85-88.
72. Teunisse, S., Bollen, A.E., van Gool, W.A., and Walstra, G.J. (1996) Dementia and subnormal levels of vitamin B12: effects of replacement therapy on dementia, *J Neurol* **243**, 522-529.
73. Carmel, R., Gott, P.S., Waters, C.H., Cairo, K., Green, R., Bondareff, W., DeGiorgio, C.M., Cummings, J.L., Jacobsen, D.W., Buckwalter, G., *et al.* (1995) The frequently low cobalamon levels in dementia usually signify treatable metabolic, neurologic and electrophysiologic abnormalities, *Eur. J. Haematol.* **54**, 245-253.
74. Joosten, E., Lesaffre, E., Riezler, R., Ghekiere, V., Dereymaeker, L., Pelemans, W., and Dejaeger, E. (1997) Is metabolic evidence for vitamin B-12 and folate deficiency more frequent in elderly patients with Alzheimer's disease? *J Gerontol* **52**, 76-79.
75. Nilsson, K., Gustafson, L., Faldt, R., Andersson, A., Brattstrom, L., Lindgren, A., and Israelsson, B. Hultberg B. (1996) Hyperhomocysteinaemia-a common finding in a psychogeriatric population, *Eur J Clin Invest* **26**, 853-859.

76. Clarke, R., Smith, A.D., Jobst, K.A., Refsum, H., Sutton, L., Ueland, P.M. (1998) Folate, vitamin B12, and serum total homocysteine levels in confirmed Alzheimer disease, *Arch. Neurol.* **55**, 1449-1455.
77. Riggs, K.M., Spiro, A. 3rd, Tucker, K., and Rush, D. (1996) Relations of vitamine B-12, vitamin B-6, folate, and homocysteine to cognitive performance in the Normative Aging Study, *Am. J. Clin. Nutr.* **63**, 306-314.
78. Kalmijn, S., Launer, L.J., Lindemans, J., Bots, M.L., Hofman, A., and Breteler, M.M.B. (1999) Total homocysteine and cognitive decline in a community-based sample of elderly subjects: the Rotterdam Study, *Am. J. Epidemiol.* **150**, 283-289.
79. Berwanger, C.S., Jeremy, J.Y., and Stansby, G. (1995) Homocysteine and vascular disease, *Br. J. Surg.* **82**, 726-731.
80. Prince, M., Cullen, M., and Mann, A. (1994) Risk factors for Alzheimer's disease and dementia: A case-control study based on the MRC elderly hypertension trial, *Neurology* **44**, 97-104.
81. Koster, T., Rosendaal, F.R., Ronde, H., Briet, E., Vandenbroucke, J.P., and Bertina, R.M. (1993) Venous thrombosis due to poor anticoagulant response to activated protein C. Leiden Thrombophilia Study, *Lancet* **342**, 1503-1506.
82. Svensson, P.J., and Dahlback, B. (1994) Resistance to activated protein C as a basis for veneous thrombosis, *N. Engl. J. Med.* **330**, 517-522.
83. Bertina, R.M., Koeleman, B.P.C., Koster, T., Rosendaal, F.R., Dirven, R.J., de Ronde, H., van der Velden, P.A., and Reitsma, P.H. (1994) Mutation in blood coagulation factor V associated with resistance to activated protein C, *Nature* **369**, 64-67.
84. Kontula, K., Ylikorlala, A., Miettinen, H., Vuorio, A., Kauppinen-Mäkelin, R., Hämäläinen, L., Palomaki, H., and Kaste, M. (1995) Arg506Gln factor V mutation in patients with ischemic cerebrovascular disease and survivors of myocardial infarction, *Thromb. Haemost.* **73**, 558-560.
85. Ridker, P.M., Hennekens, C.H., Lindpainter, K., Stampfer, M.J., Eisenberg, P.R., and Miletich, J.P. Mutation in the gene coding for coagulation facto V and the risk of myocardial infarction, stroke and venous thrombosis in apparently helthy men, *N. Engl. J. Med.* **332**, 912-917.
86. Bots, M.L., Breteler, M.M.B., van Kooten, F., Slagboom, E., Hofman, A., Haverkate, F., Meijer, P., Koudstaal, P.J., Grobbee, D.E., and Kluft, C. (1998) Response to activated protein C in subjects with and without dementia. The Dutch Vascular Factors in Dementia Study, *Haemostasis* **28**, 209-215.
87. Bots, M.L., Breteler, M.M.B., van Kooten, F., Haverkate, F., Meijer, P., Koudstal, P., Grobbee, D.E., and Kluft, C. (1998) Coagulation and fibrinolysis markers and risk of dementia, *Haemostasis* **28**, 216-222.
88. McGeer, P.J., and McGeer, E.G. (1995) The inflammatory response system of brain: implications for therapy of Alzheimer and other neurodegenerative diseases, *Brain Res Brain Res Rev* **21**, 195-218.
89. Eikelenboom, P., and Veerhuis, R. (1996) The role of complement and activated microglia in the pathogenesis of Alzheimer's disease, *Neurobiol. Aging* **17**, 673-680.
90. Andersen, K., Launer, L.J., Ott, A., Hoes, A.A.W., Breteler, M.M.B., and Hofman, A. (1995) Do nonsteroidal anti-inflammatory drugs decrease the risk for Alzheimer's disease? The Rotterdam Study, *Neurology* **45**, 1441-1445.
91. Breitner, J.C., Gau, B.A., Welsh, K.A., Plassman, B.L., McDonald, W.M., Helms, M.J., and Anthony, J.C. (1994) Inverse association of anti-inflammatory treatments and Alzheimer's disease: initial results of a co-twin control study, *Neurology* **44**, 227-232.
92. Ross, R. (1999) Atherosclerosis--an inflammatory disease, *N. Engl. J. Med.* **340**, 115-126.

93. Blennow, K., Wallin, A., Fredman, P., Karlsson, I., Gottfries, C.G., and Svennerholm, L. (1990) Blood-brain barrier disturbance in patients with Alzheimer's disease is related to vascular factors, *Acta Neurol Scand* .**81**, 323-326.
94. Van Dam, A.M., de Vries, H.E., Kuiper, J., Zijlstra, F.J., de Boer, A.G., Tilders, F.J., and Berkenbosch, F. (1996) Interleukin-1 receptors on rat brain endothelial cells: a role in neuroimmune interaction? *FASEB J.* **10**, 351-356.
95. Breteler, M.M.B., van Duijn, C.M., Chandra, V., Fratiglioni, L., Graves, A.B., Heyman, A., Jorm, A.F., Kokmen, E., Kondo, K., Mortimer, J.A., Rocca, W.A., Shalat, S.L., Soininen, H., and Hofman A. (1991) Medical history and Alzheimer's disease, *Int. J. Epidemiol.* **20**, S36-42.
96. Itzhaki, R.F., Lin, W.R., Shang, D., Wilcock, G.K., Faragher, B., and Jamieson, G.A. (1997) Herpes simplex virus type 1 in brain and risk of Alzheimer's disease, *Lancet* **349**, 241-244.
97. Fratiglioni, L., Ahlbom, A., Viitanen, M., and Winblad, B. (1993) Risk factors for late-onset Alzheimers's disease: a population-based, case-control study, *Ann. Neurol.* **33**, 258-266.
98. Saunders, P.A., Copeland, J.R., Dewey, M.E., Davidson, I.A., McWilliam, C., Sharma, V., and Sullivan, C. (1991) Heavy drinking as a risk factor for depression and dementia in elderly men. Finding from the Liverpool longitudinal study, *Br. J. Psychiatry* **159**, 213-216.
99. Orgogozo, J.M., Dartigues, J.F., Lafont, S., Letenneur, L., Commenges, D., Salamon, R., Renaud, S., and Breteler, M.M.B. (1997) Wine consumption and dementia in th elderly: a prospective community study in the Bordeaux area, *Rev. Neurol. (Paris)* **153**, 185-192.
100. Breteler, M.M.B., Mehta, K., Ott, A., Witteman, J.C.M., and Hofman, A. (1997) Alcohol consumption and dementia. The Rotterdam Study. (Abstract), *J Neurol* **244**, S47.

Chapter 4

CEREBRAL MICROVASCULAR AND MACROVASCULAR DISEASE IN THE AGING BRAIN; SIMILARITIES AND DIFFERENCES

HARRY V. VINTERS
Department of Pathology & Laboratory Medicine, Section of Neuropathology, Brain Research Institute & Neuropsychiatric Institute, UCLA Medical Center, Los Angeles,CA, USA

While the focus of this volume has been on a specific type of cerebral microvascular disease important in both 'stroke' and dementia, here we will consider non-CAA forms of vasculopathy. Specifically, similarities and differences between atherosclerosis and small artery lesions (arteriosclerosis/lipohyalinosis, CADASIL, Binswanger's subcortical leukoencephalopathy, HERNS) will be discussed in terms of their clinicopathologic manifestations and putative cellular/molecular pathogenesis. Abnormal migration, proliferation, degeneration and response to injury of vascular smooth muscle cells, endothelium, pericytes and (less likely) macrophages may play a central role in causation of many forms of vasculopathy, including CAA. *In vitro* and animal (including transgenic) models may have a part in elucidating their origins and eventual treatment.

1. INTRODUCTION

This chapter will not attempt to deal with an exhaustive consideration of all types of large vessel and small vessel disease that may affect the CNS in the course of brain aging, a subject which is tackled elsewhere in far greater detail and sophistication [1-4]. Rather it will attempt to take a clinicopathologic and mechanistic approach to understanding the pathogenesis of common types of macro- and microangiopathy (especially affecting arteries), as well as consequences of these various forms of vascular disease for the aging brain. Table 1 provides a simplistic subclassification of major types of cerebrovascular disease important in stroke and dementia. The author recognizes that subdividing cerebrovascular

M.M. Verbeek et al. (eds.),
***Cerebral Amyloid Angiopathy in Alzheimer's Disease and Related Disorders*, 59–78.**

disease into 'macro-' and 'micro-' vascular components itself represents a gross oversimplification of the real world situation. The two subtypes of vasculopathy (especially atherosclerosis, arteriosclerosis and/or CAA) often coexist in elderly individuals and almost certainly share major risk factors - not the least of which is simply aging. Co-morbidity from more than one form of angiopathy is more the rule than the exception. However, certain pathogenetic themes are worth emphasizing in trying to understand the role of cerebrovascular disease in CNS aging.

TABLE 1: Classification of large and small vessel diseases that affect the human nervous system

A. <u>MACRO</u>ANGIOPATHIES/ARTERIOPATHIES
Atherosclerosis
Including effects of hyperhomocysteinemia
Fibromuscular dysplasia (FMD)
Moyamoya disease
Arterial dissection
HIV-associated vasculopathy
Vasculitis/Angiitis
Giant cell arteritis
Takayasu's arteritis
B. <u>MICRO</u>ANGIOPATHIES/ARTERIOPATHIES
Cerebral amyloid angiopathy (CAA)
Arteriosclerosis/arteriolosclerosis (including lipohyalinosis)
Angiitis/vasculitis, including....
Primary angiitis of the CNS
Polyarteritis nodosa
Wegener's granulomatosis
Systemic lupus (variable inflammation)
Associated with dementia
Binswanger's subcortical arteriosclerotic leukoencephalopathy (BSLE)
CADASIL
Thrombotic thrombocytopenic purpura
? MELAS (Mitochondrial myopathy, encephalopathy, lactic acidosis, & stroke-like episodes)
Cerebroretinal vasculopathy/HERNS (Hereditary endotheliopathy, retinopathy, nephropathy, & stroke)

2. ATHEROSCLEROSIS & OTHER MACROANGIOPATHIES

Cardiovascular, including cerebrovascular disease, related to atherosclerosis remains the leading cause of morbidity and mortality in the USA, Europe and parts of Asia, especially among the elderly. The pathogenesis of atherosclerosis is now understood in remarkable detail, thanks to advances in the development of animal models, cell biologic/tissue culture and molecular tools that can be used to study its initiating factors and progression; key reviews have summarized a wealth of important data relevant to this form of large arterial disease [5-9]. The earliest events that antedate formation of atherosclerotic lesions have been hypothesized to include endothelial dysfunction, which manifests as increased endothelial permeability to lipoproteins and other plasma constituents (mediated by molecules such as nitric oxide [NO] and platelet-derived growth factor [PDGF]); up-regulation of leukocyte adhesion molecules, e.g. L-selectin and integrins, and endothelial adhesion molecules, e.g. E-selectin, P-selectin and intercellular adhesion molecule-1; and migration of leukocytes into the arterial wall, a process modulated by oxidized low-density lipoprotein, interleukin-8, PDGF, and numerous other factors.

Formation of advanced, complicated lesions of atherosclerosis involves development of a fibrous cap over the fatty streak; this cap covers a necrotic core that includes an admixture of leukocytes, lipid, and amorphous debris, and itself results from lipid accumulation, necrosis and proteolytic activity [5]. Unstable fibrous plaques - often resulting in symptomatic plaque rupture - result from thinning of the protective fibrous cap (this in essence shields atheromatous debris from the lumen) associated with the continuing influx and activation of macrophages, cells which are in turn thought to release metalloproteinases and other proteolytic enzymes. Hemorrhage into this unstable plaque may occur from vasa vasorum or, more accurately, fragile microvessels within the plaque itself. Finally, this may lead to mural or occlusive thrombus formation in an affected artery. Clearly, symptomatic atheroma results from the interplay of cholesterol, oxidized phospholipids, numerous growth factors, adhesion molecules, and 'leukins', combinations of which mediate migration of various cell types (including inflammatory cells) into the arterial wall. The subsequent intramural proliferation, interaction and demise of these cells (as well as smooth muscle cells that are the major component of arterial media), are crucial determinants of whether or not atherosclerosis shows clinical manifestations referrable to a specific body part - myocardial infarct, stroke or gangrene - in a given individual.

Atherosclerosis occurs primarily in large and medium-sized elastic and muscular arteries. It may affect the circle of Willis and its major branches

(Figure 1). In Western populations atherosclerosis is most severe in basal arteries, i.e. major branches of the circle of Willis, whereas in a study of Japanese patients, atherosclerosis was found to be almost as severe in small peripheral arterial branches [10]. A segment of an artery originating at the circle of Willis may undergo thrombosis with resultant symptomatic brain infarction. Commonly, however, the brain suffers because of severely stenotic or occlusive atherosclerosis in the intracervical, extracranial portions of the carotid and vertebro-basilar arterial systems [11-16]. Fragile, often ulcerated atheroma in these locations may result in artery-to-artery emboli of either platelet-fibrin or atheromatous material (Figure 2) to the cerebral circulation, producing (reversible) transient ischemic attacks or (irreversible) brain infarcts, patterns of which have been documented in meticulous necropsy studies [17,18].

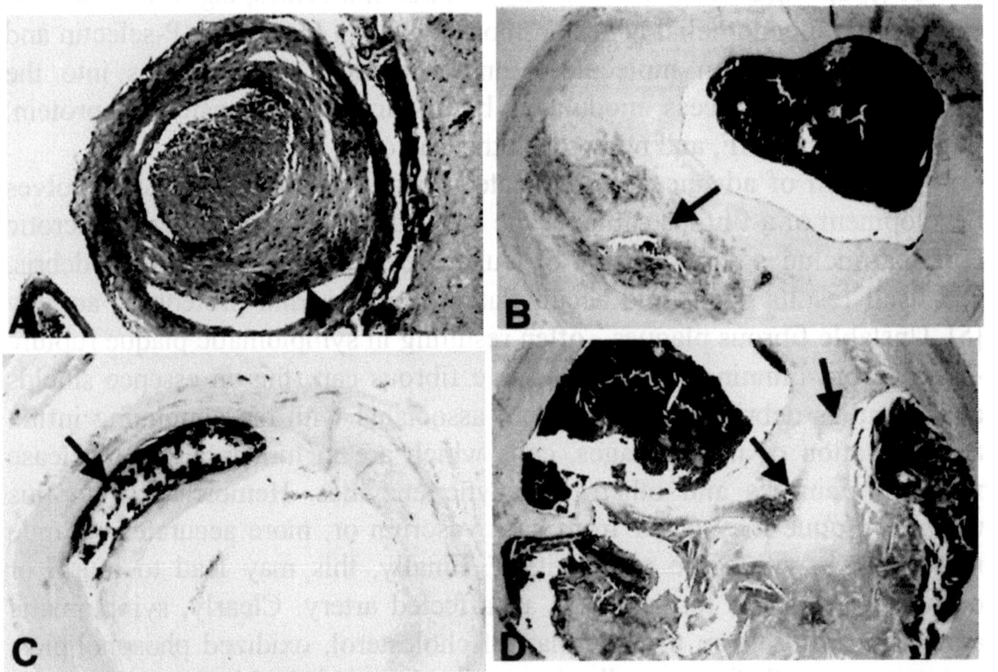

Figure 1: Meningeal and 'basal' arteries showing various manifestations and degrees of atherosclerosis. A. Small leptomeningeal artery showing fibromuscular intimal hyperplasia (arrow) only, with moderate compromise of the arterial lumen. B-D. All sections are of basal arteries representing major branches of the circle of Willis. In panel B, large eccentric fibrous plaque shows extensive old hemorrhage (arrow). In panel C, arrow indicates a large oval region of plaque calcification. In D, arrows indicate extensive acute hemorrhage into atheromatous plaque that has produced severe compromise of the arterial lumen (at upper left). Numerous cholesterol clefts are visible within the plaque itself. [All sections H & E-stained, magnifications A X 65; B-D X 25].

While atherosclerosis is by far the most common form of vasculopathy affecting large arteries and producing cerebral infarcts, several other entities may result in arterial stenosis or dissection (see Table 1, [2-4,19]).

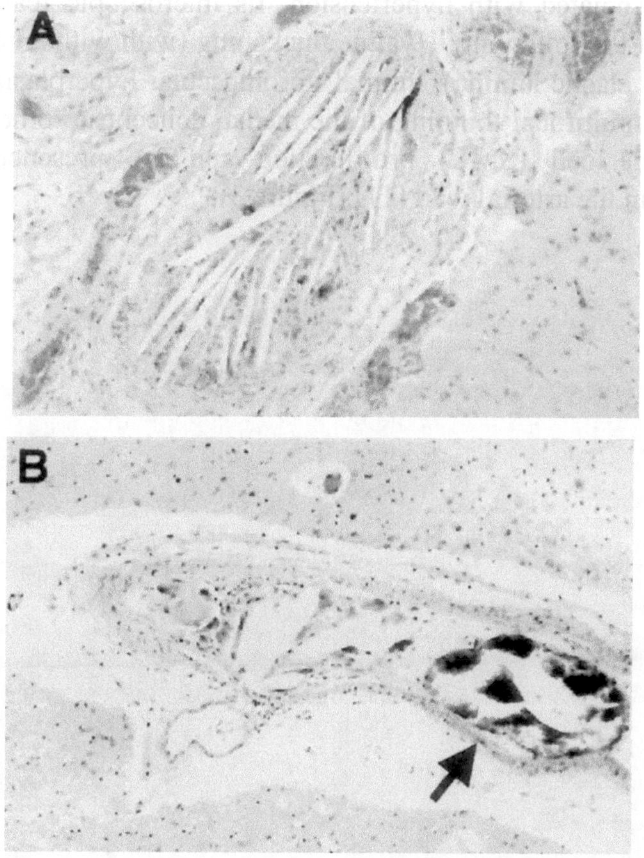

Figure 2: Atheroemboli in distal (leptomeningeal) branches of the arterial circulation. A. Occlusive embolus. B. Partly occlusive embolus, with residual lumen indicated by arrow. In both arteries, note prominent foreign body giant cell reaction surrounding cholesterol clefts. [H & E stain, both panels].

Fibromuscular dysplasia and arterial dissection, given their association with abnormalities of the arterial media, can be inferred to reflect abnormalities that derive, at some level, from lesions of the smooth muscle cells, their pattern of growth or organization, or their adjacent extracellular matrix.

3. ARTERIOSCLEROSIS

By comparison to atherosclerosis, the cellular and molecular pathogenesis of arteriosclerosis/arteriolosclerosis (AS) is poorly understood. There is not even uniform (much less universal) agreement on the microscopic brain microvascular lesions best described by the term 'AS', thus its severity is difficult to evaluate in a given case. The evolution of AS is strongly associated with hypertension. Its microscopic features include one or more of the following: Hyaline thickening (with/without degeneration of the internal elastic lamina), intimal fibromuscular hyperplasia, narrowing of the lumen, multifocal thinning of the media, concentric 'onion-skin' type smooth muscle cell (SMC) proliferation and the presence of foamy macrophages in the arterial wall [11] (Figure 3).

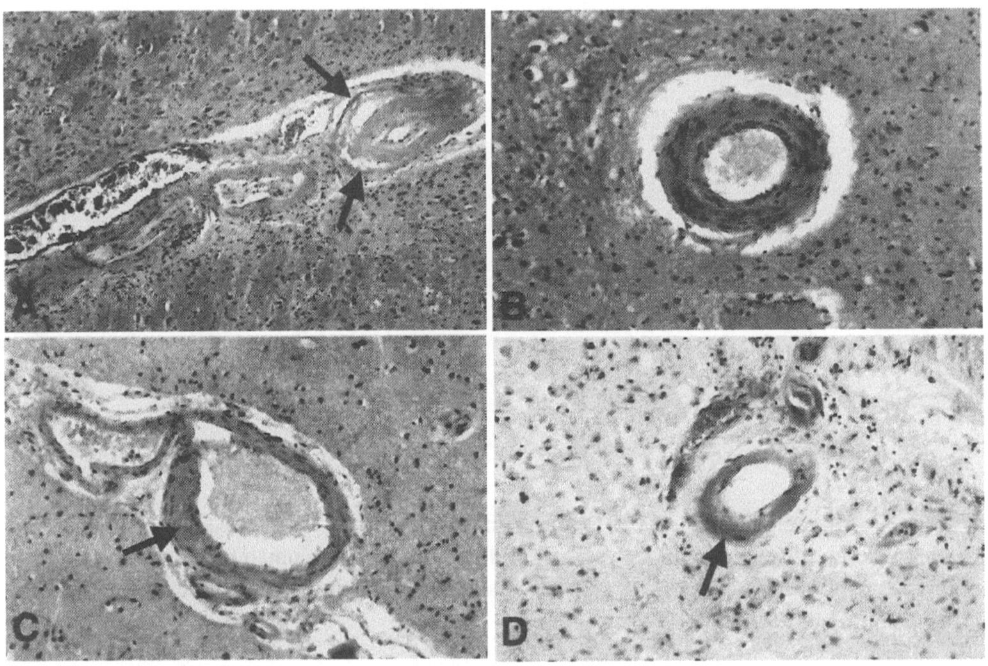

Figure 3: Arteriosclerosis/lipohyalinosis. All micrographs represent sections of brain from a 77-year-old female. A. Note hyaline thickening of small parenchymal arteriole. Arrows indicate regions in which macrophages were present within the vessel wall. B. 'Onion-skin' type thickening of a parenchymal artery, with apparent smooth muscle cell hyperplasia. C. Tortuous and ectatic parenchymal artery. Arrow indicates presence of amorphous (non-amyloid) material within thickened vessel wall. D. Fibrinoid necrosis (arrow) of a small parenchymal artery, surrounded by macrophages. [Magnifications A X 65, B-D X 130].

Ultrastructural morphometric analysis of medial injury in hypertensive patients who have experienced cerebral hemorrhage shows atrophy and loss of SMCs, and accumulation of nonfatty debris and basement membrane materials [20]. Comparable studies in occluded cerebral arteries in stroke-prone spontaneously hypertensive rats show extensive medial necrosis and exudation of plasma components (especially fibrin) within markedly thickened arterial walls [21]. Fibrinoid necrosis is said to be associated only with malignant hypertension. Apart from fibrinoid necrosis, the vascular lesions of AS thus show similarities to those in atherosclerosis. The picture becomes even more confusing when microatheroma are occasionally seen in larger meningeal (and sometimes even intraparenchymal) arteries.

3.1 Pathophysiologic Considerations

While physiology of the cerebral microcirculation is understood in considerable detail [22], pathophysiologic mechanisms important in the genesis of cerebral AS have been the subject of relatively little systematic/quantitative study. In the brain, as elsewhere, endothelium releases a substance, endothelium-derived relaxing factor (EDRF)— now known to be nitric oxide (NO)— that diffuses into vascular smooth muscle. There, EDRF/NO causes smooth muscle cell relaxation and cerebral vasodilation by activation of soluble guanylate cyclase. NO is produced in endothelial cells by the constitutive expression of nitric oxide synthase (NOS), an enzyme which is also inducible in other brain cell types. Adenylate cyclase can be activated, resulting in cAMP production, by several other molecules, including prostanoids, adenosine, vasoactive peptides, and beta-adrenergic agonists. This leads to relaxation of blood vessels, a phenomenon that may in part result from activation of K^+ channels [22]. Blood-borne elements, especially platelets and leukocytes, frequently release molecules that may serve as vasodilators (platelet-derived ADP), or vasoconstrictors (thromboxane A_2). In chronic hypertension, there may be decreased production or activity of NO and diminished activity of ATP-sensitive K^+ channels [22], and synthesis of endothelium-derived contracting factors. Hyperhomocysteinemia and diabetes mellitus both contribute to cerebral microvascular disease through complex and heterogeneous biochemical mechanisms.

3.2 Consequences of AS

Arteriosclerosis as a descriptor of cerebral microvascular (arterial/arteriolar) disease is sometimes used interchangeably with the term 'lipohyalinosis', the latter deriving from a consideration of events likely to have been important in the pathogenesis of the microangiopathy, i.e. lipid deposition and hyalinization, though not necessarily in this order. Whatever name one prefers for the the arterial lesion, it has historically been associated with two types of cerebrovascular disease or 'brain insult', viz. (a) lacunar infarcts, and (b) intracerebral (parenchymal) hemorrhage. Lacunar infarcts are small (usually less than 15 mm. in greatest dimension) regions of cystic cavitation most often found within the basal ganglia, thalamus, pons, internal capsule and deep subcortical white matter [23-25]; when found in large numbers in the subcortical white matter, they may give rise to the syndrome of Binswanger's subcortical leukoencephalopathy (BSLE; see below). Lacunar infarcts (Figure 4) are to be distinguished from enlarged perivascular spaces, which they can can mimic on gross, and sometimes even microscopic,examination of the CNS [26]. Lacunar infarcts have been associated with various stroke syndromes, many of which appear to have an excellent prognosis. Despite the historical association between lacunes, AS/lipohyalinosis and hypertension, a history of high blood pressure is absent in as many as 35-40% of patients with these small, deep encephalic infarcts [27]. Thus, it has been suggested that lacunes may result from carotid atherosclerosis — as a hemodynamic consequence or a result of atheroemboli - or even from non-atheromatous emboli [28]. Furthermore, occluded arteries of varying diame ter may be found adjacent to lacunes, depending to some extent upon the size of the infarct. When the abnormal artery has a diameter of 40-200 μm, it may show typical features of AS/lipohyalinosis, whereas an artery of 300-500 μm may show microatheroma (with especially prominent foamy histiocytes) in its wall [29,30]. Defining the clinicopathologic substrate of lacunes becomes even more problematic, given that the lesions are often asymptomatic [31]. The heterogeneity of lacunar syndromes and the underlying cerebrovascular disease that may cause them only emphasizes the importance of careful autopsy studies aimed at defining their pathogenesis.

AS/lipohyalinosis associated with hypertension is also an important cause of spontaneous (nontraumatic) intracerebral, intraparenchymal hemorrhage (IPH) [32,33]. Not surprisingly, this occurs in the same brain loci that frequently show lacunes, though both types of lesion need not be present in the same patient. Miliary microaneurysms (Charcot-Bouchard aneurysms) have been found in hypertensive patients, sometimes adjacent to IPHs, using both standard morphologic and microangiographic techniques [34,35]. More recent study of the cerebral microvasculature in autopsy brains that originate from hypertensive patients, using alkaline phosphatase cytochemistry and high-resolution microradiography, suggests that Charcot-Bouchard microaneurysms are rare and may, in some instances, represent arteriolar 'coils and twists' that are simply interpreted as aneurysms by the neuropathologist [36]. As we have seen in a discussion of CAA-associated microangiopathies, however, microaneurysms may also complicate severe CAA (see H.V. Vinters and J.-P.G. Vonsattel, this volume) and thus are probably a relatively non-specific manifestation of microvascular/arteriolar degeneration within the brain.

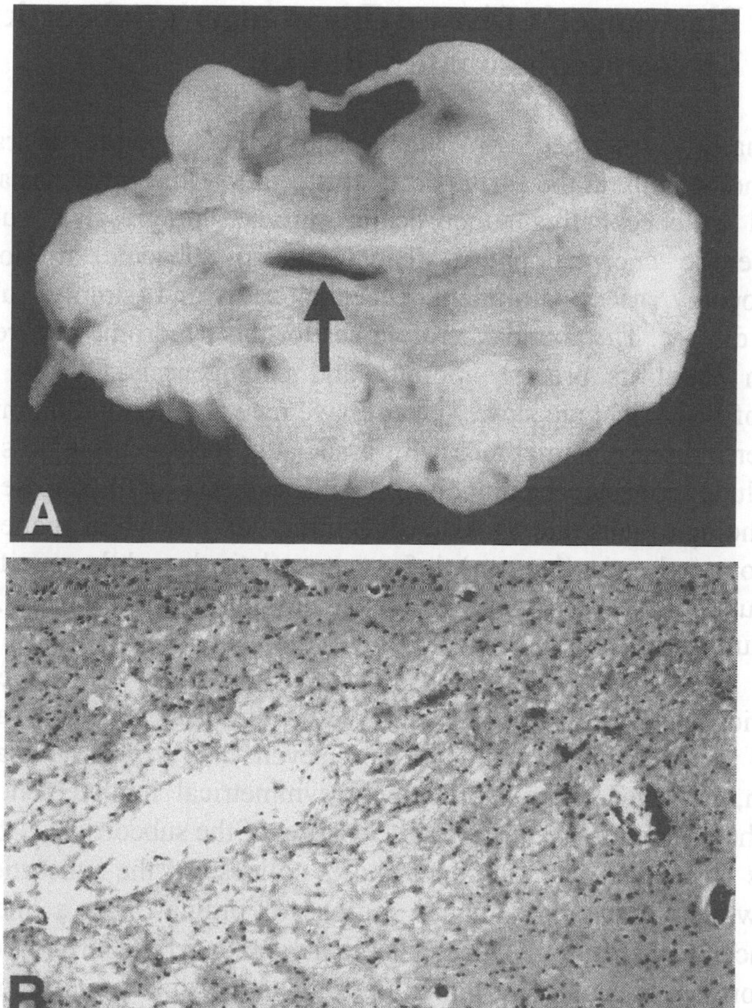

Figure 4: Lacunar infarct. A. Horizontal section of pons, showing a slit-like area of encephalomalacia in the left basis pontis (arrow). B. Microscopic section of a typical lacune, with central cavitation merging into microcystic change and astrocytic gliosis in the adjacent neuropil. An occluded vessel that may have caused the infarct is not identified in this field. [Magnification B X 65, H&E stain]

4. NON-CAA MICROANGIOPATHY ASSOCIATED WITH DEMENTIA

Clinicopathologic manifestations of CAA, including dementia, are the focus of this volume. However, other types of cerebral microangiopathy are known to produce parenchymal brain damage of a type that may result in dementia. Several of these will be reviewed below, emphasizing their similarities to, and differences from, CAA.

4.1 Binswanger's Disease (Binswanger's Subcortical Leukoencephalopathy, BSLE)

Originally described in the late 1800s and characterized neuropathologically in the early part of this century, this entity has also been described as encephalitis subcorticalis chronica progressiva, subcortical arteriosclerotic encephalopathy, (Binswanger's) subacute arteriosclerotic encephalopathy, arteriopathic leukoencephalopathy and (simply) subcortical ischemic disease. The abundance of names for this syndrome may reflect the confusion about its precise etiology and pathogenesis [37-40]. Clinical features of BSLE include slowly progressive mental deterioration mimicking Alzheimer's disease, sometimes with associated focal neurologic signs that may include aphasia, hemianopia, hemianesthesia or hemiparesis. The neuropathologic substrate is extensive injury of the white matter, which varies from ischemic (lacunar) infarcts to relatively subtle manifestations that include axonal or myelin injury with foamy macrophages. Myelin pallor/injury is, as might be expected from such a patchy disease process, variable from region to region; however, it is most severe in the periventricular regions and diminishes as one approaches the cortical ribbon. Occipital involvement is in general more severe than frontal. Overall within the brain, however, there is relatively symmetrical involvement of both cerebral hemispheres and comparative sparing of the subcortical U-fibers.

Gross inspection of brain slices shows collapse of the subcortical white matter, which frequently contains regions of pitting, with a 'moth-eaten' appearance, with resultant ventricular enlargement, but normal cortex; an impression that is usually sustained upon review of histologic sections. The corpus callosum is usually spared, possibly because of its unique neuroanatomic features [41]. Neocortical changes would be expected in such patients, if only as a 'retrograde' manifestation (within neuronal cell bodies) of axonal injury. However, to this author's knowledge such neuropathologic abnormalities - which may be subtle and thus detectable only using rigorous morphometric techniques - have not yet been proven to exist. Associated lacunar infarcts are usually found within the deep central grey structures and pons. Magnetic resonance imaging has revolutionized the antemortem study of BSLE, though has shed relatively little light on its vascular substrate(s) [42,43].

Microvascular abnormalities within white matter include changes described above as 'arteriosclerotic', i.e. there is thickening and hyalinization of small arteries and arterioles, with loss or effacement of their cellular wall components. However, amyloid is not present in the thickened vessels, though they may show enlargement of the surrounding Virchow-Robin spaces. Despite the assumption that microvascular stenosis and occlusion are the proximate cause of the extensive cerebral white matter injury, the affected arteries are usually patent [37]. Scattered perivascular lymphocytes may be seen, though BSLE is not considered to be a primarily inflammatory disorder. However, an interesting recent quantitative immmunohisto-chemical study has shown that activated white matter microglia are 2-4 times more frequent in BSLE than in controls (including patients with cortical infarcts), suggesting a role for these inflammatory cells in the impaired axonal transport (mediated by chronic ischemia) that underlies the condition [44].

Clues to the pathogenesis of BSLE have otherwise been relatively few; it has usually been the subject of correlative clinicopathologic-neuroradiologic investigations, and thoughtful speculation. Recently, the coagulation-fibrinolysis pathway has been found to be activated in BSLE patients with a 'subacute aggravation' [45], suggesting a mechanism whereby microthrombi and related microcirculatory disturbances may cause brain injury.

White matter abnormalities detectable by CT and MRI scans are now commonly recognized in elderly patients with or without dementia [42,46-50]. Such abnormalities are sometimes described by the term 'leuko-araiosis' [51], which is often – incorrectly - used as a synonym for BSLE. It has been estimated that BSLE type white matter lesions occur in as many as 60% of patients with Alzheimer's disease (AD)[39]. The 'leukoencephalopathy' of AD has been suggested to result from incomplete white matter infarction [52]. Yet, perhaps precisely because leukoencephalopathy (defined using neuroimaging criteria) is such a common occurrence in the elderly, its neuropathologic substrate(s) have been remarkably difficult to 'pin down' (for an excellent review of likely contributing etiologic factors, see [53]). White matter correlates of leuko-araiosis are believed to include demyelination, reactive astrocytic gliosis and arteriosclerosis; sometimes proportional to the degree of neuroradiographic abnormalities [54]. In an investigation of over 3300 elderly without a history of stroke or transient ischemic attack, the Cardiovascular Health Study has found that 'white matter findings' (judged by MRI) correlated with age, silent stroke, hypertension, and even income level of the patients [55]. Both white matter gliosis - hypothesized to result from blood-brain barrier breakdown - and arteriolar thickening have been documented in elderly patients (lacking dementia) with a disequilibrium syndrome and leukoencephalopathy [56,57].

4.2 CADASIL

This is an abbreviation for 'Cerebral autosomal dominant arteriopathy with subcortical infarcts and leukoencephalopathy', a disease probably recognized since the 1950s (and earlier labelled as BSLE) but more fully characterized during the past decade [58]. Initially reported among European families, CADASIL is now recognized in North American, Asian and African patients. Since it has attained a relatively high profile as a nosologic entity only recently, the condition is almost certainly underdiagnosed. The initial confusion between CADASIL and BSLE was for good reason: from both clinical and neuropathologic perspectives, they show important similarities.

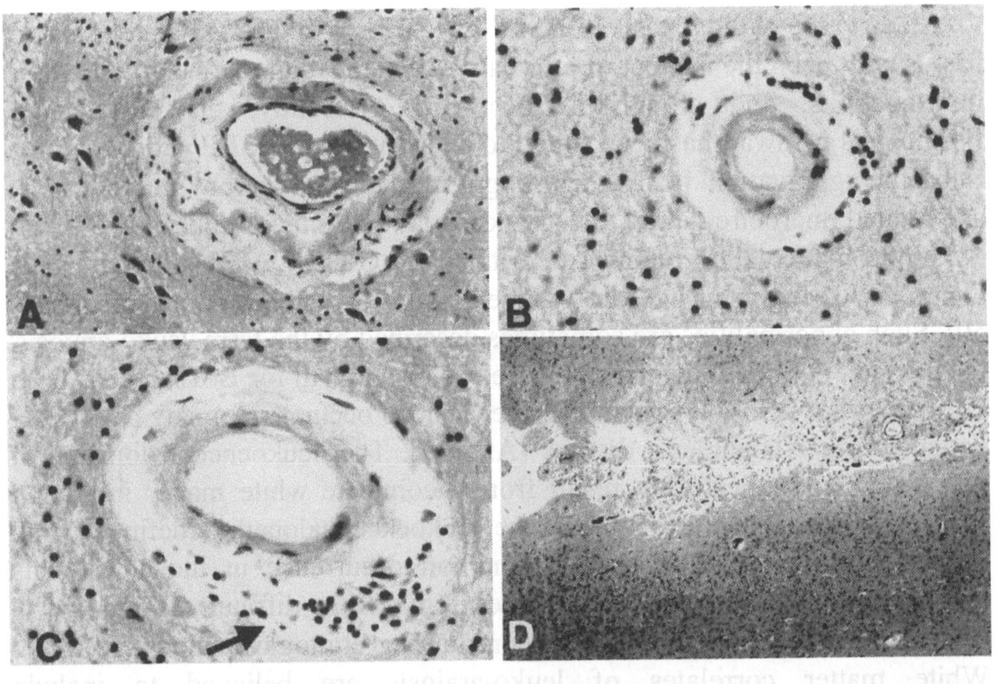

Figure 5: CADASIL. Three different arterioles (H & E-stained), showing variable degrees of intimal hyperplasia (A), granular degeneration and adventitial hyalinization (B,C), and adventitial mononuclear inflammatory cells (arrow, C). A microinfarct is shown in panel D. [Case studied by courtesy of Dr. Bruce Quinn, Northwestern University Medical Center, Chicago.]

Patients most commonly present with stroke (over 80%) involving the subcortical white matter, dementia (30-90%), migraine with aura or severe mood disturbance [58]. Onset is usually in the fifth decade of life, with subsequent survival of 10-30 yr. Dementia is usually of the frontal lobe type and similar to other 'ischemic-vascular dementias'; memory impairment is a common feature. Cognitive impairment occurs in a stepwise fashion. MRI imaging shows nodular hyperintensities, usually in a symmetrical distribution between cerebral hemispheres, and commonly involving the periventricular regions, centrum semi-ovale, basal ganglia and pons [58-60].

Neuropathologic features of CADASIL include lacune-type infarcts in the regions predicted to show abnormalities by MRI, with relative preservation of subcortical U-fibers and associated ventriculomegaly. Microscopic sections of affected brain areas demonstrate diffuse and focally prominent myelin pallor. Lacunar infarcts show typical expected histologic abnormalities, i.e. small cystic cavities containing collections of macrophages surrounded by astrocytic gliosis. Overlying neocortex may show foci of glial scarring with neuron loss [58], but Alzheimer changes are infrequently observed, except in rare cases [61].

The most intriguing aspects of this disorder are obviously (a) its underlying vascular substrate(s), and (b) given its autosomal dominant pattern of inheritance, the causal genetic abnormality. Histologic sections of affected white matter show abnormal microvasculature characterized by vessel wall thickening with 'hyaline degeneration'; the foci of degeneration are often described as showing a 'smudged' or granular substructure (Figure 5). Reduplication or fragmentation of the elastica may be seen. Medial hyaline degeneration appears to result from the disappearance of vascular SMC, which may (in what are presumed to be early stages of degeneration) show 'ballooning' [58]. Loss of SMC-specific proteins can be shown by immunohistochemistry. Less constantly observed microvascular abnormalities include fibrinoid deposits in their walls, inflammatory cells, and (using immunohistochemical detection methods), immunoglobulins. Despite vascular thickening and (sometimes) concentric intimal hyperplasia, complete arterial occlusion is infrequent. The granular material which appears to be an integral component of the degenerating media has a characteristic ultrastructural appearance: it consists of electron-dense, extracellular granular deposits lacking a filamentous component, closely apposed to (and sometimes apparently enveloped by) smooth muscle cell cytoplasm [2,58]. To date, these granular osmiophilic materials (GOMs) have not been biochemically characterized, though there is widespread suspicion that understanding their composition would greatly enhance our understanding of CADASIL itself. Of interest, small arteries in skeletal muscle and skin of patients with CADASIL can be shown to contain GOMs in their media [62], thus allowing for the definitive diagnosis of the disorder using biopsy of a non-CNS structure. CADASIL is now known to result from mutations in the *Notch3* gene on chromosome 19p13.1 [63]. A well known gene in *Drosophila*, *Notch* is known to encode a transmembrane receptor considered important in the specification of developmental cell fates in various tissues. Homologues of the gene have been identified in humans, though the precise molecular pathology of how mutations in *Notch3* produce this unique form of vascular dementia is far from understood.

4.3 Hereditary endotheliopathy with retinopathy, nephropathy, and stroke (HERNS)

Whereas virtually all of the microangiopathies considered above - and certainly CAA - appear to represent lesions of vascular SMC, HERNS defines a multi-system disorder in which cerebrovascular disease appears to result from an abnormality of endothelium and/or endothelial basement membrane(s) [64]. Not nearly as well characterized in clinico-pathologic terms as CADASIL, and clearly not linked to it genetically, HERNS is a rare autosomal dominant multi-infarct syndrome with systemic involvement that, despite its rarity, warrants further study.

4.4 Dementia Associated with Other Forms of Microangiopathy

Because cerebral arteriolar ('small vessel') disease is usually associated with hypertension, aggressive treatment of high blood pressure would be expected to lead to a decrease in the occurrence or severity of CNS arteriosclerosis. Instead, as one recent study has pointed out [65], cerebral 'hyaline arteriosclerosis' (less commonly lipohyalinosis, fibrinoid necrosis) is now seen in patients who have no demonstrable history of hypertension - suggesting that perhaps conditions which enhance small vessel permeability may contribute to the pathogenesis of this form of (non-CAA) microangiopathy. Ischemic vascular dementia (IVD) as a complication of (non-BSLE, non-CADASIL, non-HERNS) cerebral microvascular disease is described as being extremely rare in some studies [66], while other investigations claim to show that this (and not macroscopic infarction) is the major neuropathologic substrate of IVD [67]. Meticulous prospective clinicopathologic studies of well characterized patient populations thought to have IVD using best available clinical criteria [68,69] will be one way to address this issue [70].

5. APPROACHES TO STUDY PATHOGENESIS OF CEREBRAL MICROANGIOPATHY

The above discussion of atherosclerosis and various forms of small vessel disease important in stroke and dementia lead to the conclusion that abnormalities of one or more cell type in the vessel wall, or the interactions between/among them, are pivotal in understanding the pathogenesis of cerebrovascular disease: the key players here are endothelium, pericytes and perivascular microglia and astrocytes (in the case of capillaries), SMCs and (possibly) macrophages (in the case of arterioles and arteries). A discussion of cerebral microvascular endothelium - site of the blood-brain barrier (BBB) - must take into account its unique transport functions and how these may be influenced by adjacent astrocytes [71,72]. SMC proliferation and degeneration, and SMC interactions with basement membrane glycoproteins, appear to be important determinants of both atherosclerosis and cerebral microangiopathies as distinct as CAA, AS/lipohyalinosis, CADASIL, and BSLE. Pericytes (surrounding capillaries) and periendothelial (intimal) SMC may share a common origin that distinguishes them from microglia [73]. Mammalian brain pericytes may also have a contractile function similar to that of SMC [74].

Mechanistic studies aimed at understanding SMC and pericyte degeneration in the context of CAA are outlined elsewhere in this volume (see Verbeek *et al*). Such investigations take advantage of the unique opportunity to grow, manipulate and injure cerebral microvessel-derived cellular components, even derived from human brain, in tissue culture [75,76]. Both endothelium and SMC may behave differently in response to growth factors and other stimuli/toxins, depending upon the vascular bed (large artery vs. microvessel, visceral vs. brain arteriole) from which they have been isolated [77,78]. In the case of genetically determined microangiopathies (e.g. CADASIL), expression (or overexpression) of the mutant responsible gene (*Notch3* in the case of CADASIL) in cultured cells that appear to be the site of seminal pathologic changes (microvessel SMC) might be expected to reveal basic pathogenetic mechanisms. Novel molecular techniques applied to *in vitro* and animal model systems also show the remarkable versatility of vascular cells, e.g. by the observation that endothelium may differentiate into SMC-like cells [79-81]. Whether this phenomenon is important in the evolution of cerebral microvascular disease merits further exploration. Finally, knock-out and transgenic mice may provide further insights into phenomenology associated with human cerebral microangiopathies. The embryos of mice deficient in platelet-derived growth factor (PDGF)-B develop cerebral microaneurysms that are thought to result from a failure of sprouting capillaries to attract pericyte progenitor cells [82].

While this highly artificial model is unlikely to provide direct meaningful insights into the genesis of Charcot-Bouchard aneurysms, it suggests a paradigm of 'engineered' animal models that may help to unravel the more puzzling features underlying cerebral microangiopathies important in stroke and dementia.

ACKNOWLEDGEMENTS

Work in the author's laboratory has been generously supported by PHS/NIH Grants #NS26312, P30AG10123, P50AG16570, and P01AG12435. Ms. Carol Appleton assisted with preparation of illustrations, and Ms. Annetta Pierro with final manuscript preparation.

REFERENCES

1. Vinters, H.V., & Mah, V.H. (1991) Vascular diseases, in Duckett, S. (ed.), The Pathology of the Aging Human Nervous System, Lea & Febiger, Philadelphia, pp. 20-76.
2. Ellison, D., Love, S., Chimelli, L., Harding, B., Lowe, J., Roberts, G.W., & Vinters, H.V. (1998) Neuropathology. A Reference Text of CNS Pathology, Mosby, London, pp. 9.1-10.25.
3. Vinters, H.V., Farrell, M.A., Mischel, P.S., & Anders, K.H. (1998) Diagnostic Neuropathology, Marcel Dekker, New York, pp. 51-146.
4. Kalimo, H., Kaste, M., & Haltia, M. (1997) Vascular diseases, in Graham, D.I. & Lantos, P.L. (eds.), Greenfield's Neuropathology, 6th edition, Arnold, London, pp. 315-396.
5. Ross, R. (1999) Atherosclerosis--an inflammatory disease, *N. Engl. J. Med.* **340**, 115-126.
6. Gibbons, G.H., & Dzau, V.J. (1996) Molecular therapies for vascular diseases, *Science* **272**, 689-693.
7. Moore, S. (1999) Cholesterol revisited: Prime mover or a factor in the progression of atherosclerosis? *Ann. Royal Coll. Phys. Surg. Canada* **32**, 198-204.
8. Ross, R. (1993) The pathogenesis of atherosclerosis: a perspective for the 1990s, *Nature* **362**, 801-809.
9. Navab, M., Berliner, J.A., Watson, A.D., Hama, S.Y., Territo, M.C., Lusis, A.J., *et al.* (1996) The Yin and Yang of oxidation in the development of the fatty streak. A review based on the 1994 George Lyman Duff Memorial Lecture, *Arterioscler. Thromb. Vasc. Biol.* **16**, 831-842.
10. Resch, J.A., Okabe, N., Loewenson, R.B., Kimoto, K., Katsuki, S., & Baker, A.B. (1969) Pattern of vessel involvement in cerebral atherosclerosis. A comparative study between a Japanese and Minnesota population, *J. Atherosclerosis Res.* **9**, 239-250.
11. Stehbens, W.E. (1972) Pathology of the Cerebral Blood Vessels, C.V. Mosby, St. Louis.
12. Castaigne, P., Lhermitte, F., Gautier, J.C., Escourolle, R., Derouesne, C., Der Agopian, P. and Popa, C. (1973) Arterial occlusions in the vertebro-basilar system. A study of 44 patients with post-mortem data, *Brain* **96**, 133-154.
13. Fisher, C.M., Gore, I., Okabe, N., & White, P.D. (1965) Atherosclerosis of the carotid and vertebral arteries—extracranial and intracranial, *J. Neuropathol. Exp. Neurol.* **24**, 455-476.

14. Torvik, A., & Jorgensen, L. (1964) Thrombotic and embolic occlusions of the carotid arteries in an autopsy material. Part 1. Prevalence, location and associated diseases, *J. Neurol. Sci.* **1**, 24-39.
15. Torvik, A., & Jorgensen, L. (1966) Thrombotic and embolic occlusions of the carotid arteries in an autopsy series. Part 2. Cerebral lesions and clinical course, *J. Neurol. Sci.* **3**, 410-432.
16. Barnett, H.J.M., Peerless, S.J., & Kaufmann, J.C.E. (1978) "Stump" of internal carotid artery--a source for further cerebral embolic ischemia, *Stroke* **9**, 448-456.
17. Jorgensen, L., & Torvik, A. (1966) Ischaemic cerebrovascular diseases in an autopsy series. Part 1. Prevalence, location and predisposing factors in verified thrombo-embolic occlusions, and their significance in the pathogenesis of cerebral infarction, *J. Neurol. Sci.* **3**, 490-509.
18. Jorgensen, L., & Torvik, A. (1969) Ischaemic cerebrovascular diseases in an autopsy series. Part 2. Prevalence, location, pathogenesis, and clinical course of cerebral infarcts, *J. Neurol. Sci.* **9**, 285-320.
19. Vinters, H.V. (1997) Pathology of cervical and intracranial atherosclerosis and fibromuscular dysplasia, in Batjer, H.H. (ed.), Cerebrovascular Disease, Lippincott-Raven, Philadelphia, pp. 41-51.
20. Takebayashi, S. (1985) Ultrastructural morphometry of hypertensive medial damage in lenticulostriate and other arteries, *Stroke* **16**, 449-453.
21. Tagami, M., Nara, Y., Kubota, A., Sunaga, T., Maezawa, H., Fujino, H., & Yamori, Y. (1987) Ultrastructural characteristics of occluded perforating arteries in stroke-prone spontaneously hypertensive rats, *Stroke* **18**, 733-740.
22. Sobey, C.G., Faraci, F.M., & Heistad, D.D. (1998) Vascular biology of cerebral arteries, in Barnett, H.J.M., Mohr, J.P., Stein, B.M., and Yatsu, F.M. (eds.), STROKE. Pathophysiology, Diagnosis, and Management, 3d ed., Churchill Livingstone, New York, pp. 41-50.
23. Miller Fisher, C. (1965) Lacunes: Small, deep cerebral infarcts, *Neurology* **15**, 774-784.
24. Miller Fisher, C. (1978) Ataxic hemiparesis. A pathologic study, *Arch. Neurol.* **35**, 126-128.
25. Miller Fisher, C. (1965) Pure sensory stroke involving face, arm, and leg, *Neurology* **15**, 76- 80.
26. Challa, V.R., Bell, M.A., & Moody, D.M. (1990) A combined hematoxylin-eosin, alkaline phosphatase and high-resolution microradiographic study of lacunes, *Clin. Neuropathol.* **9**, 196-204.
27. Miller, V.T. (1983) Lacunar stroke. A reassessment, *Arch. Neurol.* **40**, 129-134.
28. Waterston, J.A., Brown, M.M., Butler, P., & Swash, M. (1990) Small deep cerebral infarcts associated with occlusive internal carotid artery disease. A hemodynamic phenomenon? *Arch. Neurol.* **47**, 953-957.
29. Weisberg, L.A. (1988) Diagnostic classification of stroke, especially lacunes, *Stroke* **19**, 1071-1073.
30. Bamford, J.M., & Warlow, C.P. (1988) Evolution and testing of the lacunar hypothesis, *Stroke* **19**, 1074-1082.
31. Tuszynski, M.H., Petito, C.K., & Levy, D.E. (1989) Risk factors and clinical manifestations of pathologically verified lacunar infarctions, *Stroke* **20**, 990-999.
32. Fisher, C.M. (1971) Pathological observations in hypertensive cerebral hemorrhage, *J. Neuropathol. Exp. Neurol.* **30**, 536-550.

33. Kase, C.S., Mohr, J.P., & Caplan, L.R. (1998) Intracerebral hemorrhage, in Barnett, H.J.M., Mohr, J.P., Stein, B.M., & Yatsu, F.M. (eds.), STROKE. Pathophysiology, Diagnosis, and Management, 3d ed., Churchill Livingstone, New York, pp. 649-700.
34. Cole, F.M., & Yates, P. (1967) Intracerebral microaneurysms and small cerebrovascular lesions, *Brain* **90,** 759-768.
35. Ross Russell, R.W. (1963) Observations on intracerebral aneurysms, *Brain* **86**, 425-442.
36. Challa, V.R., Moody, D.M., & Bell, M.A. (1992) The Charcot-Bouchard aneurysm controversy: Impact of a new histologic technique, *J. Neuropathol. Exp. Neurol.* **51**, 264-271.
37. Miller Fisher, C. (1989) Binswanger's encephalopathy: a review, *J. Neurol.* **236**, 65-79.
38. Babikian, V., & Ropper, A.H. (1987) Binswanger's disease: A review, *Stroke* **18**, 2-12.
39. Roman, G.C. (1999) New insight into Binswanger disease, *Arch. Neurol.* **56**, 1061-1062.
40. Dubas, F., Gray, F., Roullet, E., & Escourolle, R. (1985) Leucoencephalopathies arteriopathiques (17 cas anatomo-cliniques), *Rev. Neurol. (Paris)* **141**, 93-108.
41. Moody, D.M., Bell, M.A., & Challa, V.R. (1988) The corpus callosum, a unique white-matter tract: Anatomic features that may explain sparing in Binswanger disease and resistance to flow of fluid masses, *AJNR* **9**, 1051-1059.
42. Roman, G.C. (1996) From UBOs to Binswanger's disease. Impact of magnetic resonance imaging on vascular dementia research, *Stroke* **27**, 1269-1273.
43. Kidwell, C.S., & Saver, J.L. (1996) Causes and consequences of white matter hyperintensities: A meta-analysis, *Facts and Research in Gerontology (Supplement: Stroke),* 137-150.
44. Akiguchi, I., Tomimoto, H., Suenaga, T., Wakita, H., & Budka, H. (1997) Alterations in glia and axons in the brains of Binswanger's disease patients, *Stroke* **28**, 1423-1429.
45. Tomimoto, H., Akiguchi, I., Wakita, H., Osaki, A., Hayashi, M., & Yamamoto, Y. (1999) Coagulation activation in patients with Binswanger disease, *Arch. Neurol.* **56**, 1104-1108.
46. Hachinski, V.C., Potter, P., & Merskey, H. (1987) Leuko-Araiosis, *Arch. Neurol.* **44**, 21-23.
47. Steingart, A., Hachinski, V.C., Lau, C., Fox, A.J., Diaz, F., Cape, R., *et al.* (1987) Cognitive and neurologic findings in subjects with diffuse white matter lucencies on computed tomographic scan (Leuko-Araiosis), *Arch. Neurol.* **44**, 32-35.
48. Steingart, A., Hachinski, V.C., Lau, C., Fox, A.J., Fox, H., Lee, D., Inzitari, D., & Merskey, H. (1987) Cognitive and neurologic findings in demented patients with diffuse white matter lucencies on computed tomographic scan (Leuko-Araiosis), *Arch. Neurol.* **44**, 36-39.
49. Inzitari, D., Diaz, F., Fox, A., Hachinski, V.C., Steingart, A., Lau, C., Donald, A., *et al.* (1987) Vascular risk factors and Leuko-Araiosis, *Arch. Neurol.* **44**, 42-47.
50. Awad, I.A., Johnson, P.C., Spetzler, R.F., & Hodak, J.A. (1986) Incidental subcortical lesions identified on magnetic resonance imaging in the elderly. II. Postmortem pathological correlation, *Stroke* **17**, 1090-1097.
51. Kidwell, C.S., & Saver, J.L. (1996) Causes and consequences of leukoaraiosis: A review, *Facts and Research in Gerontology (Supplement: Stroke),* 115-135.
52. Brun, A., & Englund, E. (1986) A white matter disorder in dementia of the Alzheimer type: A pathoanatomical study, *Ann. Neurol.* **19**, 253-262.
53. Pantoni, L., & Garcia, J.H. (1997) Pathogenesis of leukoaraiosis--A review, *Stroke* **28**, 652- 659.
54. van Gijn, J. (1998) Leukoaraiosis and vascular dementia, *Neurology* **51(Suppl),** S3-S8.
55. Longstreth, W.T. Jr., Manolio, T.A., Arnold, A., Burke, G.L., Bryan, N., Jungreis, C.A., Enright, P.L, *et al.* (1996) Clinical correlates of white matter findings on cranial magnetic

resonance imaging of 3301 elderly people. The Cardiovascular Health Study, *Stroke* **27**, 1274-1282.

56. Baloh, R.W., & Vinters, H.V. (1995) White matter lesions and disequilibrium in older people. II. Clinicopathologic correlation, *Arch. Neurol.* **52**, 975-981.
57. Whitman, G.T., DiPatre, P.L., Lopez, I.A., Liu, F., Noori, N.E., Vinters, H.V., & Baloh, R.W. (1999) Neuropathology in older people with disequilibrium of unknown cause, *Neurology* **53**, 375-382.
58. Ruchoux, M-M., & Maurage, C-A. (1997) CADASIL: Cerebral autosomal dominant arteriopathy with subcortical infarcts and leukoencephalopathy, *J. Neuropathol. Exp. Neurol.* **56**, 947-964.
59. Chabriat, H., Vahedi K., Iba-Zizen, M.T., Joutel, A., Nibbio, A., Nagy, T.G., Krebs, M.O., Julien, J., *et al.* (1995) Clinical spectrum of CADASIL: a study of 7 families, *Lancet* **346**, 934-939.
60. Broe, G.A., & Bennett, H.P. (1995) Multiple subcortical infarction: CADASIL in context, *Lancet* **346**, 919.
61. Gray, F., Robert, F., Labrecque, R., Chretien, F., Baudrimont, M., Fallet-Bianco, C., Mikol, J., & Vinters, H.V. (1994) Autosomal dominant arteriopathic leuko-encephalopathy and Alzheimer's disease, *Neuropathol. Appl. Neurobiol.* **20**, 22-30.
62. Ruchoux, M-M., Guerouaou, D., Vandenhaute, B., Pruvo, J-P., Vermersch, P., & Leys, D. (1995) Systemic vascular smooth muscle cell impairment in cerebral autosomal dominant arteriopathy with subcortical infarcts and leukoencephalopathy, *Acta Neuropathol.* **89**, 500-512.
63. Joutel, A., Corpechot, C., Ducros, A., Vahedi, K., Chabriat, H., Mouton, P., Alamowitch, S., Domenga, V., *et al.* (1997) *Notch 3* mutations in cerebral autosomal dominant arteriopathy with subcortical infarcts and leukoencephalopathy (CADASIL), a Mendelian condition causing stroke and vascular dementia, *Ann. N.Y. Acad. Sci.* **826**, 213-217.
64. Jen, J., Cohen, A.H., Yue, Q., Stout, J.T., Vinters, H.V., Nelson, S., & Baloh, R.W. (1997) Hereditary endotheliopathy with retinopathy, nephropathy, and stroke (HERNS), *Neurology* **49**, 1322-1330.
65. Lammie, G.A., Brannan, F., Slattery, J., & Warlow, C. (1997) Nonhypertensive cerebral small- vessel disease. An autopsy study, *Stroke* 28, 2222-2229.
66. Hulette, C., Nochlin, D., McKeel, D., Morris, J.C., Mirra, S.S., Sumi, S.M., & Heyman, A. (1997) Clinical-neuropathologic findings in multi-infarct dementia: A report of six autopsied cases, *Neurology* **48**, 668-672.
67. Esiri, M.M., Wilcock, G.K., & Morris, J.H. (1997) Neuropathological assessment of the lesions of significance in vascular dementia, *J. Neurol. Neurosurg. Psychiatry* **63**, 749-753.
68. Chui, H.C., Victoroff, J.I., Margolin, D., Jagust, W., Shankle, R., & Katzman, R. (1992) Criteria for the diagnosis of ischemic vascular dementia proposed by the State of California Alzheimer's Disease Diagnostic and Treatment Centers, *Neurology* **42**, 473-480.
69. Roman, G.C., Tatemichi, T.K., Erkinjuntti, T., Cummings, J.L., Masdeu, J.C., Garcia, J.H., Amaducci, L., *et al.* (1993) Vascular dementia: Diagnostic criteria for research studies. Report of the NINDS-AIREN International Workshop, *Neurology* **43**, 250-260.
70. Ellis, W.G., Zarow, C., Zaias, B.W., Chui, H.C., Vinters, H.V., & Jagust, W.J. (1998) The spectrum of pathologic changes in ischemic vascular dementia, *J. Neuropathol. Exp. Neurol.* **57**, 478 [Abstract].
71. Pardridge, W.M. (1999) A morphological approach to the analysis of blood-brain barrier transport function, in Paulson, O., Knudsen, G.M., & Moos, T. (eds.), Brain Barrier Systems, Alfred Benzon Symposium 45, Munksgaard, Copenhagen, pp. 19-42.

72. Beck, D.W., Vinters, H.V., Hart, M.N., & Cancilla, P.A. (1984) Glial cells influence polarity of the blood-brain barrier, *J. Neuropathol. Exp. Neurol.* **43**, 219-224.
73. Alliot, F., Rutin, J., Leenen, P.J.M., & Pessac, B. (1999) Pericytes and periendothelial cells of brain parenchyma vessels co-express aminopeptidase N, aminopeptidase A, and nestin, *J. Neurosci. Res.* **58**, 367-378.
74. Ehler, E., Karlhuber, G., Bauer, H-C., Draeger, A. (1995) Heterogeneity of smooth muscle-associated proteins in mammalian brain microvasculature, *Cell & Tissue Res.* **279**, 393-403.
75. Vinters, H.V., Reave, S., Costello, P., Girvin, J.P., & Moore, S.A. (1987) Isolation and culture of cells derived from human cerebral microvessels, *Cell & Tissue Res.* **249**, 657-667.
76. Wang, Z., Natte, R., Berliner, J.A., van Duinen, S.G., & Vinters, H.V. (2000) Toxicity of Dutch (E22Q) and Flemish (A21G) mutant amyloid beta proteins to human cerebral microvessel and aortic smooth muscle cells, *Stroke* **31**, *in press*.
77. Vinters, H.V., Berliner, J.A., Beck, D.W., Maxwell, K., Bready, J.V., Cancilla, P.A. (1985) Insulin stimulates DNA synthesis in cerebral microvessel endothelium and smooth muscle, *Diabetes* **34**, 964-969.
78. Vinters, H.V., & Berliner, J.A. (1987) The blood vessel wall as an insulin target tissue, *Diabete & Metabolisme (Paris)* **13**, 294-300.
79. Arciniegas, E., Sutton, A.B., Allen, T.D., & Schor, A.M. (1992) Transforming growth factor beta 1 promotes the differentiation of endothelial cells into smooth muscle-like cells *in vitro*, *J. Cell Science* **103**, 521-529.
80. DeRuiter, M.C., Poelmann, R.E., VanMunsteren, J.C., Mironov, V., Markwald, R.R., & Gittenberger-de Groot, A.C. (1997) Embryonic endothelial cells transdifferentiate into mesenchymal cells expressing smooth muscle actins *in vivo* and *in vitro*, *Circ. Res.* **80**, 444- 451.
81. Majesky, M.W., & Schwartz, S.M. (1997) An origin for smooth muscle cells from endothelium? *Circ. Res.* **80**, 601-603.
82. Lindahl, P., Johansson, B.R., Leveen, P., & Betsholtz, C. (1997) Pericyte loss and microaneurysm formation in PDGF-B-deficient mice, *Science* **277**, 242-245.

SECTION II

GENETICS OF CAA

Chapter 5

APOE GENOTYPE IN RELATION TO SPORADIC AND ALZHEIMER-RELATED CAA

MARK O. McCARRON and JAMES A.R. NICOLL
University of Glasgow Department of Neuropathology, Institute of Neurological Sciences, Southern General Hospital, Glasgow, Scotland, UK

The brain is second only to the liver as a source of apolipoprotein E (*APOE* for gene; ApoE for protein), a protein which is intricately involved in lipid transport and metabolism. Following genetic linkage of late-onset familial Alzheimer's disease to chromosome 19 and the identification of ApoE in senile plaques, neurofibrillary tangles and amyloid-laden blood vessels, over-representation of the *APOE* ε4 allele was recognized in both familial late-onset and sporadic Alzheimer's disease. A dose-dependent association of the *APOE* ε4 allele with increasing cerebrovascular amyloid β-protein deposition was subsequently identified. Lobar hemorrhage is the major clinical manifestation of cerebral amyloid angiopathy (CAA) and is thought to follow the development of CAA-associated vasculopathic complications such as fibrinoid necrosis and concentric ("double barrel") blood vessel formation. There is emerging a complex relationship between *APOE* ε4, *APOE* ε2 and hemorrhage associated with CAA. For example, pathological studies have demonstrated that the *APOE* ε2 allele, the least frequent of the *APOE* alleles, is over-represented in patients with CAA-related hemorrhage. This is also true for patients with co-existing Alzheimer's disease, who otherwise have a very low ε2 allele frequency. Other forms of intracranial hemorrhage (hypertensive and subarachnoid hemorrhage) do not share the same association with *APOE* ε2. Patients with the ε2 allele and CAA-related hemorrhage are more likely to have taken anticoagulant or antiplatelet medication, have had hypertension or minor head trauma than non-ε2 carriers. In addition, there is now evidence suggesting that the ε2 allele enhances CAA-associated vasculopathic changes such as fibrinoid necrosis or concentric blood vessel formation. The role of *APOE* in other manifestations of CAA, including leukoencephalopathy and transient neurological deficits, is currently unclear. Besides being a risk factor for Alzheimer's disease, CAA and CAA-related hemorrhage, *APOE* genotype can also influence outcome following acute brain injury. By an as yet unidentified mechanism the ε4 allele is a marker of poor prognosis after head injury and intracerebral hemorrhage. Study of *APOE* has provided useful insights into the pathogenetic steps leading to CAA and CAA-related hemorrhage. Its influence on CAA and outcome from acute brain injury tantalizingly offers new prospects for future therapeutic interventions.

M.M. Verbeek et al. (eds.),
Cerebral Amyloid Angiopathy in Alzheimer's Disease and Related Disorders, 81–102.

1. INTRODUCTION

The 1990s, the decade of the brain, led to dramatic advances in our understanding of the molecular pathology of Alzheimer`s disease and amyloid-related conditions. Three genes were identified in which mutations caused autosomal dominant early-onset Alzheimer`s disease; amyloid precursor protein [1-4] (from which amyloid β-protein - a constituent of plaques and cerebral amyloid angiopathy - is derived), presenilin I [5] and presenilin II [6]. These however only make up a very small proportion of all Alzheimer cases [7]. The genetic basis of autosomal dominant forms of cerebral amyloid angiopathy (CAA) which cause cerebral hemorrhage have also been identified. These result from point mutations in the cystatin C gene in hereditary cerebral hemorrhage with amyloidosis - Icelandic type (See Ólafsson and Thorsteinsson, this volume) and in the amyloid precursor protein (APP) gene in the Dutch and Flemish forms of hereditary cerebral hemorrhage with amyloidosis (See Bornebroek *et al*, this volume). These APP mutations are different from those associated with familial Alzheimer`s disease. Other causes of CAA, including transthyretin variants [8] and prion protein disease [9] (see Ghetti *et al*, this volume) are not usually associated with cerebral hemorrhage. Despite these advances, very little was known of the genetic basis of sporadic Alzheimer`s disease, not to mention concomitant or sporadic CAA, until 1993 when the importance of apolipoprotein E (*APOE* for gene; ApoE for protein) polymorphism was first recognised.

2. THE APOLIPOPROTEIN E GENE AND PROTEIN

Human ApoE is a 299 amino acid protein encoded on chromosome 19q13.2. There are three common alleles designated ε2, ε3 and ε4, which result in the three isoforms ApoE2, ApoE3 and ApoE4. The three alleles give rise to three heterozygotes (ε2/ε3, ε3/ε4 and ε2/ε4) and three homozygotes (ε2/ε2, ε3/ε3 and ε4/ε4). Genotypes can be rapidly determined using the polymerase chain reaction to amplify a fragment of *APOE* containing both polymorphic sites, digesting the fragment with a restriction enzyme such as *Hha*I that distinguishes the allelic variation, and subsequent visualization of the bands [10] (Figure 1).

APOE ε3 is the most common allele, followed in most populations by ε4 and then ε2 [11]. *APOE* ε1, ε5, ε6 and ε7 are very rare variants. Allelic frequency varies among races. For example, the Japanese have lower ε2 and ε4 frequencies compared to Caucasians, while African Americans tend to have a higher ε4 frequency than both other races. Even among Caucasians the *APOE* allele frequencies can vary as shown by the high ε4 frequency in Finland (0.23) compared to France (0.13) [11]. In addition, *APOE* allele frequencies vary with age; *APOE* ε4 decreases and ε2 tends to rise with increasing age [12]. This has been interpreted as reflecting association of ε4 with mortality at a younger age and association of ε2 with longevity [12]. Despite the variations, average allele frequencies in Caucasians (who have been most frequently genotyped) are often quoted as: ε2 0.08, ε3 0.77 and ε4 0.15, based on summed data from 5805 individuals [11].

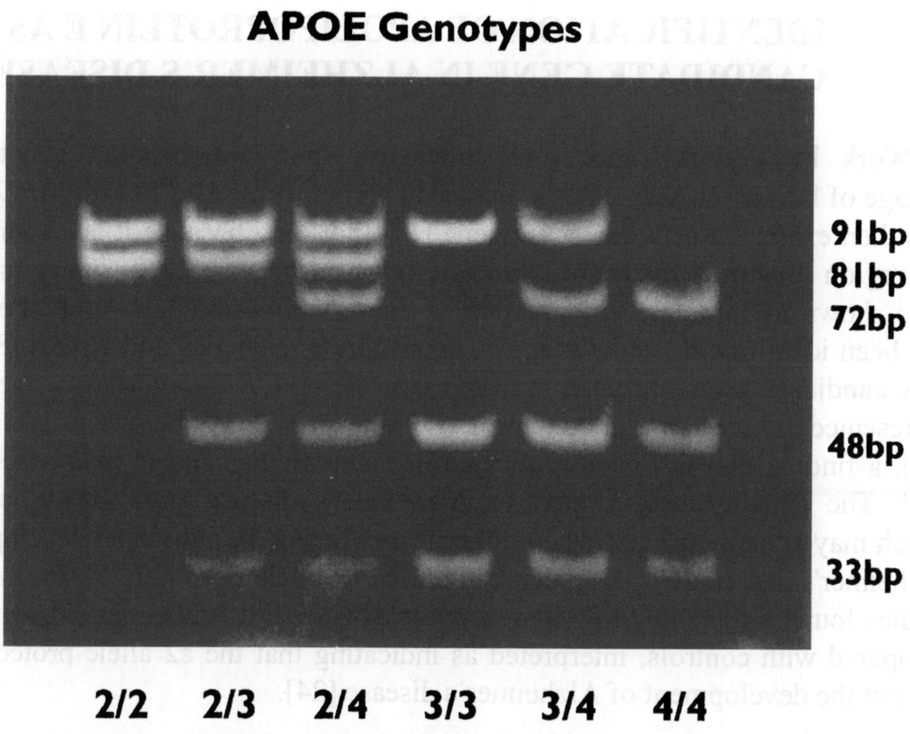

Figure 1. Gel of *APOE* genotypes. Following PCR amplification of a fragment of the *APOE* gene and digestion with the restriction enzyme, *Hha*I, the products are run on a 10% polyacrylamide gel and visualized under ultraviolet light.

The ApoE isoforms differ in cysteine and arginine contents at positions 112 and 158: ApoE3 contains a single cysteine in its structure at residue 112 and arginine at 158. Both positions have cysteine in the ApoE2 isoform and arginine in ApoE4. ApoE has two structural domains which can influence the properties of each other – a phenomenon known as domain interaction [13]. The amino-terminal (residues 1-191) is arranged in a four helical bundle and contains the lipoprotein receptor binding region (residues 136-

150). The carboxyl-terminal domain (residues 216-299) contains the major lipid binding determinants [14]. Domain interaction is thought to explain how the cysteine-arginine change at position 112 results in the preference of ApoE4 to bind very low density lipoproteins (with resulting elevated plasma cholesterol levels) compared to ApoE2 and ApoE3, which have a preference for high density lipoproteins. Although much work has focused on the role of ApoE in lipid transfer and metabolism in terms of atherosclerosis [13], it had been clear from 1985 that the brain was second only to the liver as a source of *APOE* mRNA [15]. Subsequent findings confirmed the importance of ApoE in the central nervous system.

3. IDENTIFICATION OF APOLIPOPROTEIN E AS CANDIDATE GENE IN ALZHEIMER`S DISEASE

Work by Pericak-Vance and colleagues first demonstrated genetic linkage of late-onset Alzheimer`s disease to chromosome 19q.13 [16]. *APOE* was then examined as a candidate gene [17], not only because it lay within the region identified by genetic linkage, but also because brain injury had been shown to stimulate upregulation of ApoE expression and ApoE itself had been identified in senile plaques, neurofibrillary tangles and CAA [18]. This candidate gene approach revealed that the *APOE* ε4 allele was over-represented in both familial late-onset [17] and sporadic Alzheimer`s disease [19], a finding that has been replicated in many studies and in many races [20]. The genetic susceptibility to Alzheimer`s disease from *APOE* ε4, which may actually predict when rather than whether an individual develops Alzheimer's disease [21], is dependent on the ε4 allele dose [22,23]. Further studies found a relatively low frequency of *APOE* ε2 in Alzheimer`s disease compared with controls, interpreted as indicating that the ε2 allele protects against the development of Alzheimer`s disease [24].

4. APOE ε4: DOSE–DEPENDENT RISK FACTOR FOR CAA

The identification of *APOE* ε4 as a dose-dependent risk factor for familial late-onset and sporadic Alzheimer`s disease as well as co-localisation of ApoE with amyloid β-protein (Aβ) in CAA prompted further work on cerebrovascular pathology both in patients with and without Alzheimer`s disease. Not only do 90% of Alzheimer patients have evidence of CAA [25], but nearly 50% of all individuals over 80 years of age have

CAA [26]. Both the extent and severity of CAA are a function of age. CAA is however usually asymptomatic. Polymorphism of the *APOE* gene has provided insights into factors that influence CAA severity and the uncommon clinical manifestation of intracerebral hemorrhage.

Aβ-laden blood vessels in Alzheimer patients were first examined with respect to *APOE* genotype by Schmechel and colleagues [27]. In an autopsy study of Alzheimer brains they found a strong dose-dependent association of the *APOE* ε4 allele with increased vascular amyloid deposition. Other groups confirmed this association in Alzheimer patients [28,29] and in aged patients without Alzheimer's disease [28,30]. A Japanese study showed a similar, although not statistically significant pattern in patients with and without Alzheimer's disease [31,32], which may be a reflection of a rather low frequency of the ε4 allele in the Japanese population despite CAA being a common finding [33]. In other words, the *APOE* ε4 allele may exert less influence on the severity of CAA in elderly Japanese individuals than in Caucasians. Such racial variation points to other genetic or environmental factors, and in Japanese individuals α_1-antichymotrypsin polymorphism appears to increase CAA severity [34].

5. MECHANISTIC CLUES FOR EFFECT OF APOE ε4 ON CAA

The underlying mechanism by which *APOE* ε4 has a dose-dependent influence on Alzheimer's disease and CAA has to date remained elusive. A number of potentially pertinent findings have yielded many hypotheses. In terms of CAA, the sequence of Aβ and ApoE deposition, the length of the Aβ peptide in CAA, *in vitro* interactions between ApoE and Aβ and a transgenic model of Alzheimer's disease with and without the *APOE* gene have all been examined.

In contrast to Aβ plaque formation in Down's syndrome patients, in which ApoE clearly follows the deposition of $A\beta_{42}$ [35], it has been suggested that ApoE accumulation may precede deposition of Aβ in CAA [36]. This conclusion was based on circumstantial evidence from the observation that ApoE accumulated at the junction between the arterial media and adventitia and beneath the internal elastic lamina, where Aβ deposition has been found in the early stage of CAA [37]. Non-demented individuals in their eighth and ninth decades had fine punctate ApoE immunostaining on meningeal vessels lacking Aβ when senile plaques were seen in nearby cortex [36]. In addition, a canine model of CAA revealed ApoE in leptomeningeal vessels before CAA was present [38]. A clever experiment in which *APOE* knockout mice were crossed with transgenic mice carrying a human mutant APP (V717F) demonstrated that ApoE is indeed required to enhance Aβ deposition in this animal model of Alzheimer`s disease [39].

In vitro work has examined ApoE - Aβ interactions. A stable complex forms, which has isoform-specific characteristics; ApoE4 binds more rapidly and over a narrower pH range than ApoE3 [40]. Different sources of ApoE can however alter its binding properties [41]. Nevertheless Aβ-ApoE complexes have been characterized as the principal component of amyloid in Alzheimer brains [42]. The next step in trying to identify the mechanism of ApoE`s influence in promoting Aβ deposition was to characterize the effect of ApoE on Aβ fibril formation. An electron microscopic study demonstrated that the ApoE4 isoform interacts with Aβ to produce a more complex meshwork of monofibrils than that observed with ApoE3 [43]. Other studies have similarly suggested that ApoE promotes Aβ fibril formation [44-46], although this has not been consistently demonstrated [47,48].

In terms of CAA, some of this controversy has been settled by a recent post mortem brain study, which has shown that ApoE influences the progression of CAA in vessels which have already been seeded with Aβ [49]. In CAA there is more of the shorter $A\beta_{40}$ than $A\beta_{42}$ (with the additional carboxyl residues) in contrast to plaques where $A\beta_{42}$ is more abundant [35,50-52]. Alonzo and colleagues [49] not only confirmed that advanced cases of CAA have increased deposition of $A\beta_{40}$ but that the mean amount of $A\beta_{40}$ per vessel is a function of *APOE* ε4 allele dose. In addition, because there is no apparent change in the proportion of vessels affected in mild and severe cases, they concluded that ApoE ε4 enhances $A\beta_{40}$ deposition in already established or seeded lesions [49]. A canine model of CAA had already established that after the initiation of CAA pathology, physiological concentrations of Aβ are sufficient to sustain CAA progression [38]. It has been proposed that Aβ-ApoE complexes within the cerebrospinal fluid or extracellular fluid are internalised and accumulate in smooth muscle cells [53], but this remains to be established. The determinants of seeding of Aβ remain unknown.

6. CAA-RELATED HEMORRHAGE

Intracerebral hemorrhage can be a devastating clinical manifestation of CAA (see Greenberg, this volume). Increasing severity of CAA [25,54] and secondary vessel changes termed "CAA-associated vasculopathy" [55] are thought to increase the risk of vessel rupture and cerebral hemorrhage. These microvascular changes (vessel-within-a-vessel or "double-barrel" appearance, thickened vessel walls apparently compromising lumens, dilated vessels, microaneurysms, fibrinoid necrosis, cortical microinfarcts and evidence of previous microscopic hemorrhage) are described by Vinters and Vonsattel (this volume). Work with canine CAA supports the belief that Aβ deposition leads to segmental loss of vessel wall viability and may lead to secondary vascular changes and subsequent hemorrhage [56]. Despite the common risk factor of the *APOE* ε4 allele, only 50% of CAA patients who develop cerebral hemorrhage have evidence of Alzheimer`s disease [57]. Other factors which do not enhance Alzheimer pathology, yet predispose to hemorrhage, may account for this discrepancy.

7. APOE GENOTYPE AND CAA-RELATED HEMORRHAGE

Because of the important findings associating the *APOE* ε4 allele and Aβ deposition both in cerebral blood vessels and brain parenchyma [27], *APOE* polymorphism was also examined in CAA-related hemorrhage [30,58,59]. Initial findings on a small number of patients reported an excess of the *APOE* ε4 allele [30,59]. However it became clear that confounding bias from co-existing Alzheimer`s disease and the need for pathologically verified (autopsy) or supported (biopsy) diagnoses would have to be considered. A larger series of pathologically confirmed cases of CAA-related hemorrhage also assessed for Alzheimer`s disease revealed a surprising excess of the ε2 allele [58,60], which had previously been documented to protect patients from Alzheimer`s disease [24]. Because of the association between CAA and Alzheimer`s disease it might have been predicted that patients with CAA-related hemorrhage would have a low, rather than a high ε2 frequency. The elevated ε2 frequency in CAA-related hemorrhage was, however present in patients both with and without pathological evidence of Alzheimer`s disease (Figure 2). One third of the patients had evidence of multiple cerebral hematomas and these had an even higher ε2 allele frequency (0.35 versus 0.2), reinforcing the association of CAA-related hemorrhage with this allele [60]. The CAA-related hemorrhage patients with Alzheimer`s disease had an ε4 frequency similar to controls with Alzheimer`s disease but without CAA-related hemorrhage, whereas patients with CAA-related hemorrhage but without Alzheimer`s disease had an ε4 frequency similar to controls without CAA-related hemorrhage or Alzheimer`s disease. This was interpreted as indicating that the over-representation of ε4 in this population was due to the association of Alzheimer`s disease with CAA. The over-representation of the ε2 allele in CAA-related hemorrhage appears to be specific for this pathological process and is not seen in patients with hypertension-associated deep intracerebral hemorrhages or with subarachnoid hemorrhage due to saccular aneurysms [61]. Although an important genetic risk factor, *APOE* is neither sensitive nor specific for the diagnosis of CAA [62] or CAA-related hemorrhage. As described below, however, *APOE* genotyping may have clinical significance in the management of such patients.

Table. Reported *APOE* allele frequency in patients with pathologically confirmed CAA-related hemorrhage compared to literature controls

	No. of *APOE* alleles	Allele frequencies (number) ε2	ε3	ε4
Greenberg [30]	16	0.12 (2)	0.44 (7)	0.44 (7)
Premkumar [28]	26	0.00 (0)	0.38 (10)	0.62 (16)
Nicoll [60]	72	0.25 (18)	0.57 (41)	0.18 (13)
Greenberg [64]	12	0.17 (2)	0.50 (6)	0.33 (4)
McCarron [61]	8	0.13 (1)	0.38 (3)	0.50 (4)
Unpublished*	48	0.27 (13)	0.48 (23)	0.25 (12)
Total	182	0.201 (36)	0.49 (90)	0.312 (56)
Controls	3798	0.09 (345)	0.77 (2917)	0.14 (536)

*Includes 14 patients kindly provided by SM Greenberg, Boston, Massachusetts, 7 patients from Glasgow, Scotland and 3 patients from Duke University Medical Center, North Carolina provided by Dr MJ Alberts.

By $\chi 2$: †P<0.001 and ‡P<0.0001 compared with controls.

To date there have been relatively few series of CAA-related hemorrhage reported; fewer exist with documented *APOE* genotypes. The table summarizes the genotypes in such patients, who have been pathologically diagnosed (autopsy or biopsy). Comparison with elderly controls confirms a statistically significant higher frequency of both the *APOE* ε2 and ε4 alleles in CAA-related hemorrhage patients.

These data gave rise to an emerging hypothesis that whereas the *APOE* ε4 allele predisposes to Aβ deposition in cerebral vasculature (CAA), the ε2 allele predisposes to rupture of blood vessels already laden with Aβ [58,60,63]. However, it needs to be emphasised that neither ε2 nor ε4 are necessary or sufficient for the development of CAA-related hemorrhage, which can occur in individuals with the ε3/ε3 genotype. Examination of the clinical features and microvascular pathology in CAA-related hemorrhage has provided not only some support for this explanation, but also an insight into potential pathogenic mechanisms involved in this type of cerebral hemorrhage.

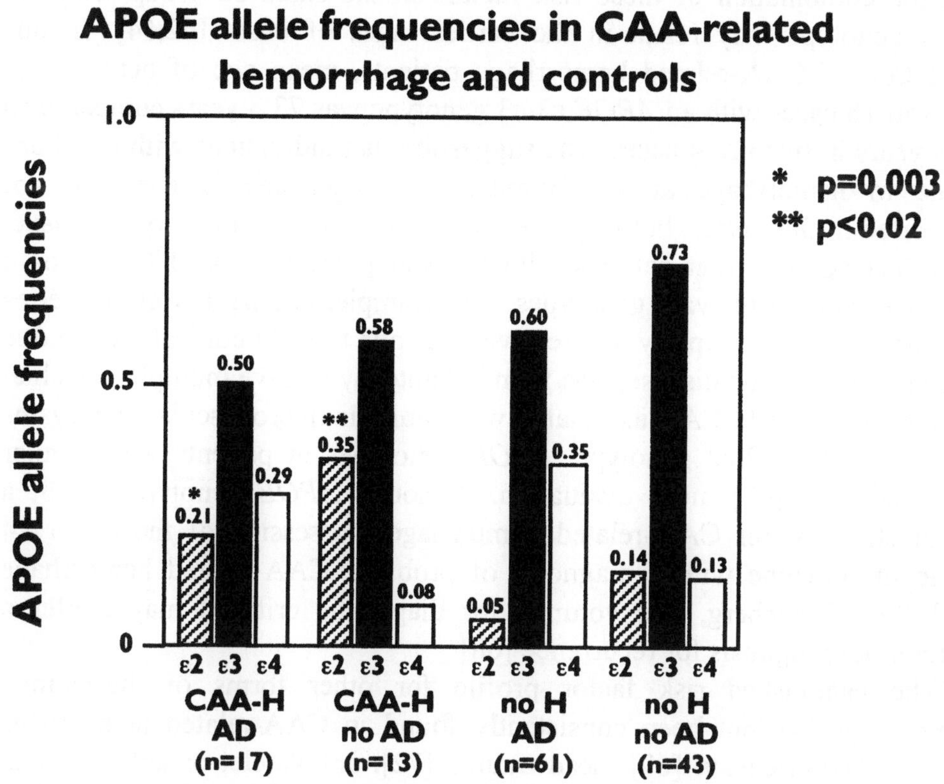

Figure 2. *APOE* allele frequency in CAA-related hemorrhage (CAA-H) with and without Alzheimer's disease (AD) compared with autopsy controls. *Comparison of ε2 frequency between patients with CAA-H/AD and AD/no H. **Comparison of ε2 frequency between patients with CAA-H/no AD and controls with no H and no AD. (Data reproduced from Nicoll, J.A., Burnett, C., Love, S., Graham, D.I., Dewar, D., Ironside, J.W., Stewart, J. & Vinters, H.V. (1997) High frequency of apolipoprotein E ε2 allele in hemorrhage due to cerebral amyloid angiopathy. *Annals of Neurology*, **41**, 716-721 with permission from Lippincott, Williams & Wilkins)

8. APOE GENOTYPE AND CLINICAL FEATURES IN CAA-RELATED HEMORRHAGE

If the *APOE* ε2 and ε4 alleles are indeed important in the pathogenesis of CAA-related hemorrhage, then patients carrying these alleles may be expected to have their first cerebral hemorrhage at a younger age than individuals without one or both of these alleles. This has been demonstrated with respect to ε2 [60,64] and ε4 [64,65], although the latter association has not been found in one of the larger pathologically diagnosed series of patients with CAA-related hemorrhage [60,66]. It has also been suggested that the combination of these risk factors in the small subgroup with the ε2/ε4 genotype may result in the earliest ages of onset [62,64]. In our collection of CAA-related hemorrhage patients, mean age of hemorrhage onset in 18 cases with an *APOE* ε3/ε3 genotype was 73.3 years compared to 70.8 years in five ε2/ε4 cases. The suggestion that individuals with ε2/ε4 are prone to hemorrhage at a substantially younger age is therefore not supported in this series, but larger numbers are required to clarify this issue. It is also possible that patients with unusual presentations of CAA could have particularly relevant genotypes. For example, one patient in our series who presented with rapidly progressive dementia thought clinically to be due to Creutzfeldt Jakob disease, was found at autopsy to have multiple petechial hemorrhages due to CAA associated with multiple microvascular aneurysms and the *APOE* ε2/ε2 genotype. *APOE* genotypes in patients with similar presentations [67] merit evaluation. Although *APOE* genotype is not a diagnostic tool for CAA-related hemorrhage, possession of the ε2 or ε4 allele in someone with a diagnosis of probable CAA-related hemorrhage [30] (See Greenberg, this volume, for diagnostic criteria) may confer a differential prognosis for recurrence [68].

The established risk factor profile for other forms of intracranial hemorrhage has not been consistently found in CAA-related hemorrhage [69-71]. Hypertension [69], head trauma [72], antiplatelet or anticoagulant medication [70,73,74] and thrombolytic therapy [75,76] have been described as potential precipitants of hemorrhage in only some individuals with CAA. Interaction of such putative clinical risk factors with *APOE* genotype could conceivably lead to CAA-related hemorrhage. Intriguingly, this has been noted in one of the pathologically-diagnosed series of CAA-related hemorrhage, in which those patients carrying an ε2 allele were more likely to have one or more clinical features implicated in other forms of intracranial hemorrhage (antiplatelet or anticoagulant medication, hypertension or minor head trauma) [66]. This requires confirmation, but tantalizingly supports the idea that ApoE ε2 renders amyloid-laden blood vessels vulnerable to rupture.

9. ASSOCIATION OF APOE GENOTYPE AND MICROVASCULAR PATHOLOGY IN CAA-RELATED HEMORRHAGE

It has been suspected for some time that CAA-associated vasculopathic complications are the predecessors to hemorrhage [25,55,57,70]. Because *APOE* genotype appears to influence the development of CAA and subsequent hemorrhage, it may exert an influence on the morphology of CAA vessels. In particular, the ε2 allele may provide insight into the critical CAA-associated vasculopathic complication or complications which lead to vessel rupture.

Two studies have systematically assessed the presence of the CAA-associated vascular complications with respect to the *APOE* ε2 allele. Greenberg *et al* [64] investigated the ε2 allele frequency in 75 brains in which amyloid had completely replaced the walls of blood vessels. They looked for brains that had both cracking of amyloid-laden vessels (a "double barrel" appearance) and evidence of microscopic paravascular blood leak (n=23). The *APOE* ε2 frequency of this group was 0.09, which was significantly greater than that found in the group without this combination of vascular complications (0.01, n=52). These results were, however based on just five ε2 alleles (0.03) from a total of 150. A further study [77] analyzed each of the individual vascular complications of CAA in patients who also demonstrated complete replacement of vessel walls with Aβ with and without macroscopic lobar hemorrhage. This study examined serial sections from each patient not only for CAA and morphological vascular complications, but also vascular immunoreactivity for ApoE, cystatin C and perivascular activated microglia with respect to the ε2 allele. Eighteen of these 62 CAA patients had an ε2 allele. The CAA-associated vascular complications were compared in patients with and without macroscopic lobar hemorrhages. Vessels with an increased ratio of wall thickness to lumen diameter (which appeared stenosed), dilated/microaneurysmal vessels and fibrinoid necrosis were more common in the group with the macroscopic lobar hemorrhages as was immunoreactivity for vascular cystatin C and perivascular activated microglia. ApoE immunoreactivity in the walls of blood vessels closely mirrored that of Aβ. Among all the CAA-associated vasculopathic changes, only fibrinoid necrosis was associated with possession of the *APOE* ε2 allele. This suggests that ε2 may predispose to hemorrhage by promoting fibrinoid necrosis in Aβ-laden blood vessels. This vulnerability to hemorrhage would also fit with how the clinical features such as antiplatelet/anticoagulant medication, minor head trauma or even hypertension appear to increase the chances of rupture of these blood vessels.

These two pathological studies support the view that CAA-associated vascular complications do indeed precede macroscopic lobar hemorrhage and are under the influence of the *APOE* ε2 allele. The vasculopathic findings of fibrinoid necrosis and cracking with paravascular blood leak may even represent different stages of the same pathology. For example, fibrinoid necrosis may resolve into the "double barrel" appearance with or without a microscopic paravascular blood leak. This is, however, speculative. Further work, including ultrastructural analysis of CAA in different *APOE* genotypes and possibly transgenic animal models (if the vascular complications can be identified) will clarify the pathogenic role of *APOE* genotype in CAA-related hemorrhage. The current evidence in this complex disease is summarised in a schematic diagram in Figure 3.

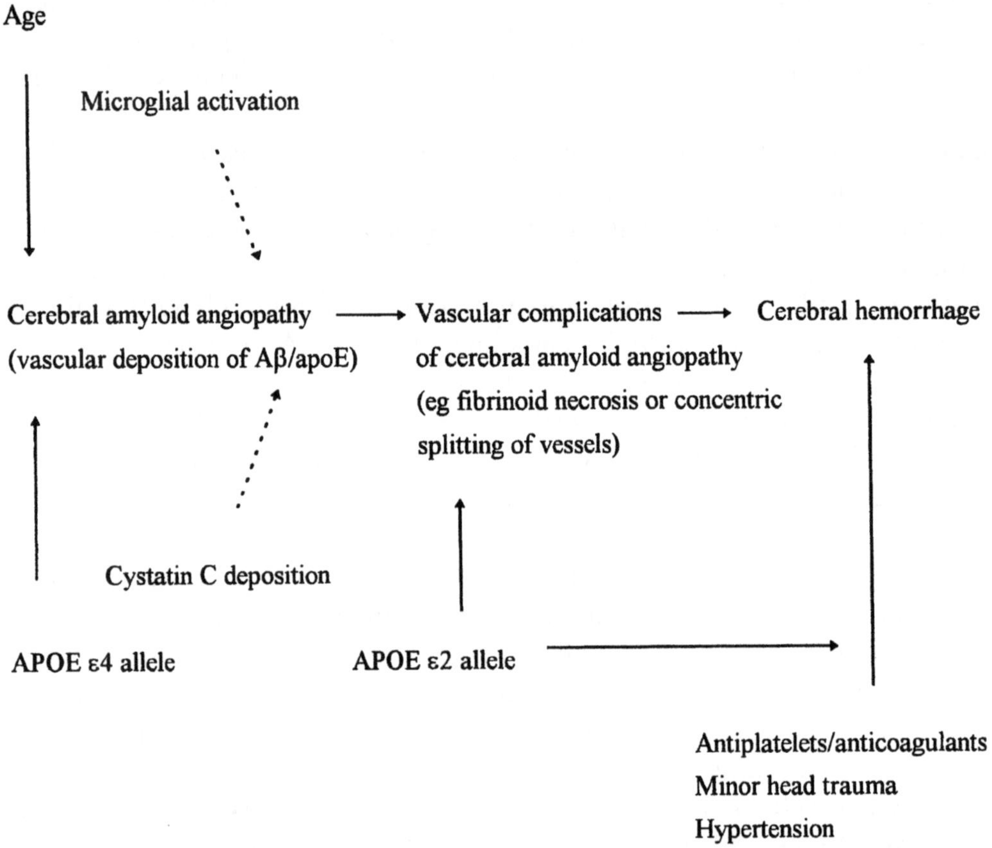

Figure 3. Schematic diagram of identified factors involved in CAA-related hemorrhage

10. TRANSIENT NEUROLOGICAL DEFICITS, LEUKOENCEPHALOPATHY AND DOWN`S SYNDROME WITH CAA

To date no conclusive associations have been reported between *APOE* polymorphism and CAA-related leukoencephalopathy [67,78-80], or transient neurological deficits [73,74,79,81], which sometimes precede cerebral hemorrhage [73,74,79], primarily because these are uncommon manifestations and series of a sufficiently large size have not been examined.

Epistatic (gene-gene) interactions of *APOE* genotype have been established in some instances. For example, *APOE* genotype influences the age of onset in patients with early-onset Alzheimer's disease due to APP mutations [82-85]. This has not been found in the Dutch and Flemish types of hereditary cerebral hemorrhage with amyloidosis [86,87] which are due to different APP mutations. *APOE* genotype does influence cognitive decline in Down's patients [88-91] in whom three copies of the APP gene on chromosome 21 have been associated with Alzheimer's disease pathology including CAA [92]. CAA-related hemorrhage has been rarely reported in Down's patients [93-96], possibly because of under-recognition or early mortality from other conditions such as hematological malignancies, pneumonia or congenital heart disease [96]. It is not clear if ApoE ε2 can interact with the excess of APP in Down's patients to lead to CAA-related hemorrhage; the *APOE* genotype of only one such patient has been reported [96]. That patient had an ε2/ε4 genotype and suffered a fatal CAA-related hemorrhage at the age of 46 years, one of the youngest patients recorded with CAA-related hemorrhage. Although this is suggestive of an epistatic interaction between APP and *APOE* alleles in CAA-related hemorrhage, a series of such patients will be required to confirm or refute this hypothesis.

11. APOE GENOTYPE AND OUTCOME AFTER ACUTE BRAIN INJURY

Besides the recognition of the role of *APOE* polymorphism in Alzheimer's disease and sporadic CAA with and without cerebral hemorrhage, a number of studies have addressed the influence of *APOE* polymorphism on clinical outcome from acute brain injury [97-102]. These followed work on *APOE* deficient or knockout mice, which displayed slower synaptic recovery following entorhinal cortex lesions than controls [103]. More severe cognitive and motor deficits following closed head injury have also been demonstrated in *APOE* deficient mice compared to controls [104]. Similarly in models of ischemia stroke infarct volumes were larger and neuronal injury more extensive in *APOE* deficient mice compared to age and sex-matched controls [105,106]. In patients, possession of the *APOE* ε4 allele has been associated with an adverse outcome following acute head injury [97,100] and intracerebral hemorrhage [99,102]. Most of the intracerebral hemorrhages in these studies were hypertensive in origin. In an autopsy study of *APOE* genotypes the ε4 allele frequencies exceeded 20% in patients with deep hypertensive intracerebral hemorrhage, CAA-related hemorrhage and subarachnoid hemorrhage from saccular aneurysms compared to 17% in controls drawn from the same population. [61]

Although the difference was not statistically significant, this finding raises the possibility that the adverse association between ε4 and outcome is a general effect found in all types of intracranial hemorrhage. However, as the ε4 allele may also have an etiological role in CAA [27,49] (Figure 3), confounding bias cannot be eliminated with certainty in this form of hemorrhage. Improving antemortem diagnosis of CAA-related hemorrhage [74] should allow future prospective studies to determine if the association of *APOE* ε4 with poor outcome following acute brain injury also applies to CAA-related hemorrhage.

The mechanism underlying the adverse association between the *APOE* ε4 allele and outcome following intracerebral hemorrhage (as well as following head injury) is not clear. Different properties of ApoE such as anti-oxidant [107], immunomodulatory [108,109], neurotrophic [110] and intracellular calcium effects [111] are possible candidates for this isoform-specific association. Confirmatory evidence that any of these effects are important in intracerebral hemorrhage is currently lacking. In addition, examination of computerized tomographic scans from patients with intracerebral hemorrhage suggests that acute hematoma volume and the extent of perihematoma edema are not associated with the *APOE* ε4 allele [112]. Identification of the mechanism may be important as a means of improving medical management of these patients, who currently suffer significant disability and a very high (30% to 50%) mortality [113].

12. CONCLUSION

Whereas many association studies have been criticised because they have shed no light on the underlying pathogenic mechanisms of complex diseases [114], the genetic findings in CAA and subsequent cerebral hemorrhage have enhanced our knowledge of this distinct pathology and produced testable hypotheses. The current view is that possession of the *APOE* ε4 allele is a risk factor for the deposition of Aβ in the walls of blood vessels whereas possession of the *APOE* ε2 allele is a risk factor for hemorrhage from amyloid-laden blood vessels by promoting specific "CAA-associated vasculopathies". This followed upon the important discovery of the association between Alzheimer`s disease and the *APOE* ε4 allele. ApoE research has demonstrated how a gene may not only predispose to more than one complex disease, but the same gene may also influence outcome from different forms of acute brain injury. Although it has kindled an exciting interest in CAA and its underlying molecular biology, it must be remembered, that like Alzheimer`s disease itself, other, as yet unidentified genetic and/or environmental factors must also influence CAA and its clinical manifestations. Investigation of other genes, particularly those which may also have associations with Alzheimer's disease such as α_2-macroglobulin [115] and the low-density lipoprotein receptor-related protein (LRP) [116] (an ApoE receptor) may further our understanding of the pathogenesis of CAA.

REFERENCES

1. Goate, A., Chartier-Harlin, M.C., Mullan, M., Brown, J., Crawford, F., Fidani, L., Giuffra, L., Haynes, A., Irving, N. & James, L. (1991) Segregation of a missense mutation in the amyloid precursor protein gene with familial Alzheimer's disease. Nature **349**, 704-706.
2. Chartier-Harlin, M.C., Crawford, F., Houlden, H., Warren, A., Hughes, D., Fidani, L., Goate, A., Rossor, M., Roques, P. & Hardy, J. (1991) Early-onset Alzheimer's disease caused by mutations at codon 717 of the beta-amyloid precursor protein gene. *Nature* **353**, 844-846.
3. Hendriks, L., van, D.C., Cras, P., Cruts, M., Van, H.W., van, H.F., Warren, A., McInnis, M.G., Antonarakis, S.E. & Martin, J.J. (1992) Presenile dementia and cerebral hemorrhage linked to a mutation at codon 692 of the beta-amyloid precursor protein gene. *Nature Genet.* **1**, 218-221.
4. Mullan, M., Crawford, F., Axelman, K., Houlden, H., Lilius, L., Winblad, B. & Lannfelt, L. (1992) A pathogenic mutation for probable Alzheimer's disease in the APP gene at the N-terminus of beta-amyloid. *Nature Genet.* **1**, 345-347.
5. Sherrington, R., Rogaev, E.I., Liang, Y., Rogaeva, E.A., Levesque, Ikeda, M., Chi, H., Lin, C., Li, G. & Holman, K. (1995) Cloning of a gene bearing missense mutations in early-onset familial Alzheimer's disease. *Nature* **375**, 754-760.
6. Levy-Lahad, E., Wasco, W., Poorkaj, P., Romano, DM, Oshima, J., Pettingell, W.H., Yu, C.E., Jondro, P.D., Schmidt, S.D., Wang, K., et & al (1995) Candidate gene for the chromosome 1 familial Alzheimer's disease locus. *Science* **269**, 973-977.
7. Lendon, C.L., Ashall, F. & Goate, A.M. (1997) Exploring the etiology of Alzheimer disease using molecular genetics. *JAMA* **277**, 825-831.
8. Benson, M.D. (1996) Leptomeningeal amyloid and variant transthyretins. *Am. J. Pathol.* **148**, 351-354.
9. Ghetti, B., Piccardo, P., Frangione, B., Bugiani, O., Giaccone, Young, K., Prelli, F., Farlow, M.R., Dlouhy, S.R. & Tagliavini, F. (1996) Prion protein amyloidosis. *Brain Pathol.* **6**, 127-145.
10. Wenham, P.R., Price, W.H. & Blandell, G. (1991) Apolipoprotein E genotyping by one-stage PCR. *Lancet*, **337** 1158-1159.
11. Davignon, J., Gregg, R.E. & Sing, C.F. (1988) Apolipoprotein E polymorphism and atherosclerosis. *Arteriosclerosis* **8**, 1-21.
12. Schachter, F., Faure-Delanef, L., Guenot, F., Rouger, H., Froguel, P., Lesueur-Ginot, L. & Cohen, D. (1994) Genetic associations with human longevity at the APOE and ACE loci. *Nature Genet.* **6**, 29-32.
13. Weisgraber, K.H. (1994) Apolipoprotein E: structure-function relationships. *Adv. Prot. Chem.* **45**, 249-302.
14. Weisgraber, K.H. & Mahley, R.W. (1996) Human apolipoprotein E: the Alzheimer's disease connection. *FASEB J.* **10**, 1485-1494.

15. Elshourbagy, N.A., Liao, W.S., Mahley, R.W. & Taylor, J.M. (1985) Apolipoprotein E mRNA is abundant in the brain and adrenals, as well as in the liver, and is present in other peripheral tissues of rats and marmosets. *Proc. Natl. Acad. Sci. USA* **82**, 203-207.
16. Pericak-Vance, M.A., Bebout, J.L., Gaskell, P.C., Jr, Yamaoka, L.H., Hung, W.Y., Alberts, M.J., Walker, A.P., Bartlett, R.J., Haynes, C.A. & Welsh, K.A. (1991) Linkage studies in familial Alzheimer disease: evidence for chromosome 19 linkage. *Am. J. Hum. Genet.* **48**, 1034-1050.
17. Strittmatter, W.J., Saunders, A.M., Schmechel, D., Pericak-Vance, M., Enghild, J., Salvesen, G.S. & Roses, A.D. (1993) Apolipoprotein E: high-avidity binding to beta-amyloid and increased frequency of type 4 allele in late-onset familial Alzheimer disease. *Proc. Natl. Acad. Sci. USA* **90**, 1977-1981.
18. Namba, Y., Tomonaga, M., Kawasaki, H., Otomo, E. & Ikeda, K. (1991) Apolipoprotein E immunoreactivity in cerebral amyloid deposits and neurofibrillary tangles in Alzheimer's disease and kuru plaque amyloid in Creutzfeldt-Jakob disease. *Brain Res.* **541**, 163-166.
19. Saunders, A.M., Strittmatter, W.J., Schmechel, D., George, H., PH, Pericak-Vance, M.A., Joo, S.H., Rosi, B.L., Gusella, J.F., Crapper, M.L., DR & Alberts, M.J. (1993) Association of apolipoprotein E allele ε4 with late-onset familial and sporadic Alzheimer's disease. *Neurology* **43**, 1467-1472.
20. Farrer, L.A., Cupples, L.A., Haines, J.L., Hyman, B., Kukull, W.A., Mayeux, R., Myers, R.H., Pericak-Vance, M.A., Risch, N. & van, D.C. (1997) Effects of age, sex, and ethnicity on the association between apolipoprotein E genotype and Alzheimer disease. A meta-analysis. APOE and Alzheimer Disease Meta Analysis Consortium. *JAMA* **278**, 1349-1356.
21. Meyer, M.R., Tschanz, J.T., Norton, M.C., Welshbohmer, K.A., Steffens, D.C., Wyse, B.W. & Breitner, J.S. (1998) APOE genotype predicts when-not whether-one is predisposed to develop Alzheimer disease. *Nature Genet.* **19**, 321-322.
22. Corder, E.H., Saunders, A.M., Strittmatter, W.J., Schmechel, D.E., Gaskell, P.C., Small, GW, Roses, A.D., Haines, J.L. & Pericak-Vance, M.A. (1993) Gene dose of apolipoprotein E type 4 allele and the risk of Alzheimer's disease in late onset families. *Science* **261**, 921-923.
23. Slooter, A.J.C., Cruts, M., Kalmijn, S., Hofman, A., Breteler, M.M.B., Vanbroeckhoven, C. & Vanduijn, C.M. (1998) Risk estimates of dementia by apolipoprotein E genotypes from a population-based incidence study: the Rotterdam study. *Arch. Neurol.* **55**, 964-968.
24. Corder, E.H., Saunders, A.M., Risch, N.J., Strittmatter, W.J., Schmechel, D.E., Gaskell, P.C.J., Rimmler, J.B., Locke, P.A., Conneally, P.M. & Schmader, K.E. (1994) Protective effect of apolipoprotein E type 2 allele for late onset Alzheimer disease. *Nature Genet.* **7**, 180-184.
25. Vonsattel, J.P., Myers, R.H., Hedley-Whyte, E.T., Ropper, A.H., Bird, ED & Richardson, E.P., Jr. (1991) Cerebral amyloid angiopathy without and with cerebral hemorrhages: a comparative histological study. *Ann. Neurol.* **30**, 637-649.
26. Vinters, H.V. & Gilbert, J.J. (1983) Cerebral amyloid angiopathy: incidence and complications in the aging brain. II. The distribution of amyloid vascular changes. *Stroke* **14**, 924-928.
27. Schmechel, D.E., Saunders, A.M., Strittmatter, W.J., Crain, B.J., Hulette, C.M., Joo, S.H., Pericak-Vance, M.A., Goldgaber, D. & Roses, A.D. (1993) Increased amyloid beta-peptide deposition in cerebral cortex as a consequence of apolipoprotein E genotype in late-onset Alzheimer disease. *Proc. Natl. Acad. Sci. USA* **90**, 9649-9653.

28. Premkumar, D.R., Cohen, D.L., Hedera, P., Friedland, R.P., Kalaria & RN (1996) Apolipoprotein E-ε4 alleles in cerebral amyloid angiopathy and cerebrovascular pathology associated with Alzheimer's disease. *Am. J. Pathol.* **148**, 2083-2095.
29. Olichney, J.M., Hansen, L.A., Galasko, D., Saitoh, T., Hofstetter, CR, Katzman, R. & Thal, L.J. (1996) The apolipoprotein E ε4 allele is associated with increased neuritic plaques and cerebral amyloid angiopathy in Alzheimer's disease and Lewy body variant. *Neurology* **47**, 190-196.
30. Greenberg, S.M., Rebeck, G.W., Vonsattel, J.P., Gomez-Isla, T. & Hyman, B.T. (1995) Apolipoprotein E ε4 and cerebral hemorrhage associated with amyloid angiopathy. *Ann. Neurol.* **38**, 254-259.
31. Itoh, Y., Yamada, M., Suematsu, N., Matsushita, M. & Otomo, E. (1996) Influence of apolipoprotein E genotype on cerebral amyloid angiopathy in the elderly. *Stroke* **27**, 216-218.
32. Yamada, M., Itoh, Y., Suematsu, N., Matsushita, M. & Otomo, E. (1996) Lack of an association between apolipoprotein E ε4 and cerebral amyloid angiopathy in elderly Japanese. *Ann. Neurol.* **39**, 683-684.
33. Yamada, M., Tsukagoshi, H., Otomo, E. & Hayakawa, M. (1987) Cerebral amyloid angiopathy in the aged. *J. Neurol.* **234**, 371-376.
34. Yamada, M., Sodeyama, N., Itoh, Y., Suematsu, N., Otomo, E., Matsushita, M. & Mizusawa, H. (1998) Association of alpha1-antichymotrypsin polymorphism with cerebral amyloid angiopathy. *Ann. Neurol.* **44**, 129-131.
35. Lemere, C.A., Blusztajn, J.K., Yamaguchi, H., Wisniewski, T., Saido, T.C. & Selkoe, D.J. (1996) Sequence of deposition of heterogeneous amyloid beta-peptides and APO E in Down syndrome: implications for initial events in amyloid plaque formation. *Neurobiol. Dis.* **3**, 16-32.
36. Yamaguchi, H., Ishiguro, K., Sugihara, S., Nakazato, Y., Kawarabayashi, T., Sun, X. & Hirai, S. (1994) Presence of apolipoprotein E on extracellular neurofibrillary tangles and on meningeal blood vessels precedes the Alzheimer beta-amyloid deposition. *Acta Neuropathol.* **88**, 413-419.
37. Yamaguchi, H., Yamazaki, T., Lemere, C.A., Frosch, M.P. & Selkoe, D.J. (1992) Beta amyloid is focally deposited within the outer basement membrane in the amyloid angiopathy of Alzheimer's disease. An immunoelectron microscopic study. *Am. J. Pathol.* **141**, 249-259.
38. Prior, R., D'Urso, D., Frank, R., Prikulis, I. & Pavlakovic, G. (1995) Experimental deposition of Alzheimer amyloid beta-protein in canine leptomeningeal vessels. *Neuroreport* **6**, 1747-1751.
39. Bales, K.R., Verina, T., Dodel, R.C., Du, Y., Altstiel, L., Bender, M., Hyslop, P., Johnstone, E.M., Little, S.P., Cummins, D.J., Piccardo, P., Ghetti, B. & Paul, S.M. (1997) Lack of apolipoprotein E dramatically reduces amyloid beta-peptide deposition. *Nature Genet.* **17**, 263-264.
40. Strittmatter, W.J., Weisgraber, K.H., Huang, D.Y., Dong, L.M., Salvesen, G.S., Pericak-Vance, M., Schmechel, D., Saunders, A.M., Goldgaber, D. & Roses, A.D. (1993) Binding of human apolipoprotein E to synthetic amyloid beta peptide: isoform-specific effects and implications for late-onset Alzheimer disease. *Proc. Natl. Acad. Sci. USA* **90**, 8098-8102.
41. LaDu, M.J., Falduto, M.T., Manelli, A.M., Reardon, C.A., Getz, G.S. & Frail, D.E. (1994) Isoform-specific binding of apolipoprotein E to beta-amyloid. *J. Biol. Chem.* **269**, 23403-23406.
42. Naslund, J., Thyberg, J., Tjernberg, L.O., Wernstedt, C., Karlstrom, A.R., Bogdanovic, N., Gandy, S.E., Lannfelt, L., Terenius, L. & Nordstedt (1995) Characterization of stable

complexes involving apolipoprotein E and the amyloid beta peptide in Alzheimer's disease brain. *Neuron* **15**, 219-228.

43. Sanan, D.A., Weisgraber, K.H., Russell, S.J., Mahley, R.W., Huang, D., Saunders, A., Schmechel, D., Wisniewski, T., Frangione, B. & Roses, A.D. (1994) Apolipoprotein E associates with beta amyloid peptide of Alzheimer's disease to form novel monofibrils. Isoform apoE4 associates more efficiently than apoE3. *J. Clin. Invest.* **94**, 860-869.
44. Ma, J., Yee, A., Brewer, H.B.J., Das, S. & Potter, H. (1994) Amyloid-associated proteins alpha 1-antichymotrypsin and apolipoprotein E promote assembly of Alzheimer beta-protein into filaments. *Nature* **372**, 92-94.
45. Castano, E.M., Prelli, F., Wisniewski, T., Golabek, A., Kumar, R.A., Soto, C. & Frangione, B. (1995) Fibrillogenesis in Alzheimer's disease of amyloid beta peptides and apolipoprotein E. *Biochem. J.* **306**, 599-604.
46. Golabek, A.A., Soto, C., Vogel, T. & Wisniewski, T. (1996) The interaction between apolipoprotein E and Alzheimer's amyloid beta-peptide is dependent on beta-peptide conformation. *J. Biol. Chem.* **271**, 10602-10606.
47. Naiki, H., Gejyo, F. & Nakakuki, K. (1997) Concentration-dependent inhibitory effects of apolipoprotein E on Alzheimer's beta-amyloid fibril formation *in vitro*. *Biochemistry* **36**, 6243-6250.
48. Beffert, U. & Poirier, J. (1998) ApoE associated with lipid has a reduced capacity to inhibit β-amyloid fibril formation. *Neuroreport* **9**, 3321-3323.
49. Alonzo, N.C., Hyman, B.T., Rebeck, G.W. & Greenberg, S.M. (1998) Progression of cerebral amyloid angiopathy - accumulation of amyloid-beta 40 in affected vessels. *J. Neuropathol. Exp. Neurol.* **57**, 353-359.
50. Iwatsubo, T., Odaka, A., Suzuki, N., Mizusawa, H., Nukina, N. & Ihara, Y. (1994) Visualization of A beta 42(43) and A beta 40 in senile plaques with end-specific A beta: evidence that an initially deposited species is A beta 42(43). *Neuron* **13**, 45-53.
51. Gravina, S.A., Ho, L., Eckman, C.B., Long, K.E., Otvos, L., Jr., Younkin, L.H., Suzuki, N. & Younkin, S.G. (1995) Amyloid beta protein (A beta) in Alzheimer's disease brain. Biochemical and immunocytochemical analysis with antibodies specific for forms ending at A beta 40 or A beta 42(43). *J. Biol. Chem.* **270**, 7013-7016.
52. Iwatsubo, T., Mann, D.M., Odaka, A., Suzuki, N. & Ihara, Y. (1995) Amyloid beta protein (A beta) deposition: A beta 42(43) precedes A beta 40 in Down syndrome. *Ann. Neurol.* **37**, 294-299.
53. Urmoneit, B., Prikulis, I., Wihl, G., D'Urso, D., Frank, R., Heeren, J., Beisiegel, U. & Prior, R. (1997) Cerebrovascular smooth muscle cells internalize Alzheimer amyloid beta protein via a lipoprotein pathway: implications for cerebral amyloid angiopathy. *Lab. Invest.* **77**, 157-166.
54. Iwamoto, N., Ishihara, T., Ito, H. & Uchino, F. (1993) Morphological evaluation of amyloid-laden arteries in leptomeninges, cortices and subcortices in cerebral amyloid angiopathy with subcortical hemorrhage. *Acta Neuropathol.* **86**, 418-421.
55. Mandybur, T.I. (1986) Cerebral amyloid angiopathy: the vascular pathology and complications. *J. Neuropathol. Exp. Neurol.* **45**, 79-90.
56. Prior, R., D'Urso, D., Frank, R., Prikulis, I. & Pavlakovic, G. (1996) Loss of vessel wall viability in cerebral amyloid angiopathy. *Neuroreport* **7**, 562-564.
57. Vinters, H.V. (1987) Cerebral amyloid angiopathy. A critical review. *Stroke* **18**, 311-324.
58. Nicoll, J.A., Burnett, C., Love, S., Graham, D.I., Ironside, J.W. & Vinters, H.V. (1996) High frequency of apolipoprotein E ε2 in patients with cerebral hemorrhage due to cerebral amyloid angiopathy. *Ann. Neurol.* **39**, 682-683.

59. Kalaria, R.N. & Premkumar, D.R. (1995) Apolipoprotein E genotype and cerebral amyloid angiopathy. *Lancet* **346**, 1424
60. Nicoll, J.A., Burnett, C., Love, S., Graham, D.I., Dewar, D., Ironside, J.W., Stewart, J. & Vinters, H.V. (1997) High frequency of apolipoprotein E ε2 allele in hemorrhage due to cerebral amyloid angiopathy. *Ann. Neurol.* **41**, 716-721.
61. McCarron, M.O. & Nicoll, J.A.R. (1998) High frequency of apolipoprotein E ε2 allele is specific for patients with cerebral amyloid angiopathy-related hemorrhage. *Neurosci. Lett.* **247**, 45-48.
62. Greenberg, S.M. (1998) Cerebral amyloid angiopathy: prospects for clinical diagnosis and treatment. *Neurology* **51**, 690-694.
63. Greenberg, S.M. & Hyman, B.T. (1997) Cerebral amyloid angiopathy and apolipoprotein E: bad news for the good allele? *Ann. Neurol.* **41**, 701-702.
64. Greenberg, S.M., Vonsattel, J.G., Segal, A.Z., Chiu, R.I., Clatworthy, A.E., Liao, A., Hyman, B.T. & Rebeck, G.W. (1998) Association of apolipoprotein E ε2 and vasculopathy in cerebral amyloid angiopathy. *Neurology* **50**, 961-965.
65. Greenberg, S.M., Briggs, M.E., Hyman, B.T., Kokoris, G.J., Takis, C., Kanter, D.S., Kase, C.S. & Pessin, M.S. (1996) Apolipoprotein E ε4 is associated with the presence and earlier onset of hemorrhage in cerebral amyloid angiopathy. *Stroke* **27**, 1333-1337.
66. McCarron, M.O., Nicoll, J.A.R., Ironside, J.W., Love, S., Alberts, M.J. & Bone, I. (1999) Cerebral amyloid angiopathy-related hemorrhage: interaction of apolipoprotein E ε2 with putative clinical risk factors. *Submitted.*
67. Silbert, P.L., Bartleson, J.D., Miller, G.M., Parisi, J.E., Goldman, M.S. & Meyer, F.B. (1995) Cortical petechial hemorrhage, leukoencephalopathy, and subacute dementia associated with seizures due to cerebral amyloid angiopathy. *Mayo Clin. Proc.* **70**, 477-480.
68. O`Donnell, H.C., Rosand, J., Chiu, R.I., Furie, K.L., Ikeda, D. & Greenberg, S.M. (1999) Apolipoprotein E genotype predicts hemorrhage recurrence in cerebral amyloid angiopathy. *Stroke*, **30** 255-255.(Abstract)
69. Jellinger, K. (1977) Cerebrovascular amyloidosis with cerebral hemorrhage. *J. Neurol.* **214**, 195-206.
70. Okazaki, H., Reagan, T.J. & Campbell, R.J. (1979) Clinicopathologic studies of primary cerebral amyloid angiopathy. *Mayo Clin. Proc.* **54**, 22-31.
71. Ferreiro, J.A., Ansbacher, L.E. & Vinters, H.V. (1989) Stroke related to cerebral amyloid angiopathy: the significance of systemic vascular disease. *J. Neurol.* **236**, 267-272.
72. Greene, G.M., Godersky, J.C., Biller, J., Hart, M.N. & Adams, H.P., Jr. (1990) Surgical experience with cerebral amyloid angiopathy. *Stroke* **21**, 1545-1549.
73. Smith, D.B., Hitchcock, M. & Philpott, P.J. (1985) Cerebral amyloid angiopathy presenting as transient ischemic attacks. Case report. *J. Neurosurg.* **63**, 963-964.
74. Anonymous (1996) Case records of the Massachusetts General Hospital. Weekly clinicopathological exercises. Case 22-1996. Cerebral hemorrhage in a 69-year-old woman receiving warfarin. *New Engl. J. Med.* **335**, 189-196.
75. Leblanc, R., Haddad, G. & Robitaille, Y. (1992) Cerebral hemorrhage from amyloid angiopathy and coronary thrombolysis. *Neurosurgery* **31**, 586-590.
76. Sloan, M.A., Price, T.R., Petito, C.K., Randall, A.M., Solomon, R.E., Terrin, M.L., Gore, Collen, D., Kleiman, N. & Feit, F. (1995) Clinical features and pathogenesis of intracerebral hemorrhage after rt-PA and heparin therapy for acute myocardial infarction: the Thrombolysis in Myocardial Infarction (TIMI) II Pilot and Randomized Clinical Trial combined experience. *Neurology* **45**, 649-658.

77. McCarron, M.O., Nicoll, J.A.R., Stewart, J., Ironside, J.W., Mann, D.M.A., Love, S., Graham, D.I. & Dewar, D. (1998) The apolipoprotein E ε2 allele and the pathological features in cerebral amyloid angiopathy-related hemorrhage. *Submitted.*
78. Gray, F., Dubas, F., Roullet, E. & Escourolle, R. (1985) Leukoencephalopathy in diffuse hemorrhagic cerebral amyloid angiopathy. *Ann. Neurol.* **18**, 54-59.
79. Greenberg, S.M., Vonsattel, J.P., Stakes, J.W., Gruber, M. & Finklestein, S.P. (1993) The clinical spectrum of cerebral amyloid angiopathy: presentations without lobar hemorrhage. *Neurology* **43**, 2073-2079.
80. Shimode, K., Kobayashi, S., Imaoka, K., Umegae, N. & Nagai, A. (1996) Leukoencephalopathy-related cerebral amyloid angiopathy with cystatin C deposition. *Stroke* **27**, 1417-1419.
81. Cosgrove, G.R., Leblanc, R., Meagher-Villemure, K. & Ethier, R. (1985) Cerebral amyloid angiopathy. *Neurology* **35**, 625-631.
82. St George-Hyslop, P., McLachlan, D.C., Tsuda, Rogaev, E., Karlinsky, H., Lippa, C.F., Pollen, D. & Tuda, T. (1994) Alzheimer's disease and possible gene interaction. *Science* **263**, 537
83. Nacmias, B., Latorraca, S., Piersanti, P., Forleo, P., Piacentini, S., Bracco, L., Amaducci, L. & Sorbi, S. (1995) ApoE genotype and familial Alzheimer's disease: a possible influence on age of onset in APP717 Val-->Ile mutated families. *Neurosci. Lett.* **183**, 1-3.
84. Sorbi, S., Nacmias, B., Forleo, P., Piacentini, S., Latorraca, S. & Amaducci, L. (1995) Epistatic effect of APP717 mutation and apolipoprotein E genotype in familial Alzheimer's disease. *Ann. Neurol.* **38**, 124-127.
85. Mann, D.M., Iwatsubo, T., Ihara, Y., Cairns, N.J., Lantos, P.L., Bogdanovic, N., Lannfelt, L., Winblad, B., Maat-Schieman, M.L. & Rossor, M.N. (1996) Predominant deposition of amyloid-beta 42(43) in plaques in cases of Alzheimer's disease and hereditary cerebral hemorrhage associated with mutations in the amyloid precursor protein gene. *Am. J. Pathol.* **148**, 1257-1266.
86. Haan, J., Van, B.C., van, D.C., Voorhoeve, E., van, H.F., van, S.J., Maat-Schieman, M.L., Roos, RA & Bakker, E. (1994) The apolipoprotein E ε4 allele does not influence the clinical expression of the amyloid precursor protein gene codon 693 or 692 mutations. *Ann. Neurol.* **36**, 434-437.
87. Haan, J., Roos, R.A. & Bakker, E. (1995) No protective effect of apolipoprotein E ε2 allele in Dutch hereditary cerebral amyloid angiopathy. *Ann. Neurol.* **37**, 282
88. Schupf, N., Kapell, D., Lee, J.H., Zigman, W., Canto, B., Tycko, B. & Mayeux, R. (1996) Onset of dementia is associated with apolipoprotein E ε4 in Down's syndrome. *Ann. Neurol.* **40**, 799-801.
89. Alexander, G.E., Saunders, A.M., Szczepanik, J., Strassburger, T.L., Pietrini, P., Dani, A., Furey, M.L., Mentis, M.J., Roses, A.D., Rapoport, S.I. & Schapiro, M.B. (1997) Relation of age and apolipoprotein E to cognitive function in Down syndrome adults. *Neuroreport* **8**, 1835-1840.
90. Del, B.R., Comi, G.P., Bresolin, N., Castelli, E., Conti, E., Degiuli, A., Ausenda, CD & Scarlato, G. (1997) The apolipoprotein E ε4 allele causes a faster decline of cognitive performances in Down's syndrome subjects. *J. Neurol. Sci.* **145**, 87-91.
91. Schupf, N., Kapell, D., Nightingale, B., Rodriguez, A., Tycko, B. & Mayeux, R. (1998) Earlier onset of Alzheimer's disease in men with Down syndrome. *Neurology* **50**, 991-995.
92. Prasher, V.P., Farrer, M.J., Kessling, A.M., Fisher, E.M., West, R.J., Barber, P.C. & Butler, A.C. (1998) Molecular mapping of Alzheimer-type dementia in Down's syndrome. *Ann. Neurol.* **43**, 380-383.

93. Belza, M.G. & Urich, H. (1986) Cerebral amyloid angiopathy in Down's syndrome. *Clin. Neuropathol.* **5**, 257-260.
94. Anonymous (1990) Congophilic angiopathy. *New York State J. Med.* **90**, 64-68.
95. Donahue, J.E., Khurana, J.S. & Adelman, L.S. (1998) Intracerebral hemorrhage in two patients with Down's syndrome and cerebral amyloid angiopathy. *Acta Neuropathol.* **95**, 213-216.
96. McCarron, M.O., Nicoll, J.R. & Graham, D.I. (1998) A quartet of Down's syndrome, Alzheimer's disease, cerebral amyloid angiopathy, and cerebral hemorrhage - interacting genetic risk factors. *J. Neurol. Neurosurg. Psych.* **65**, 405-406.
97. Nicoll, J.A., Roberts, G.W. & Graham, D.I. (1995) Apolipoprotein E ε4 allele is associated with deposition of amyloid beta-protein following head injury . *Nat. Med.* **1**, 135-137.
98. Sorbi, S., Nacmias, N., Piacentini, S., Repice, A., Latorraca, S., Forleo, P. & Amaducci, L. (1995) ApoE as a prognostic factor for post-traumatic coma. *Nat. Med.* **1**, 852
99. Alberts, M.J., Graffagnino, C., McClenny, C., DeLong, D., Strittmatter, W., Saunders, A.M. & Roses, A.D. (1995) ApoE genotype and survival from intracerebral hemorrhage. *Lancet* **346**, 575
100. Teasdale, G.M., Nicoll, J.R., Murray, G. & Fiddes, M. (1997) Association of apolipoprotein E polymorphism with outcome after head injury. *Lancet* **350**, 1069-1071.
101. Laskowitz, D.T., Horsburgh, K. & Roses, A.D. (1998) Apolipoprotein E and the CNS response to injury. *J. Cerebr. Blood F. Met.* **18**, 465-471.
102. McCarron, M.O., Muir, K.W., Weir, C.J., Dyker, A.G., Bone, I., Nicoll, J.A.R . & Lees, K.R. (1998) The apolipoprotein E ε4 allele and outcome in cerebrovascular disease. *Stroke* **29**, 1882-1887.
103. Masliah, E., Mallory, M., Alford, M., Ge, N. & Mucke, L. (1995) Abnormal synaptic regeneration in hAPP695 transgenic and APOE knockout mice, in K. Iqbal, J.A. Mortimer, B. Winblad, & H.M. Wisniewski (eds.), *Research advances in Alzheimer's disease and related disorders*, John Wiley & Sons, Chichester, New York, Brisbane, Toronto, Singapore, pp. 405-414.
104. Chen, Y., Lomnitski, L., Michaelson, D.M. & Shohami, E. (1997) Motor and cognitive deficits in apolipoprotein E-deficient mice after closed head injury. *Neuroscience* **80**, 1255-1262.
105. Laskowitz, D.T., Sheng, H., Bart, R.D., Joyner, K.A., Roses, A.D. & Warner, D.S. (1997) Apolipoprotein E-deficient mice have increased susceptibility to focal cerebral ischemia. *J. Cerebr. Blood F. Met.* **17**, 753-758.
106. Horsburgh, K., Kelly, S., McCulloch, J., Higgins, G.A., Roses A.D., & Nicoll, J.A.R. (1999) Increased neuronal damage in apolipoprotein E deficient mice following global ischaemia. *Neuroreport (in press).*
107. Miyata, M. & Smith, J.D. (1996) Apolipoprotein E allele-specific antioxidant activity and effects on cytotoxicity by oxidative insults and beta-amyloid peptides. *Nat. Gen.* **14**, 55-61.
108. Laskowitz, D.T., Matthew, W.D., Bennett, E.R., Schmechel, D., Herbstreith, M.H., Goel, S. & Mcmillian, M.K. (1998) Endogenous apolipoprotein E suppresses LPS-stimulated microglial nitric oxide production. *Neuroreport* **9**, 615-618.
109. Laskowitz, D.T., Goel, S., Bennett, E.R. & Matthew, W.D. (1997) Apolipoprotein E suppresses glial cell secretion of TNF-alpha. *J. Neuroimmunol.* **76**, 70-74.
110. Nathan, B.P., Bellosta, S., Sanan, D.A., Weisgraber, K.H., Mahley, RW & Pitas, R.E. (1994) Differential effects of apolipoproteins E3 and E4 on neuronal growth *in vitro*. *Science* **264**, 850-852.

111. Muller, W., Meske, V., Berlin, K., Scharnagl, H., Marz, W., Ohm & TG (1998) Apolipoprotein E isoforms increase intracellular CA^{2+} differentially through a ω-agatoxin IVa-sensitive CA^{2+}-channel. *Brain Pathol.* **8**, 641-653.

112. McCarron, M.O., Hoffmann, K.L., DeLong, D.M., Gray, L., Saunders, A.M. & Alberts, M.J. (1998) Intracerebral hemorrhage outcome: APOE, haematoma and oedema volumes. *Submitted.*

113. Thrift, A.G., Donnan, G.A. & McNeil, J.J. (1995) Epidemiology of intracerebral hemorrhage. *Epidemiol. Rev.* **17**, 361-381.

114. Rosenthal, N. & Schwartz, R.S. (1998) In search of perverse polymorphisms. *New Engl. J. Med.* **338**, 122-124.

115. Blacker, D., Wilcox, M.A., Laird, N.M., Rodes, L., Horvath, S.M., Go, RCP, Perry, R., Watson, B., Bassett, S.S., McInnis, M.G., Albert, M.S., Hyman, B.T. & Tanzi, R.E. (1998) Alpha-2 macroglobulin is genetically associated with Alzheimer disease. *Nat. Genet.* **19**, 357-360.

116. Kang, D.E., Saitoh, T., Chen, X., Xia, Y., Masliah, E., Hansen, L.A., Thomas, R.G., Thal, L.J. & Katzman, R. (1997) Genetic association of the low-density lipoprotein receptor-related protein gene (LRP), an apolipoprotein E receptor, with late-onset Alzheimer's disease. *Neurology* **49**, 56-61.

Chapter 6

CLINICAL AND GENETIC ASPECTS OF HEREDITARY CEREBRAL HEMORRHAGE WITH AMYLOIDOSIS DUTCH TYPE (HCHWA-D)

MARJOLIJN BORNEBROEK, JOOST HAAN, EGBERT BAKKER and RAYMUND A.C. ROOS
Department of Neurology and Clinical Genetics, Leiden University, Medical Center, Leiden, The Netherlands

Hereditary cerebral hemorrhage with amyloidosis-Dutch type (HCHWA-D) is a rare autosomal dominant disorder clinically characterised by recurrent strokes and dementia. Symptoms occur after the age of 40 years and the mean age at first stroke is about 50 years. One third of the patients die as a consequence of the first stroke and those who survive suffer recurrent strokes and become severely disabled. The mean age at death is about 60 years. The second most frequent symptom is dementia that can develop in relation with hemorrhagic strokes or can sometimes be found as the first or only symptom. Cerebral magnetic resonance imaging (MRI) scans demonstrate diffuse white matter hyperintensities in symptomatic and asymptomatic HCHWA-D mutation carriers. Moreover, MRI scans show focal lesions which are mostly hemorrhagic lesions but sometimes infarctions. HCHWA-D is caused by a G to C mutation at codon 693 of the amyloid precursor protein (APP) gene located at chromosome 21. Two other mutation have been described in the Aβ part of the APP gene causing recurrent strokes and dementia. Five different mutations have been described outside the amyloid β protein (Aβ) part and are linked to familial early onset AD. Other loci for AD were identified on chromosome 14 and 1, named presenilin (PS) -1 and presenilin-2. All APP mutations causing AD have in common that they alter APP metabolism in a way that either more Aβ or more of the longer $A\beta_{42(43)}$ is formed. Mutations in PS-1 and PS-2 also result in a selective two-fold increase of $A\beta_{42(43)}$. Remarkably, in HCHWA-D, *in vitro* studies demonstrated a decrease of $A\beta_{40}$ and $A\beta_{42(43)}$, with a larger decrease of the latter resulting in a lower $A\beta_{42(43)/40}$ ratio. Also *in vivo*, in the plasma of HCHWA-D patients the concentration of $A\beta_{42(43)}$ was decreased.

M.M. Verbeek et al. (eds.),
***Cerebral Amyloid Angiopathy in Alzheimer's Disease and Related Disorders*, 103–119.**

1. INTRODUCTION

In cerebral β-amyloid angiopathy (CAA) congophilic material deposits in the media of cerebral vessels. Of all kinds of amyloid that exist, Aβ is most frequently found in the brain [1]. CAA is commonly found in brains of elderly and frequently remains clinically asymptomatic but can cause intracerebral hemorrhage. In about 10%-15% of this population, CAA is the cause of the spontaneous cerebral hemorrhages [2,3]. Hemorrhages are typically superficial or lobar, and can be multiple. The location of CAA hemorrhages differs from that of hemorrhages associated with hypertension, which are in the basal ganglia. The mechanism of rupture of blood vessels in CAA is not clearly understood but may be due to reduced compliance and weakening of the vessel walls as a consequence of the amyloid deposition. Aβ in CAA is similar to Aβ found in senile plaques in Alzheimer's disease (AD). The so-called cerebral Aβ-amyloidosis are grouped as: (familial) Alzheimer's disease (AD), Down syndrome, sporadic CAA, hereditary cerebral hemorrhage with amyloidosis-Dutch type (HCHWA-D) and normal aging. Although HCHWA-D is a rare disease, the detailed knowledge of the clinical and pathological data and the possibility of DNA testing make HCHWA-D a useful model to study Aβ-amyloidosis *in vivo*. In this chapter we will describe the clinical and genetic aspects of HCHWA-D and other cerebral Aβ amyloidoses related to HCHWA-D.

2. CLINICAL ASPECTS

2.1 History

HCHWA-D was first described as the familial occurrence of intracerebral hemorrhages by Luyendijk and Schoen in 1964 [4]. About twenty years later CAA was shown to be the pathological substrate of the disease [5]. Initially, a relation between hereditary cerebral hemorrhage with amyloidosis-Icelandic type (HCHWA-I), which is another autosomal dominant disease clinically characterized by recurrent strokes, was hypothesized [6]. However, in 1987 the main component of amyloid in HCHWA-D was described to be similar to Aβ. In 1990 a C to G mutation in the Aβ precursor protein (APP) gene, located on chromosome 21, was identified [7,8]. A variant of cystatin C was shown to be the amyloid component in HCHWA-I, and an A to G point mutation in the cystatin C gene on chromosome 20 was found to cause this disease [9,10]. Next to HCHWA-D two other familial forms of recurrent cerebral hemorrhage due to CAA have been identified. The so-called

Flemish APP mutation causes early onset dementia and recurrent hemorrhagic strokes [11]. Recently an Italian family with recurrent hemorrhagic strokes and dementia due a APP mutation was described [12,13].

2.2 Epidemiology and genealogy

HCHWA-D is a rare autosomal dominant disease afflicting members of three large families that originate from two Dutch coastal villages called Katwijk and Scheveningen. The majority of the patients still live in these villages although some of them moved to other parts of the Netherlands, South-Africa, Australia, and the United States [14]. No common founder of the three HCHWA-D families was identified, although all patients carry the same APP mutation and the ancestors of the three families had lived in the same Dutch area. The three families are traced back now to the beginning of the seventeenth century but further ancestral connection is impossible to prove. Haplotype analysis to prove that these families are related has not been performed.

The exact prevalence of HCHWA-D is not known. About 50 patients who suffered one or more strokes are visiting our out-patient clinic regularly. Information about another 150 patients who died as a consequence of HCHWA-D is available. About 400 members of the families are estimated to have 50% risk of developing HCHWA-D.

2.3 Clinical symptoms and radiological aspects of HCHWA-D

Recurrent strokes and dementia dominate the clinical picture of HCHWA-D. The strokes are mainly hemorrhagic, but sometimes ischemic lesions are also seen. It is not known whether all hemorrhagic strokes are primary cerebral hemorrhages or if some of them are secondary hemorrhagic infarctions [15].

The first stroke usually occurs around the age of 50 years [16]. The youngest patient was 39 years old at which age she suffered her first stroke. The oldest patient was asymptomatic till he suffered a stroke at the age of 76 years. Almost one third of the patients die of their first stroke. Those who survive the first stroke often suffer recurrent strokes leading to severe disablement [17]. The mean survival after the first stroke is about 10 years and the mean number of recurrent strokes is two. A wide variation in number of recurrences exists (one to 10) and the interval between them can vary from days to many years. The longest recorded survival is 28 years. This patient suffered a first stroke at the age of 53 years and one more till she died at the age of 81 years. The mean age at death is about 60 years, but again the variation is wide [18]. The youngest patient died a few days after a first stroke, at the age of 39 years. The oldest patient survived to the age of nearly 88 years.

Dementia is the second most frequent symptom occurring in about 75% of the patients [19]. Dementia occurs in relation to the hemorrhagic strokes but is encountered as the first or only symptom of HCHWA-D in a few patients. Dementia in HCHWA-D can develop in several ways. First, it can progress in a step-wise fashion in relation to strokes, as in vascular dementia. In about one third of the demented HCHWA-D patients dementia develops in this way. Secondly, dementia can develop in a slowly progressive way and apparently unrelated to the strokes, like in AD. Slowly progressive dementia is also found in one third of the HCHWA-D patients. Progressive dementia can occur between the strokes, but a few patients developed dementia before the first stroke [20]. It is, however, possible that "silent" strokes cause this deterioration. Routinely performed CT or MRI scans of patients often displayed small cerebral hemorrhages that had remained clinically asymptomatic. A combination of the step-wise and slowly progressive course is observed in the remaining one third of the patients. The exact cause of dementia in HCHWA-D is unclear but it is probably a combination of vascular lesions and chronic progressive parenchymal AD-like pathology [15,19,20].

Imaging studies (CT- and MRI-scan) of the brain of HCHWA-D mutation carriers show focal lesions and diffuse white matter damage [16]. The focal lesions represent recent and old strokes that can be hemorrhagic (85%) or ischemic (15%). Lobar hemorrhages are preponderantly localized in the temporal, parietal and occipital lobes. Both hemispheres are affected equally.

Diffuse white matter damage is an early radiological sign of HCHWA-D, mostly seen as white matter hyperintensities (WMHs) on MRI. WMHs are located both periventricularly and subcortically and are present on all MRI scans of symptomatic patients (Figure 1). In nearly all asymptomatic mutation carriers aged over 40 years WMHs are present on MRI, and therefore can be seen as a first indicator of the disease (Figure 2) [20]. The amount of WMHs increases with age and probably reflects chronic progressive ischemic damage caused by amyloid angiopathy, probably by stenosis of the long perforating arterioles. Neither the amount of WMHs nor the number of focal lesions correlates with the degree of cognitive deterioration in HCHWA-D patients [19].

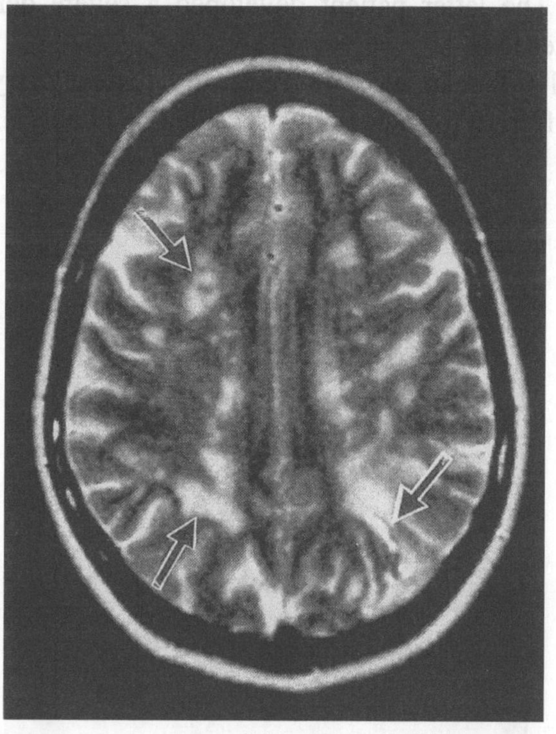

Figure 1. T2 weighted MRI images of a symptomatic HCHWA-D patient demonstrating a hemorrhage at the right parietal lobe and diffuse white matter hyperintensities in both hemispheres.

2.4 Clinical aspects of the Flemish mutation

The Flemish mutation has been found in a Dutch family (family 1302), but it was named the "Flemish mutation" because it has been found and described by a Flemish research group in Antwerp. In contrast to the clinical picture of HCHWA-D that is dominated by recurrent strokes, this Dutch family was firstly grouped among families with familial AD [11]. The Flemish mutation has been described in a four-generations family. Seven symptomatic family members are alive nowadays and 10 family members died as a consequence of hemorrhagic stroke or AD.

Clinical symptoms are described in only six mutation carriers [11,21]. In three mutation carriers the clinical diagnosis AD was confirmed. These patients suffered from dementia in the absence of cerebral hemorrhages, with an age at onset of cognitive deterioration between 45 and 50 years. Medical records of six other patients in which no mutation analysis was performed, revealed that they had died of presenile dementia in psychiatric hospitals. The remaining three mutation carriers suffered recurrent hemorrhagic strokes. One of them suffered a stroke at the age of 35 years and another at the age of 42. The latter patient developed a severe dementia syndrome several years later [11].

The cerebral CT scan of a 48 years old patient carrying the Flemish mutation and suffering from dementia revealed no abnormalities.

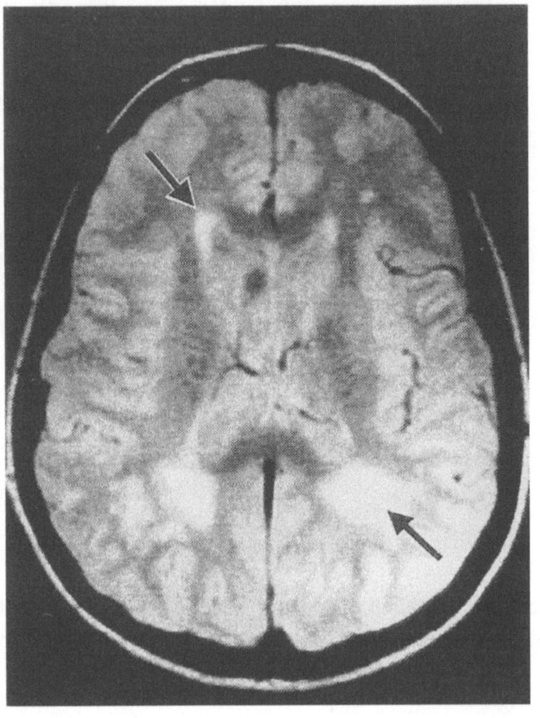

Figure 2. T2-weighted MRI image of an asymptomatic HCHWA-D point-mutation carrier showing extensive diffuse white matter hyperintensities in both hemispheres.

2.5 Symptoms of an Italian type of HCHWA

Recently, three unrelated Italian families were described in which the patients suffered hemorrhagic strokes. The patients died between the age of 60 and 75 years after a 10-20 years history of mild cognitive decline, recurrent strokes and, sometimes, epilepsy.

Six members of one family, spanning four generations, were affected. One patient suffered the first stroke at the age of 45 years and had recurrent strokes till his death 15 years later. Neuropsychological evaluation did not show signs of cognitive deterioration. Four members of three generations in another family had a history of recurrent strokes and mild cognitive decline in 10-20 years before they died at the age of 65-75 years [12,13].

3. GENETICS

3.1 APP gene

Down's syndrome patients carry an extra copy of chromosome 21 and almost invariably develop AD-like pathology at middle age. This observation led to the hypothesis that overexpression of a gene on chromosome 21 might lead to AD. This idea was further supported by the report of genetic linkage between DNA markers on chromosome 21 and AD in several kindreds. The isolation of Aβ from cerebral blood vessels and later also from senile plaques in AD and Down syndrome resulted in the identification of the APP gene, which maps on chromosome 21. The APP gene is located on chromosome 21 at 21q21.2 and is encoded by 18 exons of which exon 16 and 17 encode in part for Aβ [22,23]. Three major splice variants containing Aβ have been identified: APP695, APP751 and APP770. The two larger isoforms include exon 7 which encodes a domain similar to a Kunitz protease inhibitor (KPI) [22,24]. APP is an integral transmembrane glycoprotein. Although its primary structure suggests that APP might serve as a cell surface receptor [25] its physiological function is still unknown. A secreted form of APP comprising the KPI domain has been shown to be identical to the serine protease inhibitor, protease nexin 2 (PN2) [26]. APP(PN2) is a potent inhibitor of intrinsic blood coagulation factor IXa and XIa, by binding to this factors [27-30]. Binding between factor XIa and APP has been investigated in plasma of HCHWA-D patients [31]. A significantly higher concentration of factor XIa-APP complexes was found in HCHWA-D. It can be hypothesized that local binding of APP to factor XIa in cerebral amyloid-laden vessels can favor hemorrhage [31].

APP is processed through different proteolytic pathways. "Normal" secretory processing of APP involves cleavage of the molecule at the cell surface within the extracellular portion of Aβ, between residues 16 and 17 (APP Lys 687 and Leu 688), producing a soluble APP fraction and a membrane bound 13 kD fragment containing only part of Aβ [32]. The cleavage is thought to be made by an enzyme, or group of enzymes, referred to as α-secretase. Since this pathway does not produce intact Aβ, it is considered non-amyloidogenic. Potentially amyloidogenic pathways release intact Aβ from APP by cleavage through β- and γ-secretases. The sequence-specific β-secretase cleaves APP between Met671 and Asp672 releasing a truncated APP and leaving a larger fragment in the membrane [33]. A candidate enzyme with specific β-secretase activity has recently been cloned [34]. This fragment is further processed into Aβ by γ-secretase.

Nowadays, four different mutations are known that are located within the Aβ part of the APP gene. One of them at codon APP692 (Flemish) and two at codon APP693 (Dutch and Italian) result in recurrent cerebral hemorrhages and dementia [8,11,12,35]. The first mutation discovered in the APP gene was the HCHWA-D mutation at codon 693. This is a G to C base mutation responsible for a glutamic acid to glutamine substitution at position 22 of the Aβ protein. Another mutation at codon 693 is reported in the Italian families described above and is an G to A mutation predicting a lysine for glutamine aminoacid substitution. The Flemish mutation is a C to G mutation at codon 692. The fourth mutation at codon 693 ("Arctic" mutation) results in the substitution of a glycine for a glutamic acid, but this mutation is not associated with cerebral hemorrhage [36]. All these mutations are located near the α-secretase cleavage site.

Subsequently, five different APP mutations were identified in approximately 20 early-onset AD families worldwide. In these families with autosomal dominant AD the mutations were all found outside the Aβ region of the APP gene. Three mutations are described at APP717 and called the "London" mutation, the mutation at APP716 is named the "Florida" mutation and a double mutation at APP671/670 is found in Swedish AD families [37-41].

3.2 Other genes involved

Mutations in the APP gene are estimated to be responsible for about 5% of all familial early-onset AD cases and therefore are a rare cause of AD. Since the mutations in the APP gene explain the development of AD only in a small number of cases, a genome-wide search in families segregating with AD was started. A second locus for early-onset AD was subsequently identified on chromosome 14 and the presenilin-1 (PS-1) gene was isolated [42,43]. Over 40 different mutations have been detected in the PS-1 gene in a large number of families of different ethnic origin contributing to 18-50% of all early onset familial AD cases [44,45]. The age at onset of AD is generally slightly lower in individuals carrying PS-1 mutations (35-55 years) than in APP mutation carriers [46].

Soon after the identification of PS-1, screening of sequence databases revealed the presence of a homologous gene on chromosome 1, presenilin-2 (PS-2) [47]. Simultaneously, genetic linkage was found between PS-2 markers and AD in a group of families [48]. PS-1 and PS-2 share 67% overall homology and all mutations in PS-2 are located in regions that are conserved between PS-1 and PS-2. Only three mutations in PS-2 in a small number of families have been detected, suggesting that PS-2 mutations are a rare cause of AD. PS-2 mutations give rise to a broad range of onset of AD symptoms with incomplete penetrance [44]. Recently, a PS-1 polymorphism that is biallelic and located at intron 8 of the chromosome 14 PS-1 gene was hypothesized to be a risk factor for AD. Homozygosity for allele 1 nearly doubled the risk for late onset AD, although the results were not consistent [49]. The PS-1 2/2 genotype was not associated with CAA in Japanese elderly, whether they had AD or not [50]. In HCHWA-D, this polymorphism, did not correlate with clinical parameters (such as age at first stroke, age at death and presence of dementia) and radiological parameters [51].

Linkage analysis in families with late onset AD revealed a locus on chromosome 19 [52]. The *APOE* gene was one of the genes located in the chromosomal region that co-segregated with the disease and became immediately a candidate gene since ApoE-immunoreactivity was found in senile plaques, neurofibrillary tangles and CAA. Three common alleles of the gene encode three isoforms that differ from each other in one or two amino acids at position 112 and 158. Since the *APOE* ε4 allele was found with a higher frequency in late-onset AD patients than in control population, it was identified as a genetic risk factor for late onset AD [53]. The *APOE* ε4 risk effect has been extended to early onset and sporadic AD cases and confirmed by many other groups [54,55]. There is evidence that the risk associated with the *APOE* ε4 allele is dose-dependent, reflecting in the age at onset of the disease. Not only in AD patients, but also in sporadic CAA, several investigators have found an association between the *APOE* ε4 allele and sporadic CAA-related hemorrhage (see McCarron and Nicoll, this volume). The *APOE* ε4 allele is associated with an earlier age at onset and poor neurologic outcome after an intracerebral hemorrhage [56]. Histopathologically, both in AD and in sporadic CAA *APOE* ε4 is associated with a larger number of plaques and more severe amyloid angiopathy, although not all studies could confirm this finding [53,57-59]. The *APOE* genotype further modulates the age at onset of AD in APP717 and APP670/671 families [60]. However, for the Flemish APP692 and the Dutch APP693 mutation no effect on the age at onset, occurrence of dementia or the number of strokes was found [61,62]. Also, the severity of amyloid-related structural lesions, radiologically visible as WMHs and focal lesions on MRI scans or pathological lesions as CAA and plaques, appeared not to correlate with the *APOE* genotype [63]. Likewise, no modification of age at onset was observed for PS-1 and PS-2 mutation carriers [44].

An protective effect in developing AD has been reported for the *APOE* ε2 allele in some studies [64], but other studies contradict with these results [65]. In sporadic CAA it was found that the possession of an *APOE* ε2 allele was a risk factor for developing cerebral hemorrhages (see McCarron and Nicoll, this volume) [66]. In HCHWA-D the *APOE* ε2 allele neither protects nor favors the occurrence of cerebral hemorrhages or dementia [62].

Female gender is described to be an independent risk factor for AD and may play a role in its pathogenesis. *In vitro* studies showed that estrogen influences the secretase metabolism of APP by stimulating the α-secretase [67]. In HCHWA-D patients the mortality rate was found nearly two-fold higher in female than in male patients [18]. Age at onset did not differ between the two sexes. The interpretation of these findings is uncertain but an effect of hormonal factors or additional genetic factors (eg. X-chromosome) is possible. Paternal transmission of HCHWA-D was correlated with a higher mortality rate than maternal transmission, but these differences were marginally significant [18]. Genetic imprinting is known in diseases with unstable trinucleotide repeats, but this was the first time that such a phenomenon has been observed in a disease caused by a point mutation.

4. DISCUSSION

CAA remains clinically asymptomatic in the majority of the elderly but can cause (recurrent) hemorrhages both in sporadic and familial forms of CAA such as HCHWA-D. Moreover, CAA is often found in AD and is probably related to dementia in HCHWA-D. Many studies have been performed to elucidate the mechanism of amyloid deposition.

Aβ includes 39-43 amino acids, 28 residues of the extracellular domain and 11-15 residues of the transmembrane domain [68]. The length of the peptides, due to heterogeneity of the C-terminal is probably of great clinical and pathological relevance. Synthetic peptides varying in length between 1-28 and 1-42(43) can all form fibrils *in vitro*. However $A\beta_{1\text{-}42(43)}$ is less soluble and forms fibrils faster than the shorter isoforms. It has been suggested that amyloidosis may be seeded by trace amounts of amyloid fibrils. Small seeds of $A\beta_{1\text{-}42(43)}$ fibrils may grow fast by subsequent deposition of $A\beta_{1\text{-}40}$ [69,70].

All APP mutations have in common that they alter the APP metabolism in ways that either more or longer $A\beta_{1\text{-}42(43)}$ is formed, suggesting a direct link to Aβ deposition in the brain. The Swedish mutation is located near the β-cleavage site and results in a 3- to 8- fold overproduction of Aβ in transfected neuronal cells and cultured skin fibroblasts of AD patients with this type of mutation [71,72]. The London and Florida mutations are located near the cleavage site of α-secretase and lead to an increase of the $A\beta_{1\text{-}42(43)/1\text{-}40}$ ratio *in vitro* and *in vivo* [41,73,74]. The APP692 and the APP693 mutations are located within the Aβ part of APP near the α-secretase cleavage site. Transfection studies with mutant APP692 DNA demonstrated a 2-fold increase in both secreted $A\beta_{1-40}$ and $A\beta_{1-42}$ [75].

Remarkably, recently performed studies with mutant APP693 transfectants demonstrated a decrease of $A\beta_{1-40}$ and $A\beta_{1-42(43)}$, with a larger decrease of the latter isoform, resulting in a lower $A\beta_{1-40/1-42(43)}$ ratio. In plasma of HCHWA-D patients the concentration of $A\beta_{1-42(43)}$ was also decreased in mutation carriers compared to family members that did not carry the mutation, whereas the concentration of $A\beta_{1-40}$ did not differ between the groups (unpublished results). It has been shown *in vitro* that Aβ peptides with the HCHWA-D mutation adopt an altered conformation compared to wild-type Aβ peptides [76]. This suggests that the APP693 mutation has an effect on the processing of Aβ that differs from that of early-onset AD APP gene mutations. In HCHWA-D not the concentration of $A\beta_{1-42(43)}$ is increased but the conformation of Aβ is altered, resulting in a peptide that is more prone to aggregation. In cerebrospinal fluid of HCHWA-D patients the level of APP is markedly decreased compared to age-matched normal subjects [77]. The decreased levels may result from increased proteolysis of mutated and/or abnormally produced precursor proteins.

PS1 and PS-2 mutations are shown to selectively increase the production of $A\beta_{1-42(43)}$ about two-fold [78,79]. There is no evidence that the expression of APP or its cleavage by α- and β-secretase is altered by PS mutations. There are currently three hypotheses about how presenilins may regulate APP proteolysis. First, presenilin could facilitate α-secretase cleavage and regulate the access of the protease [80]. Second, presenilins regulate the trafficking of α-secretase or APP [81]. Third, presenilins may actually function as γ-secretases themselves [82].

The mechanism underlying the interaction between Aβ and ApoE is not clear. *In vitro* studies demonstrated that purified ApoE4 was shown to bind synthetic Aβ more avidly than ApoE3. This preferential binding was shown to result in more rapid formation of monofibrils [83,84]. However in its lipidated form, isolated from plasma, ApoE3 binds more efficiently to Aβ than ApoE4, which suggests that it protects Aβ from fibril formation [85]. In HCHWA-D, where no association between the *APOE* genotype and CAA was found, it has been hypothesized that these findings can be explained by a lower binding affinity of ApoE for Aβ when the Dutch APP693 mutation is present. Another explanation could be that the mutated Aβ forms fibrils so fast that ApoE can not influence this process anymore.

Therapeutic options preventing CAA deposition or progression are not available now, but are likely to focus on blocking the deposition or inhibition of β- or γ-secretase. Clinical trials with γ-secretase inhibitors will start in the very near future. Until that moment the only options are to withdraw anticoagulants or antiplatelet agents, and rehabilitation and physical therapy for those who have suffered a stroke.

REFERENCES

1. Castano EM, Frangione B. (1988) Human amyloidosis, Alzheimer's disease and related disorders. *Lab. Invest.* **58**, 122-132.
2. Greenberg SM. (1998) Cerebral amyloid angiopathy: prospects for clinical diagnosis and treatment. *Neurology* **51**, 690-694.
3. Greenberg SM, Vonsattel JPG, Stakes JW, Gruber M, Finklestein SP. (1993) The clinical spectrum of cerebral amyloid angiopathy: presentations without lobar hemorrhage. *Neurology* **43**, 2073-2079.
4. Luyendijk W, Schoen JHR. (1964) Intracerebral haematomas: a clinical study of 40 surgical cases. *Psychiat. Neurol. Neurochir.* **67**, 445-468.
5. Luyendijk W, Bots GTAM, Vegter-van der Vlis M, Went LN, Frangione B. (1988) Hereditary cerebral hemorrage caused by cortical amyloid angiopathy. *J. Neurol. Sci.* **85**, 267-280.
6. Haan J, Roos RAC. (1992) Comparison betwewen the Icelandic and Dutch forms of hereditary cerebral amyloid angiopathy. *Clin. Neurol. Neurosurg.* **94 (suppl)**, S82-S83.
7. Van Duinen SG, Castano EM, Prelli F, Bots GTAM, Luyendijk W, Frangione B. (1987) Hereditary cerebral hemorrhage with amyloidosis in patients of Dutch origin is related to Alzheimer's disease. *Proc. Natl. Acad. Sci.USA* **84**, 5991-5994.
8. Van Broeckhoven C, Haan J, Bakker E, Hardy JA, Vanhul W, Wehnert A, *et al.* (1990) Amyloid β-protein precursor gene and hereditary cerebral hemorrhage with amyloidosis (Dutch). *Science* **248**, 1120-1122.
9. Jensson O, Gudmundsson G, Arnason A, Blondal H, Petursdottir I, Thorsteinsson L, *et al.* (1987) Hereditary cystatin C (gamma trace) amyloid angiopathy of the CNS causing cerebral hemorrhage. *Acta Neurol. Scand.* **76**, 102-114.
10. Abrahamson M, Islam MQ, Szpirer J, Szpirer C, Levan G. (1989) The human cystatin C gene (CST3), mutated in hereditary cystatin C amyloid angiopathy, is located on chromosome 20. *Hum. Genet.* **82**, 223-226.
11. Hendriks L, van Duijn CM, Cras P, Cruts M, Van Hul W, Van Harskamp F, *et al.* (1992) Presenile dementia and cerebral hemorrhage linked to a mutation at codon 692 of the β-amyloid precursor protein gene. *Nat. Genet.* **1**, 218-221.
12. Bugiani O, Padovani A, Magoni M, Andora G, Sgarzi M, Savoiardo A, *et al.* (1998) An Italian type of HCHWA. *Neurobiol. Aging* S238.(Abstract)
13. Tagliavini F, Rossi G, Padovani A, Magoni M, Andora G, Sgarzi M, *et al.* (1999) A new βPP mutation related to hereditary cerebral hemorrhage. *Alzheimer's Reports* **2 (suppl)**, S28.
14. Fernandez-Madrid Y, Levy E, Marder K, Frangione B. (1991) Codon 618 variant of Alzheimer's amyloid gene associated with inherited cerebral hemorrhage. *Ann. Neurol.* **30**, 730-733.
15. Haan J, Algra P, Roos RAC. (1990) Hereditary cerebral hemorrhage with amyloidosis - Dutch type: clinical and computed tomographic analysis of 24 cases. *Arch. Neurol.* **47**, 649-653.

16. Bornebroek M, Haan J, Maat-Schieman MLC, Van Duinen SG, Roos RAC. (1996a) Hereditary cerebral hemorrhage with amyloidosis-Dutch type (HCHWA-D): I - A review of the clinical, radiologic and genetic aspects. *Brain Pathol.* **6**, 111-114.
17. Wattendorf AR, Frangione B, Luyendijk W, Bots GTAM. (1995) Hereditary cerebral hemorrhage with amyloidosis, Dutch type (HCHWA-D) : clinicopathological studies. *J. Neurol. Neurosurg. Psychiatry.* **58**, 699-705.
18. Bornebroek M, Westendorp R, Haan J, Bakker E, Timmers WF, Roos RAC. (1997c) Mortality from hereditary cerebral hemorrhage with amyloidosis - Dutch type: Impact of sex, year of birth and line of transmission. *Brain* **120**, 2243-2249.
19. Bornebroek M, van Buchem MA, Haan J, Brand J, Lanser JKB, Bruïne FT, *et al.* (1996c) Hereditary cerebral hemorrhage with amyloidosis - Dutch type: better correlation of cognitive deterioration with advancing age than with the number of focal lesions or white matter hyperintensities. *Alz. Dis. Assoc. Dis.* **10**, 224-231.
20. Bornebroek M, Haan J, van Buchem MA, Lanser JKB, de Vries-vd Weerd MAC, Zoetewey M, *et al.* (1996b) White matter lesions and cognitive deterioration in presymptomatic carriers of the amyloid precursor protein gene codon 693 mutation. *Arch. Neurol.* **53**, 43-48.
21. Van Harskamp, F., Cras, P., Hendriks, L., Kros, J.M., Martin, J.J., Hofman, A., Van Broeckhoven, C. & van Duijn, C.M. (1997) in Alzheimer's disease: Biology, Diagnosis and Therapeutics (Iqbal, K., Winblad, B., Nishimura, T., Takeda, M. & Wisniewski, H.M., eds.), pp. 155-159, John Wiley and sons.
22. Tanzi RE, Gusella JF, Watkins PC, Bruns GAP, St George-Hyslop P, Van Keuren ML, *et al.* (1987) Amyloid β - gene: cDNA, mRNA distribution, and genetic linkage near Alzheimer locus. *Science* **235**, 880-884.
23. Yoshikai S, Sasaki H, Dohura K, Furuya H, Sakaki Y. (1990) Genomic organisation of the human amyloid β-protein precursor gene. *Gene* **87**, 257-263.
24. Ponte P, Gonzalez-DeWhitt P, Schilling J, Miller J, Hsu D, Greenberg B, *et al.* (1988) A new A4 amyloid mRNA contains a domain homologues to serine protease inhibitors. *Nature* **331**, 525-532.
25. Kang J, Lemaire H, Unterbeck A, Salbaum JM, Masters CL, Grzeschik K, *et al.* (1987) The precursor of Alzheimer's disease amyloid A4 protein resembles a cell surface receptor. *Nature* **325**, 733-736.
26. Van Nostrand WE, Wagner SL, Suzuki M, Choi BH, Farrow JS, Geddes JW, *et al.* (1989) Protease nexin-II, a potent antichymotrypsin, shows identity to amyloid beta-protein precursor. *Nature* **341**, 546-549.
27. Schmaier AH, Dahl LD, Rozemuller AJ, Roos RAC, Wagner SL, Chung R, *et al.* (1993) Protease nexin-2 / amyloid β protein precursor: a tight-binding inhibitor of coagulation factor IXa. *J. Clin. Invest.* **92**, 2540-2545.
28. Smith RP, Higuchi DA, Broze J. (1990) Platelet coagulation factor XIa-inhibitor, a form of Alzheimer amyloid precursor protein. *Science* **248**, 1126-1128.
29. Van Nostrand WE, Schmaier AH, Wagner SL. (1992a) Potential role of protease nexin-2/amyloid β-protein precursor as a cerebral anticoagulant. *Ann. N .Y. Acad Sci* **674**, 243-252.
30. Van Nostrand WE, Wagner SL, Farrow JS, Cunnigham DD. (1990) Immunopurification and protease inhibitory properties of protease nexin-2/amyloid beta-protein precursor. *J. Biol. Chem.* **265**, 9591-9594.
31. Bornebroek M, von dem Borne PAKr, Haan J, Meijers JCM, Van Nostrand WE, Roos RAC. (1998) Binding of β precursor protein to coagulation factor XIa in patients with hereditary cerebral hemorrhage with amyloidosis Dutch type. *J. Neurol.* **245**, 111-115.

32. Esch FS, Keim PS>, Beattie EC, Blacher RW, Culwell AR, Olterdorf T, *et al.* (1990) Cleavage of amyloid β-protein during constitutive processing of its precursor. *Science* **248**, 122-1124.
33. Citron M, Teplow DB, Selkoe DJ. (1995) Generation of amyloid β protein from its precursor is sequence specific. *Neuron* **14**, 661-670.
34. Vassar R, Bennett BD, Babu-Kahn S, Kahn S, Mendiaz EA, Denis P, Teplow DB, Ross S, Amarante P, Loeloff R, Luo Y, Fisher S, Fuller J, Edenson S, Lile J, Jarosinski MA, Leona Biere A, Curran E, Burgess T, Louis J-C, Collins F, Treanor J, Rogers G, Citron M. (1999) β-Secretase cleavage of Alzheimer's amyloid precursor protein by the transmembrane aspartic protease BACE. *Science* **286**, 735-741.
35. Levy E, Carman MD, Fernandez-Madrid Y, Power MD, Lieberburg I, Van Duinen SG, *et al.* (1990) Mutation of the Alzheimer's disease amyloid gene in hereditary cerebral hemorrhage, Dutch type. *Science* **248**, 1124-1126.
36. Nilsberth J, Forsell C, Axelman K, Gustafsson C, Luthman J, Näslund J, Lannfelt L (1999) A novel APP mutation (E693G) – The Arctic mutation, causing Alzheimer's disease with vascular symptoms. *Soc. Neurosci. Abstr.* **25,** p. 297.
37. Elleder M, Christomanou H, Kustermann-Kuhn B, Harzer K. (1994) Leptomeningeal lipid storage patterns in Fabry disease. *Acta Neuropathol. (Berl)* **88**, 579-582.
38. Murrell J, Farlow M, Ghetti B, Benson MD. (1991) A mutation in the amyloid precursor protein gene associated with hereditary Alzheimer's disease. *Science* **254**, 97-99.
39. Goate A, Chartier-Harlin MC, Mullan M, Brown J, Crawford F, Fidani L, *et al.* (1991) Segregation of a missense mutation in the amyloid precursor protein gene with familial Alzheimer's disease. *Nature* **349**, 728-730.
40. Mullan M, Crawford F, Axelman K, Houlden H, Lilius L, Winblad B, *et al.* (1992) A pathogenic mutation for probable Alzheimer's disease in the APP gene at the N-terminus of β-amyloid. *Nat. Genet.* **1**, 345-347.
41. Hardy J. (1997) Amyloid, presenilins and Alzheimer's disease. *Trends Neurosci.* **20**, 154-159.
42. Schellenberg GD, Bird TD, Wijsman EM, *et al.*. (1992) Genetic linkage evidence for a familial form of Alzheimer's disease locus on chromosome 14. *Science* **258**, 668-671.
43. Van Broeckhoven C, Backhovens H, Cruts M, *et al.*. (1992) Mapping of a gene on predisposing to early-onset Alzheimer's disease on chromosome 14q24.3. *Nat. Genet.* **2**, 335-339.
44. Cruts M, Van Broeckhoven C. (1998a) Presenilin mutations in Alzheimer's disease. *Hum. Mutat.* **11**, 183-190.
45. Cruts M, Vanduijn CM, Backhovens H, *et al.*. (1998b) Estimation of the genetic contribution of presenilin-1 and -2 mutations in a population-based study of presenile Alzheimer's disease. *Hum. Mol. Genet.* **7**, 43-51.
46. Van Broeckhoven C. (1995) Presenilins and Alzheimer disease. *Nat. Genet.* **11**, 230-232.
47. Rogaev EI, Sherrington R, Rogaev EA, Levasque G, Ikeda M, Liang Y, *et al.* (1995) Familial Alzheimer's disease in kindreds with missense mutations in a gene on chromosome 1 related to Alzheimer's disease type 3 gene. *Nature* **376**, 775-778.
48. Levy-Lahad E, Wijsman EM, Nemens E, Anderson L, Goddard KA, Weber JL, *et al.* (1995) A familial Alzheimer's disease locus on chromosome 1 [see comments]. *Science* **269**, 970-973.
49. Wragg M, Hutton M, Talbot C, Busfield F, Han SW, Lendon C, *et al.* (1996) Genetic association between intronic polymorphism in presenilin-1 gene and late-onset Alzheimer's disease. *Lancet* **347**, 509-512.

50. Yamada M, Sodeyama N, Itoh Y, Suematsu N, Otomo E, Matsushita M, *et al.* (1997) Association of presenilin-1 polymorfism with cerebral amyloid angiopathy in the elderly. *Stroke* **28**, 2219-2221.
51. Bornebroek M, Haan J, Backhovens H, Deutz P, van Buchem MA, van den Broeck M, *et al.* (1997a) Presenilin 1 polymorphism and hereditary cerebral hemorrhage with amyloidosis Dutch type. *Ann. Neurol.* **42**, 108-110.
52. Pericak-Vance MA, Bebout JL, Gaskell PC, *et al.* . (1991) Linkage studies in Alzheimer's disease: evidence for chromosome 19 linkage. *Am. J. Hum. Genet.* **48**, 1034-1050.
53. Strittmatter WJ, Saunders AM, Schmechel DE, Pericak-Vance MA, Enghild J, Salvesen GS, *et al.* (1993a) Apolipoprotein E: high-avidity binding to beta-amyloid and increased frequency of type 4 allele in late-onset familial Alzheimer disease. *Proc. Natl. Acad. Sci. USA* **90**, 1977-1981.
54. van Duijn CM, de Knijff P, Cruts M, *et al.* . (1994) Apolipoprotein E4 allele in a population-based study of early-onset Alzheimer's disease. *Nat. Genet.* **7**, 74-78.
55. Hardy J. (1995) Apolipoprotein E in the genetics and epidemilology of Alzheimer's disease. *Am. J. Hum. Genet.* **60**, 456-460.
56. Greenberg SM, Briggs ME, Hyman BT, Kokoris GJ, Takis C, Kanter DS, *et al.* (1996) Apolipoprotein E ε4 is associated with the presence and earlier onset of hemorrhage in cerebral amyloid angiopathy. *Stroke* **27**, 1333-1337.
57. Schmechel DE, Saunders AM, Strittmatter WJ, Crain BJ, Hulette CM, Joo SH, *et al.* (1993) Increased amyloid beta-peptide deposition in cerebral cortex as a consequence of apolipoprotein E genotype in late-onset Alzheimer disease. *Proc. Natl. Acad. Sci. USA* **90**, 9649-9653.
58. Polvikoski T, Sulkava R, Haltia M, Kainulainen K, Vuorio A, Verkkoniemi A, *et al.* (1995) Apolipoprotein E, dementia, and cortical deposition of β-amyloid protein. *N. Engl. J. Med.* **333**, 1242-1247.
59. Greenberg SM, Rebeck GW, Vonsattel JPG, Gomez-Isla T, Hyman BT. (1995) Apolipoprotein E epsilon 4 and cerebral hemorrhage associated with amyloid angiopathy. *Ann. Neurol.* **38**, 254-259.
60. Alzheimer's Disease Collaborative Group . (1993) ApoliproteinE genotype and Alzheimer's disease. *Lancet* **342**, 737-738.
61. Haan J, Van Broeckhoven C, van Duijn CM, Voorhoeve E, Van Harskamp F, Van Swieten JC, *et al.* (1994) The apolipoprotein E ε4 allele does not influence the clinical expression of the amyloid precursor protein gene codon 693 or 692 mutations. *Ann. Neurol.* **36**, 434-437.
62. Haan J, Roos RAC, Bakker E. (1995) No protective effect of apolipoprotein E ε2 allele in Dutch hereditary cerebral amyloid angiopathy [letter]. *Ann. Neurol.* **37**, 282.
63. Bornebroek M, Haan J, Van Duinen SG, Maat-Schieman MLC, van Buchem MA, Bakker E, *et al.* (1997b) Dutch hereditary cerebral amyloid angiopathy: structural lesions and *APOE* genotype. *Ann. Neurol.* **41**, 695-698.
64. Corder EH, Saunders AM, Risch NJ, Strittmatter WJ, Schmechel DE, Gaskell PC, *et al.* (1994) Protective effect of apolipoprotein E type 2 for late onset alzheimer's disease. *Nat. Genet.* **7**, 180-184.
65. van Duijn CM, de Knijff P, Wehnert A, De Voecht J, Bronzova JB, Havekes LM, *et al.* (1995) The apolipoprotein E ε2 allele is associated with an increased risk of early-onset Alzheimer's disease and a reduced survival. *Ann. Neurol.* **37**, 605-610.
66. Nicoll JAR, Burnett C, Love S, Graham DI, Ironside JW, Vinters HV. (1996) High frequency of apolipoprotein E epsilon 2 in patients with cerebral hemorrhage due to cerebral amyloid angiopathy. *Ann. Neurol.* **39**, 682.

67. Jaffe AB, Toran-Allerand CD, Greengard P, Gandy SE. (1994) Estrogen regulates metabolism of Alzheimer amyloid beta precursor protein. *J. Biol. Chem.* **269**, 13065-13068.
68. Masters CL, Simms G, Weinman NA, Multhaup G, Beyreuther K. (1985) Amyloid plaque core protein in Alzheimer's disease and Down syndrome. *Proc. Natl. Acad. Sci. USA* **82**, 4245-4249.
69. Jarrett JT, Berger EP, Lansbury PT. (1993) The carboxy terminus of β amyloid is critical for the seeding of amyloid formation: Implications for the pathogenesis of Alzheimer's disease. *Biochemistry* **32**, 4693-4697.
70. Iwatsubo T, Odaka A, Suzuki N, Mizusawa H, Nukina N, Ihara Y. (1994) Visualization of A beta 42(43) and A beta 40 in senile plaques with end-specific A beta monoclonals: evidence that an initially deposited species is A beta 42(43). *Neuron* **13**, 45-53.
71. Citron M, Oltersdorf T, Haass C, McConlogue L, Hung AY, Seubert P, *et al.* (1992) Mutation of the β-amyloid precursor protein in familial Alzheimer's disease increases β-protein production. *Nature* **360**, 672-674.
72. Felsenstein KM, Hunihan L, Roberts SB. (1994) Altered cleavage and secretion of a recombinant β-APP bearing the swedish familial Alzheimer's disease mutation. *Nat. Genet.* **6**, 251-256.
73. Suzuki N, Cheung TT, Cai X, Odaka A, Otvos JrL, Eckman C, *et al.* (1994) An increased percentage of long amyloid β protein secreted by familial amyloid β-protein precursor (βAPP717) mutants. *Science* **264**, 1336-1340.
74. Tamaoka A, Odaka A, Ishibashi Y, Usami M, Sahara N, Suzuki N, *et al.* (1994) APP 717 missense mutation affects the ratio of amyloid β-protein species (Aβ1-42/43 and $A\beta_{1\text{-}40}$) in familial Alzheimer's disease brain. *J. Biol.Chem.* **269**, 32721-32724.
75. De Jonghe C, Zehr C, Yager D, Prada C, Younkin S, Hendriks L, *et al.* (1998) Flemisch and Dutch mutations in amyloid β precursor protein have different effects on amyloid β secretion. *Neurobiol. Dis.* **4**, 281-286.
76. Sorimachi K, Craik DJ. (1994) Structure determination of extracellular fragments of amyloid proteins involved in Alzheimer's disease and Dutch-type hereditary cerebral hemorrhage with amyloidosis. *Eur. J. Biochem.* **219**, 237-251.
77. Van Nostrand WE, Wagner SL, Haan J, Bakker E, Roos RAC. (1992b) Alzheimer's disease and hereditary cerebral hemorrhage with amyloidosis-Dutch type share a decrease in cerebrospinal fluid levels of amyloid β-protein precursor. *Ann. Neurol.* **32**, 215-218.
78. Scheuner D, Eckman C, Jensen M, Song X, Citron M, Suzuki N, *et al.* (1996) Secreted amyloid β-protein similar to that in the senile plaques of Alzheimer's disease is increased *in vivo* by the presenilin 1 and 2 and APP mutations linked to familial Alzheimer's disease. *Nat. Med.* **2**, 864-870.
79. Mann DMA, Iwatsubo T, Cairns NJ, Lantos PL, Nochlin D, Sumi SM, *et al.* (1996) Amyloid β protein Aβ deposition in chromosome 14-linked Alzheimer's disease: predominancy of Aβ42(43). *Ann. Neurol.* **40**, 149-156.
80. Xia Y, Zhang XH, Koo EH, Selkoe DJ. (1997) Interaction between amyloid precursor protein and presenilins in mammalian cells: Implications for the pathogenesis of Alzheimer's disease. *Proc. Natl. Acad. Sci. USA* **94**, 8208-8213.
81. Thinakaran G, Borchelt DR, Lee MK, Slunt HH, Spitzer L, Kim G, *et al.* (1996) Endoproteolysis of presenilin 1 and accumulation of processed derivatives *in vivo*. *Neuron* **17**, 181-190.
82. Wolfe MS, Xia W, Ostaszewski BL, Diehl TS, Taylor Kimberley W, Selkoe DJ. (1999) Two transmembrane aspartates in presenilin-1 required for presenilin endoproteolysis and γ-secretase activity. *Nature*, **398**, 513-517.

83. Strittmatter WJ, Weisgraber KH, Huang DY, Dong L-M, Salvesen GS, Pericakvance M, *et al.* (1993b) Binding of human apolipoprotein E to synthetic amyloid L pepetide: isoform-specific effects and implications for late-onset Alzheimer's disease. *Proc. Natl. Acad. Sci. USA* **90**, 8098-8102.
84. Wisniewski T, Castano EM, Golabek AA, Vogel.T. , Frangione B. (1994) Acceleration of Alzheimer's fibril formation by Apolipoprotein E *in vitro*. *Am. J. Pathol.* **145**, 1030-1035.
85. Ladu MJ, Pederson TM, Prail DE, Reardon CA, Getz GS, falduto MT. (1995) Purification of apolipoprotein E attenuates isoform-specific binding to β-amyloid. *J. Biol. Chem.* **270**, 9039-9042.

Chapter 7

GENETICS AND NEUROPATHOLOGY OF HEREDITARY CYSTATIN C AMYLOID ANGIOPATHY (HCCAA)

ÍSLEIFUR ÓLAFSSON* and LEIFUR THORSTEINSSON*[†]
**Department of Clinical Biohemistry, Reykjavík Hospital and [†]deCODE Genetics Inc., Reykjavík, Iceland*

Hereditary cystatin C amyloid angiopathy (HCCAA) is an autosomal dominant disorder causing fatal brain hemorrhages in normotensive young adults. Nucleotide substitution in the gene encoding the cysteine proteinase inhibitor, cystatin C, gives rise to a structurally unstable variant protein with increased tendency to dimerize, aggregate and form amyloid depositions, primarily in the walls of small arteries and arterioles of the brain. In the following chapter the epidemiology, clinical aspects, pathology, molecular biology and pathophysiology of HCCAA are reviewed.

1. INTRODUCTION

Cerebral amyloid angiopathy (CAA) is an important feature of senile dementia and Alzheimers disease and is thought to play an important role in the pathophysiology of transient and chronic neurologic symptoms and cerebral hemorrhage [1,2,3]. This type of angiopathy has also been identified in hereditary conditions caused by mutations in the amyloid precursor protein gene (APP), transthyretin and cystatin C. Hereditary cerebral hemorrhage with amyloidosis-Dutch type (HCHWA-D) is an autosomal dominant disorder caused by a mutation in the APP gene (E693Q) [4,5]. It is characterized by cerebral hemorrhages occurring between 45 and 65 years of age and development of dementia in a stepwise fashion (see Bornebroek *et al*, this volume). In one Dutch family (family 1302) a different missense mutation in the APP gene (A692G APP, "Flemish mutation") causes early

M.M. Verbeek et al. (eds.),
Cerebral Amyloid Angiopathy in Alzheimer's Disease and Related Disorders, 121–134.

onset Alzheimer's disease and in some cases cerebral hemorrhages [6]. CAA has also been described in patients with transthyretin polyneuropathy but in this condition affected individuals do not have symptoms from the central nervous system [7]. Hereditary cystatin C amyloid angiopathy (HCCAA) (also called hereditary cerebral hemorrhage with amyloidosis-Icelandic type - HCHWA-I) is an autosomal dominant disease due to a mutation in the cystatin C gene [8]. It can be classified as a primary systemic amyloidosis because amyloid depositions are not limited to the central nervous system, but can be found in various other tissues. Knowledge about the pathophysiologic mechanisms causing HCCAA has rapidly increased in the last decade and the aim of this chapter is to provide an overview of the current state.

1.1 Epidemiology and clinical aspects

A hereditary disorder causing cerebral hemorrhage between the age of 20 and 45 was initially described in 1935 by Árnason [9]. The families all originated from the same geographic area on the west coast of Iceland, but later a separate family from the south of Iceland was described [10]. The disorder has been identified in about 140 members in nine families and can be traced back at least nine to twelve generations. Family studies clearly demonstrate the classical characteristics of an autosomal dominant genetic disorder [9,11].

A total population survey on stroke amongst persons 35 years and younger in Iceland revealed that 17% of strokes in this group of patients were due to hereditary cerebral hemorrhage. During the period 1960 to 1994 53 patients were identified in the total Icelandic population of about 250.000 [12]. The disease has hitherto been thought to be confined to the Icelandic population, but recently a sporadic case with the same mutation as in HCCAA was reported in an American patient of Croatic and British origin [13].

The clinical manifestations of HCCAA are the classical symptoms and signs of an intracerebral hemorrhage. A majority of the patients have their first attack before 30 years of age and die before the age of 50 [11]. The earliest signs are often headaches, nausea, vomiting and blurred speech. Some patients have died following the first stroke, but most of the patients have lived for many years disabled by multiple strokes causing progressive paralysis, dementia and behavioural changes. In some cases the disease seems to be stable for periods of up to several years and occasionally affected family members reach a normal lifespan [14].

2. PATHOLOGICAL FINDINGS IN HCCAA

Postmortem macroscopic examination of HCCAA patients usually shows multiple hemorrhagic lesions of different sizes in both cortex and white matter of all lobes of the brain and in the basal ganglia region [15]. Remnants of old hemorrhages are found in most cases. Macroscopic lesions are not found in the brainstem, cerebellum and spinal cord.

Microscopic examination shows a widespread hyalinization and thickenings in the walls of arteries and arterioles throughout the brain and spinal cord which concentrically narrows the vessels to a variable extent [11,15]. Some vessels are completely occluded. Splitting of media of the arterial wall as well as perivascular fibrosis is found. Thrombi, both fresh and organized, can be observed and lesions from recent or older insults are present. Veins and capillaries are minimally or not affected.

Congo red staining of brain tissue sections, viewed under polarized light, reveals the classic red-green birefringence of amyloid in the wall of arteries and arterioles throughout the central nervous system (CNS) (Figure 1A) [15,16]. In some cases amyloid deposits can be present in the nervous tissue, especially in the basal ganglia and the hippocampus. The congophilic hyalin material can also be found in the leptomeninges, preferably in the spinal pia and pial septa of the optic nerve. Occasionally the vessel walls show amorphous eosinophilia without green birefringence. Neither senile plaques nor neurofibrillary tangles are found.

Immunohistochemical studies of tissue sections from the CNS using polyclonal and monoclonal antibodies against human cystatin C have shown its amyloid deposits in practically all arteries and arterioles, but capillaries, veins and venules are minimally or not affected (Figure 1B) [16,17]. In the arterioles massive deposits are found primarily in the media of the vessel wall and in these severely affected vessels, smooth muscle cells are often almost obliterated [18]. These findings indicate severe degeneration and weakening of the arterial walls ultimately leading to cerebral hemorrhage. Occasionally cells, probably of monocytic origin, are seen inside the amyloid deposits which are completely negative when stained with an antibody against cystatin C. In the leptomeningeal vessels the adventitia is mainly affected. Massive cystatin C amyloid deposits are also present in interstitial tissue of the basal ganglia and the hippocampus, especially perivascularly. Spinal pia mater shows positive immunostaining as do frequently the arachnoid and arachnoid granulations.

Using immunohistochemical techniques amyloid deposits have been shown to be present not only in the CNS but also in other tissues such as lymph nodes, spleen and salivary glands [17,18,19]. Arterioles of the seminal vesicle, adrenal cortex and skin are also positive.

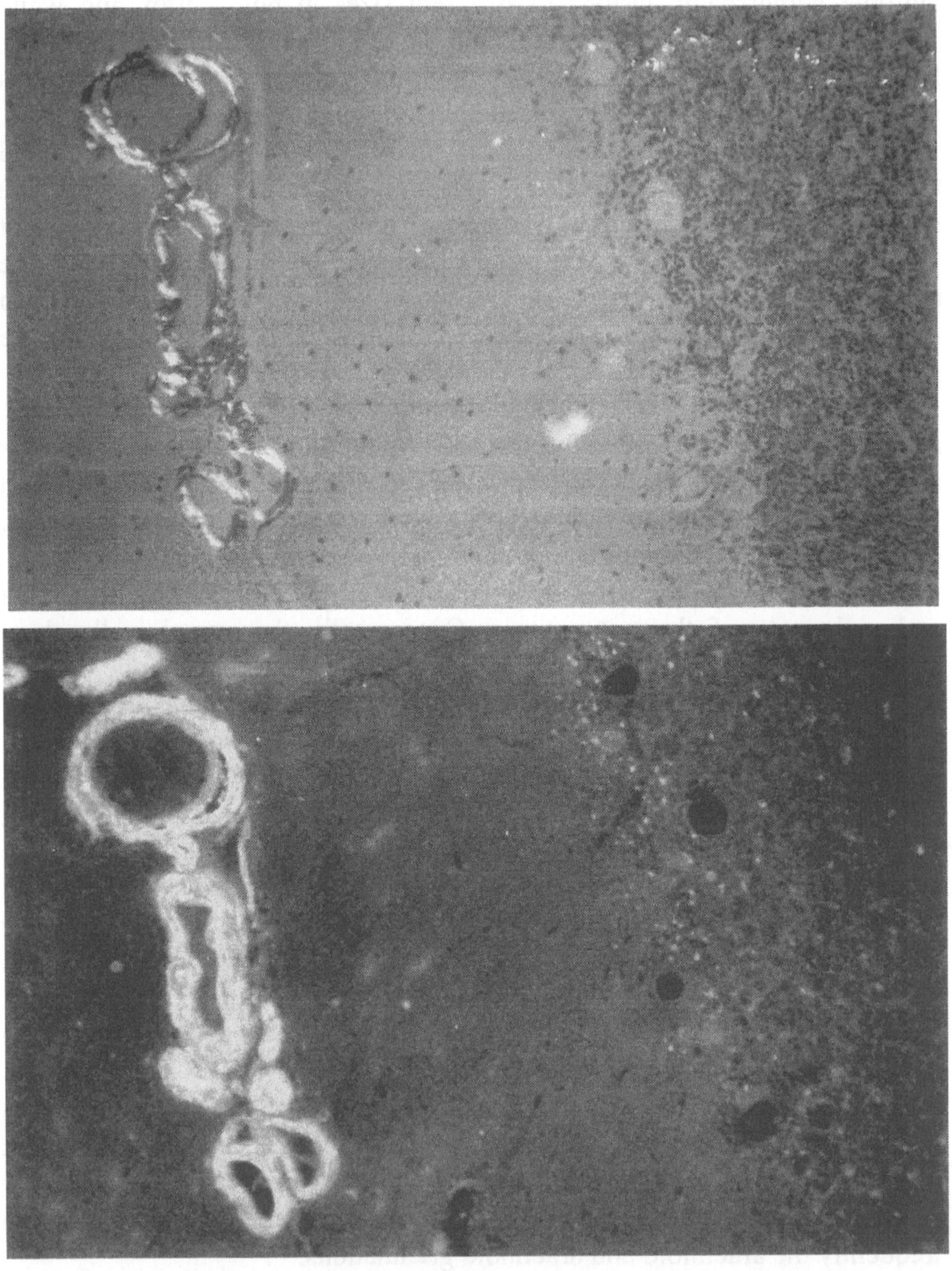

Figure 1. Cystatin C amyloid deposits in cerebellar arteries. A) A section stained with Congo red viewed under polarized light. Note yellow-green birefringence of congophilic substance representing CAA. B) Immunostaining using polyclonal rabbit anti-cystatin C and indirect immunofluorecence technique. Note intense immunofluorescence of the vessel walls. Capillaries are completely negative.

Electron microscopy of brain arterioles shows dense fibrillar material throughout the media [20,21]. The randomly arranged nonbranching fibrils are interspersed with pentagonal particles identical in size and appearance to tissue amyloid P-component [20]. The presence of protein amyloid P component in the cerebral amyloid has also been demonstrated by immunohistochemical techniques.

Interestingly, immunohistochemical examinations have demonstrated the presence of cystatin C in amyloid β-protein associated cerebrovascular amyloid deposits, in patients with Alzheimer's disease, sporadic CAA and HCHWA-D [22,23,24,25,26]. The clinical and pathophysiological significance of these findings has not been elucidated.

3. BIOCHEMISTRY AND MOLECULAR BIOLOGY OF HCCAA

The amyloid protein was originally isolated from pooled leptomeningeal tissue and meningeal blood vessels obtained at autopsy from HCCAA patients by extraction and chromatographic methods. The amino terminus of the protein was sequenced by Edman degradation and by comparison to other known protein sequences it was found that the amyloid protein was cystatin C, previously called γ-trace, starting at residue 11 [27].

Cystatin C is an alkaline protein comprised of one nonglycosylated 120-residue polypeptide chain [28]. It is an effective inhibitor of cysteine proteinases, such as cathepsins B, H and L and belongs to Family 2 of the homologous cystatin superfamily of proteins, together with cystatin D, S, SN, SA, E/M and F [29,30,31,32,33,34]. Other human cystatins are the Family 1 cystatins, cystatin A and B, and the Family 3 cystatins, the kininogens [30]. The inhibitory activity, concentrations in extracellular fluids and other physicochemical properties of cystatin C strongly suggest that it is one of the most important extracellular cysteine proteinase inhibitors in the human body [29]. It is present in all body fluids but the highest concentrations are found in the cerebrospinal fluid (CSF) and seminal plasma.

Measurements of cystatin C levels in CSF from HCCAA and control patients showed significantly lower levels in the HCCAA patients indicating abnormal metabolism of cystatin C [35]. Determination of cystatin C levels in CSF was therefore used as a diagnostic test for HCCAA with relatively high predictive value. No signs of truncation could be detected in cystatin C from the CSF of the patients, either by isoelectric focusing or N-terminal sequence determination [36].

The human cystatin C gene, together with eight or nine other genes encoding family 2 cystatins, has been mapped to a 905 kb fragment on chromosome 20p11.2 [37,38]. It has the overall size of about 6.5 kb and contains three exons and two introns [39]. The gene has several transcription initiation sites and functional analysis of the promoter region revealed a sequence with a strong positive effect on transcription of the gene [40]. This sequence contains three tandemly arranged potential Sp1 binding sites. Studies on cystatin C mRNA levels in different tissues show that the gene is expressed to a varying degree in all tissues with highest expression levels in pancreas, testis and brain [41]. Transforming growth factor β1 (TGFβ1) and dexamethasone increase cystatin C gene expression while bacterial lipopolysaccharides, interferon-γ and cigarette smoke have been shown to decrease it [42,43,44,45,46].

Subsequent total amino acid sequencing of cystatin C from amyloid material isolated from six HCCAA patients revealed an amino acid substitution of glutamine for leucine at position 68 in the protein (L68Q) [28,47]. Cloning and characterization of the cDNA encoding cystatin C showed that a point mutation in the coding sequence of the cystatin C gene could be caused either by a single base substitution at the position coding for amino acid 68 (from CTG to CAG) or by two base substitution (from CTG to CAA) [48]. Either of these base substitutions would abolish an *Alu*I restriction site in the gene. The loss of an *Alu*I restriction site in the cystatin C gene in HCCAA patients was demonstrated by restriction fragment length polymorphism studies [49]. The *Alu*I polymorphism was shown to be present in all 22 patients from 8 families suffering from HCCAA but not in healthy relatives. This polymorphism was used to develop a PCR based method for rapid and simple diagnosis of HCCAA [50]. Nucleotide sequencing showed that the disease was caused by a single T to A substitution in the codon for amino acid residue 68 of cystatin C [50, 51].

4. PATHOPHYSIOLOGY OF AMYLOID FORMATION IN HCCAA

The pathophysiologic mechanisms of amyloid deposition and decreased concentration of cystatin C in CSF in HCCAA patients are not known in details, but recent advances have increased our understanding of the basic defect. Analysis of cell homogenates and media from cultivated human and mouse cell lines overexpressing wild type human cystatin C and the L68Q mutant demonstrated that the clones expressing L68Q cystatin C secreted slightly lower levels than clones expressing the wild type cystatin C [52, 53]. Also, immunofluorescence cytochemistry and western blotting experiments showed an increased intracellular accumulation of cystatin C in cells expressing the L68Q cystatin C mutant gene compared with cells expressing the wild type gene [52]. The predominant localisation of the intracellular L68Q cystatin C accumulation was in the endoplasmic reticulum (ER). This confirms results from studies on the cystatin C secretion of cultivated peripheral blood monocytes isolated from HCCAA-patients and controls that indicated pertubation in cystatin C secretory pathway and intracellular accumulation of cystatin C aggregates [54]. The L68Q cystatin C secreted into the culture media was also shown to exhibit an increased susceptibility to proteolytic cleavage by a serine protease, as shown by the absence of degradaton when serine protease inhibitor was added, thus possibly explaining the low levels of the protein in the CSF [53].

The presence of intracellular inactive wild type cystatin C dimers was demonstrated by gel filtration of cultivated cell lysates [55]. In contrast, no cystatin C dimers could be found in the cell media, indicating that monomerization and reactivation is achieved at or prior to secretion. Cells, in which export from ER was blocked with brefeldin A, contained the dimers, strongly suggesting that the cystatin C dimerization occured in the ER [55]. NMR spectroscopy of cystatin C dimers showed that the dimerization involves cystatin C in its native folded fold and that it procedes via the reactive site of the inhibitor, which leads to loss of its inhibitory acitvity [56, 57]. This dimerization process in the secretory pathway of cystatin C could be of major relevance for the amyloid deposition of L68Q mutant (Figure 2).

According to the three-dimensional structure of the highly homologous cysteine proteinase inhibitor, chicken cystatin, the Leu-68 residue of cystatin C lies buried in the hydrophobic core of the protein molecule, suggesting that the L68Q substitution should not cause any major structural rearrangement [58,59]. Indeed, NMR analysis of the L68Q cystatin C mutant monomer produced in an *E. coli* expression system showed that it is structurally very similar to the wild type protein [60]. Studies on the functional and physicochemical properties of cystatin C L68Q mutant expressed in *E.coli* revealed that the mutant protein had normal inhibitory activity when compared to wild type cystatin C but clearly an increased tendency to form dimers and aggregates [61]. The tendency to form dimers and aggregates was shown to be temperature-dependent with much higher aggregation rate at 40 °C than at 37 °C [61]. Studies of dimerisation at different pH values, NMR studies and measurements of circular dichroism indicate that L68Q can form aggregating folding intermediates while the monomer/dimer equilibrium of wild type cystatin C is not affected by pH in the same range [60]. Thus, structural instability and state of denaturation seems to be critical for cystatin C amyloid formation.

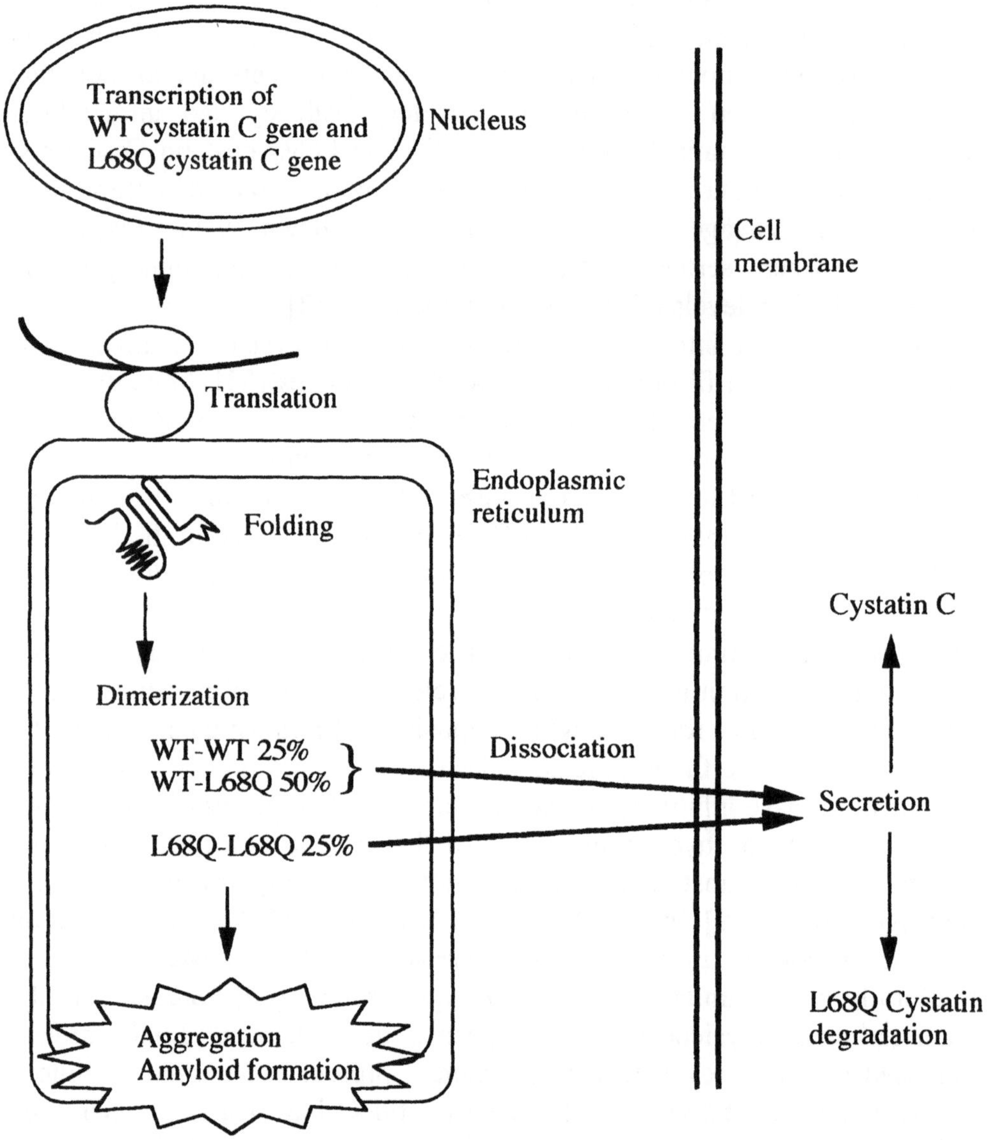

Figure 2. Proposed model for L68Q cystatin C amyloid formation. Heterozygous patients transcribe both wild type (WT) and mutated L68Q cystatin C gene. Ribosomes translate cystatin C mRNA in the endoplasmic reticulum (ER) and transfer the polypeptide chain across the ER membrane. In the ER cystatin C polypeptide chain folds through molten globule intermediates (possibly in conjuction with a molecular chaperone) and forms inactive dimers. Three different types of dimers are formed. Majority of cystatin C dimers dissociate to monomers in the secretion pathway. As L68Q-L68Q cystatin C dimers are more resistant to dissociation and have increased tendency to aggregate, they susequently form fibrillar amyloid structures. Due to structural instabilty L68Q cystatin C monomers degrade rapidly extracellularly.

Oligomerisation of nascent protein conformational intermediates and their interaction with chaperones, that assist in protein folding, is a well known process that takes place in the ER. Mutations causing aberrant conformational intermediates can either lead to accelerated degradation of the defective protein or its accumulation. It is therefore postulated that HCCAA may be one of the diseases of disorders in protein conformation affecting ER processing [62].

To summarize, it is apparent that structural instability of the L68Q variant of cystatin C causes increased tendency to dimerisation, aggregation, accumulation and amyloid fibril formation. Proteolytic enzymes of cells fail to degrade the inert amyloid fibrils with the exception of the flexible amino-terminal end of cystatin C, which is cleaved off. The degree of amyloid deposition in tissues seems to be dependent on the rate of cystatin C synthesis and the efficiency of cells to remove protein aggregates. The high degree of amyloid deposition in the cerebral arteries and arterioles leads to degeneration and weakening of the vascular wall and subsequently cerebral hemorrhage.

5. POSSIBILITIES FOR THERAPEUTIC INTERVENTION

At present no therapeutic intervention has been found for HCCAA. The results of extensive studies on the biology of cystatin C and pathophysiology of HCCAA provide at least two possible ways of treating HCCAA. One is to interfere or minimize the dimerization and aggregation of cystatin C by some therapeutic agent or by abortion of clinical situations that might lead to increased amyloid formation. The fact that cystatin C dimerization rate increases by 150% by raising temperature from 37°C to 40°C indicates harmful effects of increased body temperature [55]. Inflammation and tissue damage causing release of cellular mediators that increase cystatin C gene expression could also increase amyloid formation [41]. Long term treatment with glucocorticoids could have similar effects [43]. An alternative way of treating HCCAA would be to downregulate cystatin C production by administration of a therapeutic agent but no such nontoxic agent has been identified. Gene therapy using DNA vectors or antisense oligonucleotides is also a future therapeutic possibility.

ACKNOWLEDGEMENTS

We wish to express our sincere gratitude to our collaborators, especially at the Blood Bank and Department of Neurology, National University Hospital, Reykjavik, Iceland and at the Department of Clinical Chemistry, University Hospital, Lund, Sweden. We are also grateful to Heilavernd Society and other supporters of this project.

REFERENCES

1. Vinters HV. (1987) Cerebral amyloid angiopathy. A critical review. *Stroke* **18**, 311-324.
2. Haan J, Roos RAC. (1990) Amyloid in central nervous system disease. *Clin. Neurol. Neurosurg* **92**, 305-310.
3. Haan J, Maat-Schieman MLC, Roos RAC. (1994) Clinical effects of cerebral amyloid angiopathy. *Dementia* **5**, 210-213.
4. Van Broeckhoven C, Haan J, Bakker E, Hardy JA, Van Hul W, Wehnert A, Vegter-Van Der Vlis M, Roos RAC. (1990) Amyloid β protein precursor gene and hereditary cerebral hemorrhage with amyloidosis (Dutch). *Science* **248**, 1120-1122.
5. Levy E, Carman MD, Fernandez-Madrid I, Lieberburg I, Power MD, van Duinen SG, Bots GTAM, Luyendijk W, Frangione B (1990) Mutation of the Alzheimer's disease amyloid gene in hereditary cerebral hemorrhage, Dutch type. *Science* **248**, 1124-1126.
6. Hendriks L, Van Duijn CM, Cras P, Cruts M, Van Hul W, Van Harskamp F, Warren A, McInnes MG, Anatonarakis SE, Martin JJ, Hofman A, Van Broeckhoven C. (1992) Presenile dementia and cerebral hemorrhage linked to a mutation at codon 692 of beta-amyloid precursor protein gene. *Nat. Genet.* **1**, 218-221.
7. Kametani F, Ikeda S, Yangisawa N, Ishi T, Hanyu N. (1992) Characterization of a transthyretin-related amyloid fibril protein from cerebral amyloid angiopathy type I familial amyloid polyneuropathy. *J. Neurol. Sci.* **108**, 178-183.
8. Olafsson I, Thorsteinsson L, Jensson O. (1996) The molecular pathology of hereditary cystatin C amyloid angiopathy causing brain hemorrhage. *Brain Pathol.* **6**, 121-126.
9. Arnason A. (1935) Apoplexie und ihre vererbung (thesis). *Acta Psychiat. Neurol.*, suppl VII.
10. Jensson O, Thorsteinsson L, Palsdottir A, Gudmundsson G, Arnasson A, Blöndal H, Abrahamson M, Grubb A, Olafsson I, Lundwall Å. (1988) An isolate of families with hereditary cystatin C amyloid angiopathy and cerebral hemorrhage in the south of Iceland. In Isobe T, Araki S, Uchino F, Kito S, Tsubura E (eds), *Amyloid and Amyloidosis.* Plenum Publishing Corp, New York, pp. 579-584.
11. Jensson Ó, Gudmundsson G, Árnason A, Blöndal H, Pétursdóttir I, Thorsteinsson L, Grubb A, Löfberg H, Cohen D, Frangione B. (1987) Hereditary cystatin C (γ-trace) amyloid angiopathy of the CNS causing cerebral hemorrhage. *Acta Neurol. Scand.* **76**, 102-114.
12. Gudmundsson G, Olafsson E, Blöndal H, Arnasson A, Heuser WA. (1998). Hereditary cerebral hemorrhage in Iceland. Incidence for the 35 year period 1960-1994. *Icelandic Med. J.* suppl **37**, 76.

13. Graffagnino C, Herbstreith MH, Schmechel DE, Levy E, Roses AD, Alberts MJ. (1995) Cystatin C mutation in an elderly men with sporadic amyloid angiopathy and intracerebral hemorrhage. *Stroke* **25**, 2190-2193.
14. Sveinbjornsdottir S, Blondal H, Gudmundsson G, Kjartansson O, Jonsdottir S, Gudmundsson G. (1996) Progressive dementia and leucoencephalopathy as the initial presentation of late onset hereditary cystatin-C amyloidosis. Clinicopathological presentation of two cases. *J. Neurol. Sci.* **140**, 101-108.
15. Gudmundsson G, Hallgrímsson J, Jónasson TA, Bjarnason Ó. (1972) Hereditary cerebral hæmorrhage with amyloidosis. *Brain* **95**, 387-404.
16. Löfberg H, Grubb AO, Nilsson EK, Jensson Ó, Gudmundsson G, Blöndal H, Árnasson A, Thorsteinsson L. (1987) bImmunohistochemical characterization of the amyloid deposits and quantitation of pertinent cerebrospinal proteins in hereditary cerebral hemorrhage with amyloidosis. *Stroke* **18**, 431-440.
17. Thorsteinsson L, Blöndal H, Jensson Ó, Gudmundsson G. (1988) Distribution of cystatin C amyloid deposits in the Icelandic patients with hereditary cystatin C amyloid angiopathy. In Isobe T, Araki S, Uchino F, Kito S, Tsubura E (eds.), *Amyloid and Amyloidosis*, Plenum Publishing Corp, New Tork, pp. 585-590.
18. Wang ZZ, Jensson O, Thorsteinsson L, Vinters HV. (1997) Microvascular degeneration in hereditary cystatin C amyloid angiopathy of the brain. *APMIS* **105**, 41-47.
19 Benedikz E, Blöndal H, Gudmundsson G. (1990) Skin deposits in hereditary cystatin C amyloidosis. *Virchows Archiv. A* **417**, 325-331.
20. Jensson O, Gudmundsson G, Arnasson A, Blöndal H, Grubb A, Löfberg H. (1986) Hereditary central nervous system γ-trace amyloid angiopathy and stroke in Icelandic families. In Glenner GG, Ossserman EF, Benditt EP, Calkins E, Cohen AS, Zucker-Franklin D. (eds.), *Amyloidosis*, Plenum Publishing Corp, New Tork, pp. 585-590.
21. Rowe IF, Jensson O, Lewis PD, Candy J, Tennent GA, Pepys MB. (1984) Immunohistochemical demonstration of amyloid P component in cerebrovascular amyloidosis. *Neuropathol. Appl. Neurobiol.* **10**, 53-61.
22. Haan J, Maat-Schieman MLC, Van Duinen SG, Jensson Ó, Thorsteinsson L, Roos RAC. (1994) Colocalization of A4 and cystatin C in cortical blood vessels in Dutch but not in Icelandic hereditary cerebral angiopathy with amyloidosis. *Acta Neurol. Scand.* **89**, 367-371.
23. Shimode K, Kobayashi S, Imoka K, Umage N, Nagai A. (1996). Leukoencephalopathy-related cerebral amyloid angiopathy with cystatin C deposition. *Stroke* **27**, 1417-1419.
24 Vinters HV, Secor DL, Pardridge WM, Gray F. (1990) Immunohistochemical study of cerebral amyloid angiopathy. III. Widespread Alzheimer A4 peptide in cerebral microvessel walls colocalizes with gamma trace in patients with leukoencephalopathy. *Ann. Neurol.* **28**, 34-42.
25. Maruyama K, Ikeda S, Ishihara T, Allsop D, Yanagiswa N. (1990) Immunohistochemical characterization of cerebrovascular amyloid in 46 autopsied cases using antibodies to β protein and cystatin C. *Stroke* **21**, 397-403.
26 Anders KH, Wang ZZ, Kornfeld M, Gray F, Soontornniyomkij V, Reed LA, Hart MN, Menchine M, Secor DL, Vinters HV. (1997) Giant cell arteritis in association with cerebral amyloid angiopathy: immunohistochemical and molecular studies. *Hum. Pathol.* **28**, 1237-1246.
27. Cohen DH, Feiner H, Jensson Ó, Frangione B. (1983) Amyloid fibril in hereditary cerebral hemorrhage with amyloidosis (HCHWA) is related to the gastroentero-pancreatic neuroendocrine protein, gamma trace. *J. Exp. Med.* **158**, 623-628.

28. Grubb A, Löfberg H. (1982) Human γ-trace, a basic microprotein: Amino acid sequence and presence in the adenohypophysis. *Proc. Natl. Acad. Sci. USA* **79**, 3024-3027.
29. Abrahamson M, Barrett AJ, Salvesen G, Grubb A. (1986) Isolation of six cysteine proteinase inhibitors from human urine. Their physicochemical and enzyme kinetic properties and concentrations in biological fluids. *J. Biol. Chem.* **261**, 11281-11289.
30. Rawlings ND, Barrett AJ. (1990) Evolution of proteins of the cystatin superfamily. *J. Mol. Evol.* **30**, 60-71.
31. Freije JP, Abrahamson M, Olafsson I, Velasco G, Grubb A, Lopez-Otin C. (1991) Structure and expression of the gene encoding cystatin D, a novel human cysteine proteinase inhibitor. *J. Biol. Chem.* **266**, 20538-20543.
32. Sotiropoulou G, Anisowicz A, Sager R. (1997) Identification, cloning, and characterization of cystatin M, a novel cysteine proteinase inhibitor, down-regulated in breast cancer. *J. Biol. Chem.* **272**, 903-910.
33. Ni J, Abrahamson, Zhang M, Fernandez MA, Grubb A, Su J, Yu G-L, Li Y, Parmelee D, Xing L, Coleman TA, Gentz S, Thotakura R, Nguyen N, Hesselberg M, Gentz R. (1997) Cystatin E is a novel human cysteine proteinase inhibitor with structural resemblance to family 2 cystatins. *J. Biol. Chem.* **272**, 10853-10858.
34. Ni J, Fernandez MA, Danielsson L, Chillakuru RA, Zhang J, Grubb A, Su J, Gentz R, Abrahamson M. (1998) Cystatin F is a glycosylated human low molecular weight cysteine proteinase inhibitor. *J. Biol. Chem.* **273**, 24797-24804.
35 Grubb A, Jensson Ó, Gudmundsson G, Árnason A, Löfberg H, Malm J. (1984) Abnormal metabolism of γ-trace alkaline microprotein: The basic defect in hereditary cerebral hemorrhage with amyloidosis. *New. Engl. J. Med.* **311**, 1547-1549.
36. Ólafsson Í, Gudmundsson G, Abrahamson M, Jensson Ó, Grubb A. (1990) The amino terminal portion of cerebrospinal fluid cystatin C in hereditary cystatin C amyloid angiopathy is not truncated: direct sequence analysis from agarose gel electropherograms. *Scand. J. Clin. Lab. Invest.* **50**, 85-93.
37. Abrahamson M, Islam MQ, Szpirer J, Szpirer C, Levan G. (1989) The human cystatin C gene (CST3) mutated in hereditary cystatin C amyloid angiopathy, is located on chromosome 20. *Hum. Genet.* **82**, 223-226.
38. Schnittger S, Gopal Rao VVN, Abrahamson M, Hansmann I. (1993) Cystatin C (CST3), the candidate gene for hereditary cystatin C amyloid angiopathy (HCCAA), and other members of the cystatin gene family are clustered on chromosome 20p11.2. *Genomics* **16**, 50-55.
39. Abrahamson M, Ólafsson Í, Pálsdóttir Á, Ulvsbäck M, Lundwall Å, Jensson Ó, Grubb A. (1990) Structure and expression of the human cystatin C gene. *Biochem. J.* **268**, 287-294.
40. Ólafsson Í. (1995) The human cystatin C gene promoter: functional analysis and identification of heterogenous mRNA. *Scand. J. Clin. Lab. Invest.* **55**, 597-609.
41. Emilsson V, Thorsteinsson L, Jensson Ó, Gudmundsson G. (1996) Human cystatin C gene expression and regulation by TGFβ1: Implications for the pathogenesis of hereditary cystatin C amyloid angiopathy causing brain hemorrhage. *Amyloid: Int. J. Exp. Clin. Invest.* **3**, 110-118.
42 Solem M, Rawson C, Lindburg K, Barnes D. (1990) Transforming growth factor beta regulates cystatin C in serum free mouse embryo (SFME) cells. *Biochem. Biophys. Res. Commun.* **172**, 945-951.
43. Bjarnadóttir M, Grubb A, Ólafsson Í. (1995) Promoter-mediated, dexamethasone-induced increase in cystatin C production by HeLa cells. *Scand. J. Clin. Lab. Invest.* **55**, 617-623.

44. Warfel AH, Zucker-Franklin D, Frangione B, Ghiso J. (1987) Constitutive secretion of cystatin C (γ-trace) by monocytes and macrophages and its downregulation after stimulation. *J. Exp. Med.* **166**, 1912-1917.
45. Warfel AH, Cardozo C, Yoo OH, Zucker-Franklin D. (1991) Cystatin C and cathepsin B production by alveolar macrophages from smokers and nonsmokers. *J. Leukocyte Biol.* **49**, 41-47.
46. Chapman HA, Reilly JJ, Yee R, Grubb A. (1990) Identification of cystatin C, a cysteine proteinase inhibitor, as a major secretory product of human alveolar macrophages *in vitro*. *Am. Rev. Respir. Dis.* **141**, 698-705.
47. Ghiso J, Jensson Ó, Frangione B. (1986) Amyloid fibrils in hereditary cerebral hemorrhage with amyloidosis of Icelandic type is a variant of γ-trace basic protein (cystatin C). *Proc. Natl. Acad. Sci. USA* **83**, 2974-2978.
48. Abrahamson M, Grubb A, Ólafsson Í, Lundwall Å. (1987) Molecular cloning and sequence analysis of a cDNA coding for the precursor of the human cysteine proteinase inhibitor cystatin C. *FEBS Lett.* **216**, 229-233.
49. Pálsdóttir Á, Abrahamson M, Thorsteinsson L, Árnason A, Ólafsson Í, Grubb A, Jensson Ó. (1988) Mutation in cystatin C gene causes hereditary brain hemorrhage. *Lancet* ii, 603-604.
50. Abrahamson M, Jónsdóttir S, Ólafsson Í, Jensson Ó, Grubb A (1992) Hereditary cystatin C amyloid angiopathy: Identification of the disease-causing mutation and specific diagnosis by polymerase chain reaction based analysis. *Hum. Genet.* **89**, 377-380.
51. Levy E, Lopez-Otin C, Ghiso J, Geltner D, Frangione B. (1989) Stroke in Icelandic patients with hereditary amyloid angiopathy is related to a mutation in the cystatin C gene, an inhibitor of cysteine proteinases. *J. Exp. Med.* **169**, 1771-1778.
52. Bjarnadottir M, Wulff BS, Sameni M, Sloane BF, Keppler D, Grubb A, Abrahamson M. (1998) Intracellular accumulation of the amyloidogenic L68Q variant of human cystatin C in NIH/3T3 cells. *Mol. Pathol.* **51**, 317-326.
53. Wei L, Berman Y, Castano EM, Cadene M, Beavis RC, Devi L, Levy E. (1998) Instability of the amyloidogenic cystatin C variant of hereditary cerebral hemorrhage with amyloidosis, Icelandic type. *J. Biol. Chem.* **273**, 11806-11814.
54. Thorsteinsson L, Georgsson G, Ásgeirsson B, Bjarnadóttir M, Ólafsson Í, Jensson Ó, Gudmundsson G. (1992) On the role of monocytes/macro-phages in the pathogenesis of central nervous system lesion in hereditary cystatin C amyloid angiopathy. J. Neurol. Sci. **108**, 121-128.
55. Merz GS, Benedikz E, Schwenk V, Johansen TE, Vogel LK, Rushbrook JI, Wisniewski HM. (1997) Human cystatin C forms an inactive dimer during intracellular trafficking in transfected CHO cells. *J Cell Physiol* **173**, 423-432.
56. Ekiel I, Abrahamson M. (1996) Folding-related dimerization of human cystatin C. *J. Biol. Chem.* **271**, 1314-1321.
57. Ekiel I, Abrahamson M, Fulton DB, Lindahl P, Storer AC, Levadoux W, Lafrance M, Labelle S, Pomerleau Y, Groleau D, LeSauteur L, Gehring K. (1997) NMR structural studies of human cystatin C dimers and monomers. *J. Mol. Biol.* **271**, 266-277.
58. Bode W, Engh R, Musil D, Thiele U, Huber R, Karashikov A, Brzin J, Kos J, Turk V. (1988) The 2.0 Å X-ray crystal structure of chicken egg white cystatin and its possible mode of interaction with cysteine proteinases. *EMBO* **7**, 2593-2599.
59. Dieckman T, Mitschang L, Hofmann M, Kos J, Turk V, Auerswald EA, Jaenicke R, Oschkinat H. (1993) The structures of native phosphorylated chicken cystatin and of a recombinant unphosphorylated variant in a solution. *J. Mol. Biol.* **234**, 1048-1059.

60. Gerhartz B, Ekiel I, Abrahamson M. (1998) Two stable unfolding intermediates of the disease-causing L68Q variant of human cystatin C. *Biochemistry* **37**, 17309-17317.
61. Abrahamson M, Grubb A. (1994) Increased body temperature accelerates aggregation of (Leu-68|Gln)cystatin C, the amyloid forming protein in hereditary cystatin C amyloid angiopathy. *Proc. Natl. Acad. Sci. USA* **91**, 1416-1420.
62. Aridor M, Balch WE. (1999) Integration of endoplasmic reticulum signaling in health and disease. *Nat. Med.* **5**, 745-751.

SECTION III

CELLULAR AND MOLECULAR PATHOLOGY OF CAA

Chapter 8

NEUROPATHOLOGIC FEATURES AND GRADING OF ALZHEIMER-RELATED AND SPORADIC CAA

HARRY V. VINTERS* and JEAN-PAUL G. VONSATTEL[†]
*Department of Pathology & Laboratory Medicine, Brain Research Institute & Neuropsychiatric Institute, UCLA Medical Center, Los Angeles, CA, USA; [†]Department of Pathology and Neuroscience Center, Massachusetts General Hospital and Harvard Medical School, Charlestown, MA, USA

This chapter will consider neuropathologic features of CAA, an approach to grading the severity of CAA using straightforward morphologic criteria, and consequences of CAA in the brain; especially cerebral hemorrhage. Specifically, we will evaluate the relationship between sporadic CAA and that related to Alzheimer disease, unique neuropathologic features of CAA-related brain hemorrhage, and introduce the concept of CAA-associated microangiopathies (CAA-AM, e.g. microaneurysms, secondary angiitis, etc.); lesions that may well be the proximate cause of brain hemorrhage. Approaches to studying CAA pathogenesis and CAA-AM will be discussed in relation to the basic biology of the cerebral microvessel wall and its components, aspects of which are taken up elsewhere in this volume.

1. INTRODUCTION

This chapter will describe clinicopathologic features of lesions that commonly result from, or are associated with CAA, viz. stroke resulting from cerebral (intraparenchymal) hemorrhage, and other microscopic stigmata of Alzheimer disease (AD)/senile dementia of Alzheimer type (SDAT). It will focus on approaches to grading the *neuropathologic* severity of CAA, evaluating 'secondary' microvascular abnormalities through which CAA may cause cerebral bleeds and non-hemorrhagic brain lesions, and placing CAA in the context of other structural abnormalities in AD brain, specifically senile (diffuse and neuritic) plaques and neurofibrillary tangles.

M.M. Verbeek et al. (eds.),
Cerebral Amyloid Angiopathy in Alzheimer's Disease and Related Disorders, 137–155.

2. RELATIONSHIP OF AD-RELATED TO SPORADIC CAA

In most reviews of the neuropathologic features and pathophysiology of AD/SDAT, CAA is included (with variable emphasis) as one of the key microscopic lesions which, when found 'in excess' in the central nervous system (CNS), provide morphologic support for the clinical diagnosis [1-5]. The pivotal role of CAA in AD was highlighted by Mandybur's observation of slight to severe CAA in 13/15 AD brains [6], then Glenner & Wong's subsequent isolation of the Aβ protein from meningeal CAA microvascular lesions, not from amyloid-laden senile plaques (SPs) (as is sometimes mistakenly claimed) [7-9]. Despite this, one of the most widely used systems for neuropathological stageing of AD barely mentions CAA [10,11], while others give it less prominence than the finding of parenchymal SPs and neurofibrillary tangles (NFTs) and are unclear on how CAA is best quantified in a given brain--should the neuropathologist focus on *percentage* of microvessels affected by CAA per unit area of brain, the *severity or extent* of arterial wall destruction in a given vessel, or try to integrate both parameters [12,13]? Confusion over the significance of CAA with or without *other* AD brain changes is compounded by the occurrence of non-Aβ-linked forms of CAA, e.g. HCHWA-I/HCCAA (see Ólafsson and Thorsteinsson, this volume), and familial syndromes of relatively *pure* Abeta CAA associated with specific APP mutations, such as the Dutch form of CAA (see Bornebroek *et al,* this volume). Estimates of the relative significance of CAA in causing stroke also vary, depending on whether the relevant studies have been carried out at a 'stroke center' or one focusing on Alzheimer disease and related dementias.

Taking the brains of patients with autopsy-confirmed AD/SDAT as the starting point, 83% show at least a mild degree of CAA, and approximately 25% show moderate to severe CAA, involving cerebral vessels in one or more cortical regions [14]. Of interest, brains with moderate to severe CAA also show a significantly higher frequency of hemorrhagic or ischemic lesions than do brains with negligible CAA. In one comparative study of AD vs age-matched controls, CAA was present in 86% of AD cases, or more than twice as frequently as in controls [15]. If one considers an unselected autopsy population of the elderly, estimates of CAA depend upon how carefully (and using which microscopic techniques, including immunohistochemistry) this microvascular lesion is sought. In one study, moderate to severe CAA showed a range from 2.3% in the age group 65-74 yr, to 8% (age range 75-84 yr), to 12.1% (age 85 yr or older); consecutive autopsies on AD patients from the same center showed 25.6% with moderate to severe CAA and over 5% with CAA-related cerebral hemorrhage [16,17].

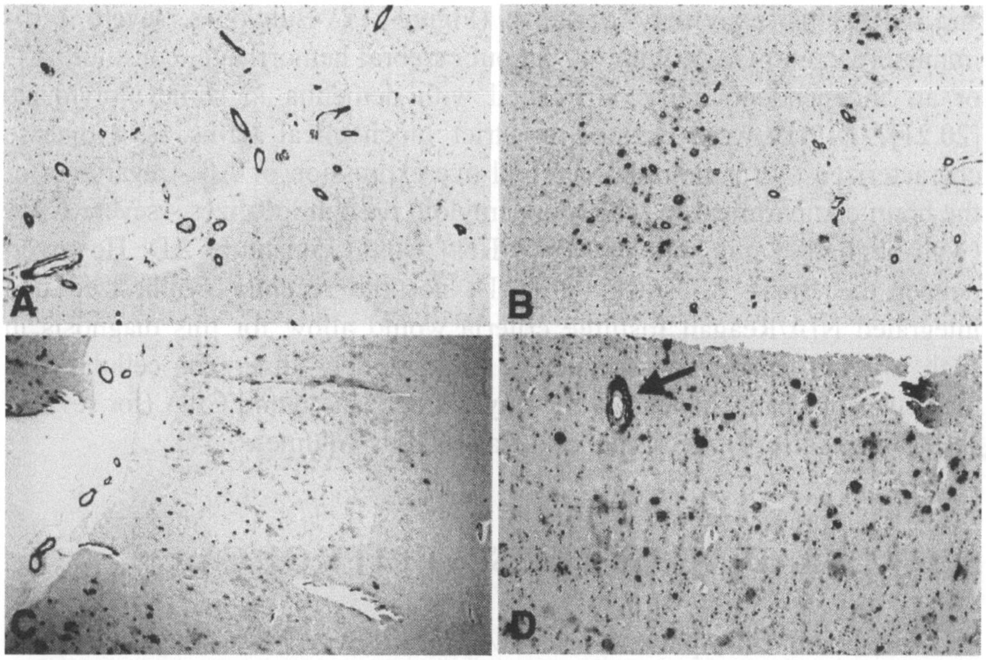

Figure 1. Aβ-immunostained sections of brain from different patients, showing variable severity of CAA in relation to senile plaques (SPs). Panel **A** shows a section from the brain of a patient with CAA-related hemorrhage occurring after a many year history of dementia. Aβ immunostaining shows overwhelming CAA in the presence of negligible SP amyloid. **B.** Section (comparable magnification) shows Aβ-immunoreactive CAA, but with abundant immunoreactive SPs. **C,D.** Brain sections from an elderly individual with longstanding dementia. Note both arteriolar and SP Aβ immunoreactivity within cortex. Immunoreactive arterioles and arteries are seen in both the subarachnoid space (C) and cortex (arrow, D). (Magnifications: A,B,C X 25; D X 65).

Clearly, CAA becomes more severe and widespread with advancing age, and significantly worse CAA is noted in AD than non-AD patients [18,19]. All neocortical regions and their overlying meninges may be affected by CAA, there being minor variations in distribution of affected microvessels among the lobes [18,19]; deep central grey matter, subcortical white matter and brainstem are usually devoid of CAA, even in otherwise severe cases, while the cerebellum shows varying degrees of CAA severity within its parenchyma or the overlying meninges. CAA may affect primarily arterioles, or manifest as capillary infiltration by amyloid with extension of amyloid fibrils into adjacent brain parenchyma, a phenomenon some describe as 'dyshoric angiopathy'.

A subset of patients who present with either dementia or cerebral hemorrhage (sometimes *without* dementia) show CAA as the major manifestation of CNS Aβ deposition. Considering all patients with AD type dementia, the extent and severity of CAA (in relation to parenchymal SPs and NFTs) is extremely variable (Figure 1). But does severe (Aβ-immunoreactive) CAA, with or without cerebral hemorrhage, constitute AD or an 'independent entity' associated with dementia in some individuals [20,21]? If AD were defined in strict biochemical terms as a disease characterized by (pathologic) cerebral overexpression of Aβ - regardless of the brain compartment in which the amyloid protein localizes - severe CAA (with negligible numbers of SPs, NFTs) would constitute AD. However, neither the Braak & Braak, CERAD, nor the recently synthesized and integrated NIA/Reagan Institute criteria would allow for this diagnosis in such a brain examined at autopsy [22]. Widely divergent cellular and molecular mechanisms lead to SP amyloid deposition and CAA (for review, see [23]; see also Verbeek *et al* and Kalaria, this volume).

3. GRADING OF THE SEVERITY OF CAA

Most brains from patients with CAA are free of hemorrhage; however, there is a relationship between the severity of CAA and the occurrence of cerebral (especially lobar) hemorrhages, initially established in Icelandic patients with HCHWA-I/HCCAA [18,24-27]. Thus, assessment of the severity of CAA is an important step in elucidating the cause of cerebral hemorrhage when amyloid angiopathy and lobar hemorrhage coexist.

There are no universally accepted criteria to grade the extent of amyloid-related vasculopathy. CAA may be absent in one brain region, while many amyloid-laden arterioles may be present in another. A segment of a vessel may be devoid of amyloid, while the media of an adjacent segment may be replaced by amyloid [28]. Within a single microscopic field, there may be a vessel whose media is extensively replaced by amyloid next to an apparently normal vessel. Usually the transition between affected and preserved segments is abrupt. With increasing amyloid accumulation, the lengths of segments that are apparently free of amyloid become shorter and their number decreases. The regional presence or absence of CAA should be considered in sampling tissue for microscopic evaluation, or for planning a biopsy. The segmental deposition of amyloid within a single vessel should be kept in mind in grading the severity of CAA.

Assessment of the severity of CAA can be determined along two paired axes: 1. By evaluating the extent and involvement of individual vessels, and 2. By evaluating the density of amyloid-laden microvessels. Usually, the severity of amyloid-laden vessels within brain has a surprisingly narrow spectrum; however, the higher the density of amyloid-laden vessels the higher the probability of encountering a severely involved vessel.

It is now accepted that the extent of amyloid deposition within vessel walls correlates with increasing risk of hemorrhage. Thus, in the absence of other cause(s), lobar cerebral hemorrhage is most likely due to CAA when the vessels are severely involved. For example, microaneurysms have not been found in isolated vessels showing moderate amyloid deposition, whereas vessels with severe amyloid deposition frequently contain microaneurysms [28]. Rupture of cerebral microaneurysms with severe CAA are believed to be a proximate cause of parenchymal hemorrhage, just as Charcot-Bouchard type microaneurysms are frequently found in the brains of patients with hypertensive cerebral hemorrhage [29,30]. It may thus be preferrable to estimate the degree of severity of CAA by determining the extent of involvement of vessel walls rather than determining the density of amyloid-laden vessels.

The grading scale developed by one of us (JPV) and used in daily practice by both of us, scores the most advanced degree of CAA present within available specimens, derived from an extensive sampling of the autopsy brain, with examination of multiple areas of neocortex, the hippocampi, striatum, thalamus, cerebellum and brainstem. Routine sections are supplemented with phosphotungstic acid hematoxylin (PTAH) for fibrin and fibrinoid. To detect amyloid routinely, 12 um thick (approx. twice normal thickness) paraffin sections are stained for Congo red and examined with and without polarization microscopy.

A brain is considered to manifest CAA when it shows at least one leptomeningeal or cortical congophilic, salmon-pink vessel with yellow-green birefringence under polarized light. Severity is graded as follows [28] (see also figure 2).

Grade 1 (mild involvement): Amyloid restricted to a congophilic rim around apparently normal or atrophic smooth muscle cells in the media of otherwise normal vessels. Smooth muscle cells may be absent, leaving an optically empty vacuole surrounded by amyloid. Deposition of amyloid appears to occur initially within the outer portion of the media. The gradual accumulation of amyloid correlates inversely with the loss of smooth muscle cell nuclei.

Grade 2 (moderate): The media is replaced by amyloid and is thicker than normal (Figure 2). At this stage, the media has lost most of its smooth muscle cell component and consists of a band of amorphous material that may have a reticular or radial structure. There is no evidence of remote or recent blood leakage.

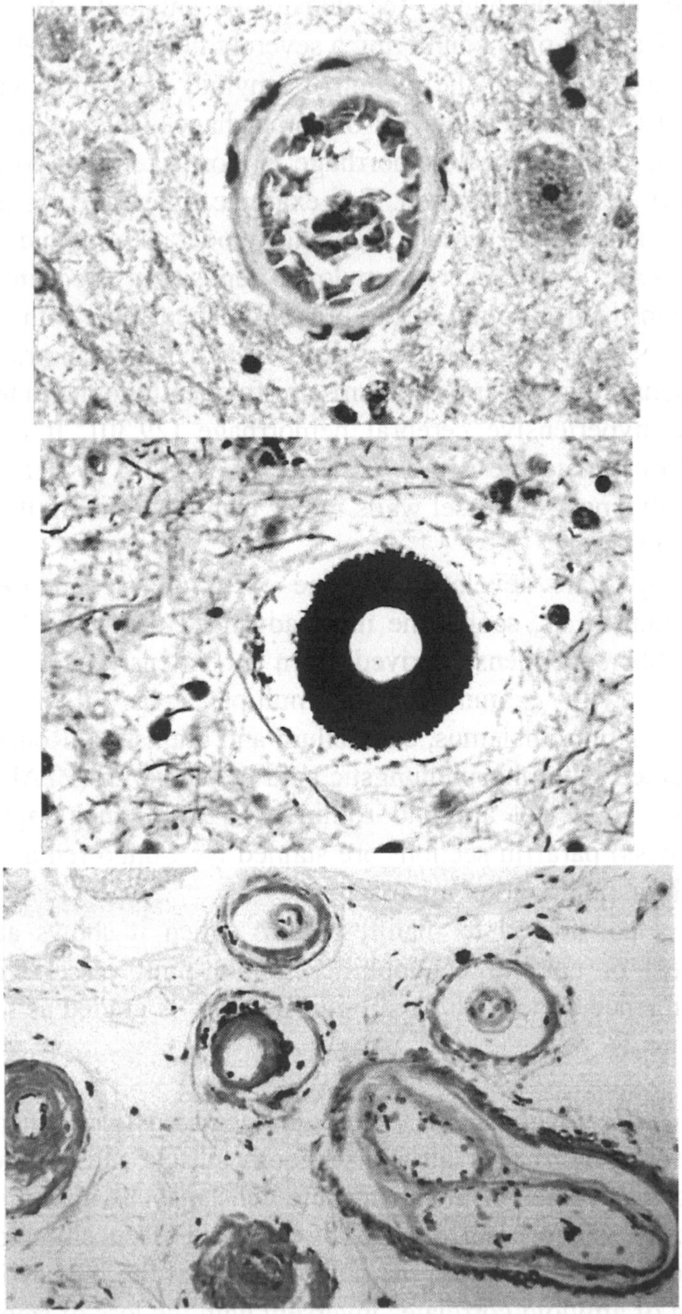

Figure 2. Grading of CAA severity. **A,B**. Moderately severe (Grade 2) CAA. A (H & E-stain) shows smooth muscle cell media to be replaced by amyloid; the vessel wall is also thickened. B (Bielschowsky stain) shows thickening of an amyloid-laden artery, but without extension of amyloid into surrounding brain parenchyma. **C**. Severe CAA (Grade 3). Arteries show thick, amyloid infiltrated walls; hemosiderin pigment, indicative of remote blood leakage from the vessels, is noted in their adventitia (H & E stain). (Magnifications: A,B X 500; C X 250).

Grade 3 (severe): There is extensive amyloid deposition with 'cracks' or focal fragmentation of vessel walls, including 'vessel-within-a-vessel' appearance and at least one focus of perivascular (acute or chronic) leakage of blood, as evidenced by the presence of erythrocytes of hemosiderin or both (Figure 2). Severe amyloid deposits may be associated with microaneurysm formation or fibrinoid necrosis, CAA-associated microangiopathies (see below).

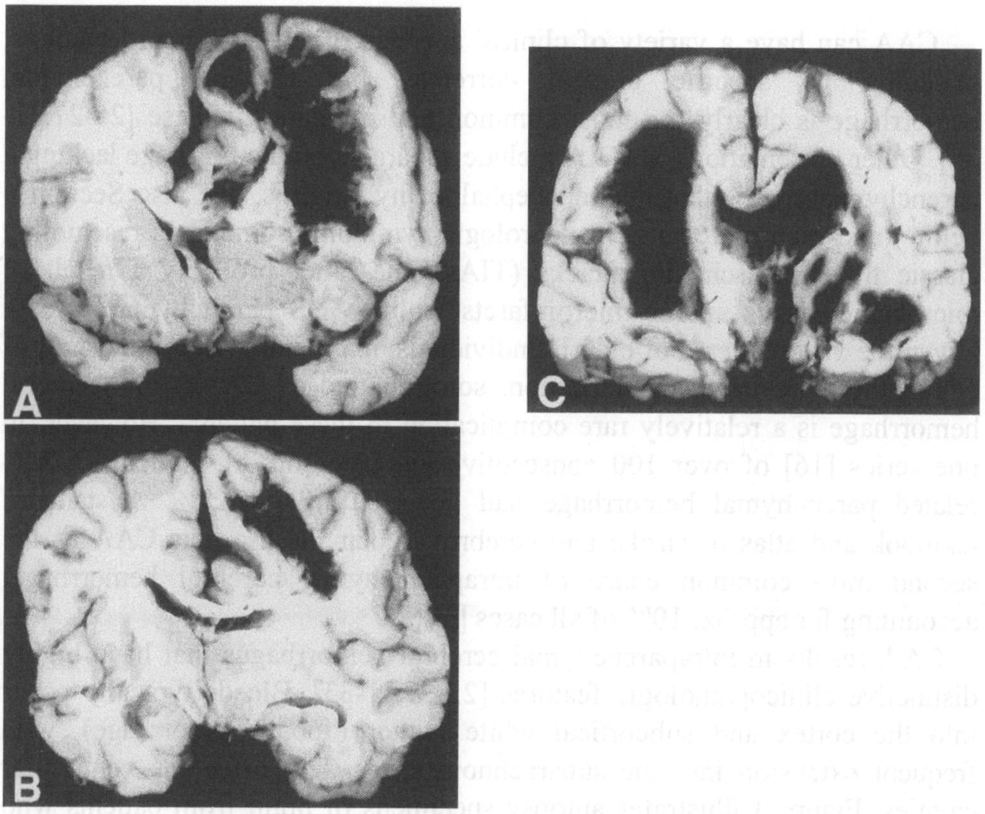

Figure 3. Examples of CAA-related cerebral hemorrhage. A,B. Massive right fronto-parietal hematoma in a 81-yr-old man who presented with rapidly evolving left hemiparesis (no previous well documented history of dementia). Clot is present in the right fronto-parietal white matter, with marked right to left shift of midline structures. C. Section of brain from a 68-yr-old man with a long history of dementia, and strong family history of same. Note bihemispheric (left fronto-parietal, right temporal) hematomas.

The presence or absence of CAA was looked for in a series of 1607 brains collected at autopsy by one of us (JPV) between 1993 and 1999; mean age at death was 69.7 +/- 16.2 yr. Up to 90% of brains examined were from subjects with neurologic or psychiatric illnesses, the rest from neurologically 'intact' individuals. Among 382 brains with CAA, 130 were categorized as grade 1 (mean age 80 +/- 10.3 yr), 185 as grade 2 (80 +/- 8.4), and 67 as grade 3 (81 +/- 7.6 yr). Of the 'grade 3' brains, seven showed fibrinoid necrosis in association with CAA and four of those seven brains had lobar hemorrhage(s) without any apparent cause other than CAA.

4. NEUROPATHOLOGIC FEATURES OF CAA-RELATED CEREBRAL HEMORRHAGE:

CAA can have a variety of clinical and neuroradiologic manifestations, including no specific clinical correlate, but cerebral parenchymal hemorrhage is clearly the most common and dramatic of these [25-27,31-35]. Other presentations of CAA include subarachnoid hemorrhage lacking a parenchymal component, leukoencephalopathy, angiitis (see also Section 6 below) and recurrent transient neurologic symptoms sometimes resembling classic transient ischemic attacks (TIAs), the latter probably a result of microhemorrhages and/or microinfarcts within CNS parenchyma [36-39]. Given the vast universe of elderly individuals, many with AD - almost all of whom therefore have, by definition, some degree of CAA - intracerebral hemorrhage is a relatively rare complication in these patients. However, in one series [16] of over 100 consecutive autopsies on AD patients, CAA-related parenchymal hemorrhage had occurred in over 5%. A standard textbook and atlas of stroke and cerebrovascular disease lists CAA as the second most common cause of intraparenchymal cerebral hemorrhage, accounting for approx. 10% of all cases [40].

CAA results in intraparenchymal cerebral hemorrhages that have highly distinctive clinicopathologic features [25-27,31-33]. Bleeds typically occur into the cortex and subcortical white matter ('lobar' hemorrhage), with frequent extension into the subarachnoid space, less often the ventricular cavities. Figure 3 illustrates autopsy specimens of brain from patients who suffered fatal CAA-related hemorrhage. This anatomic predilection reflects the topography of CAA, which is primarily a meningeal and neocortical microangiopathy [18]. Awareness of CAA as a likely etiology for cerebral lobar hemorrhage is also important, however, for the surgical pathologist--CAA can readily be detected in brain parenchyma adjacent to an evacuated hematoma [41], provided the examiner's index of suspicion is appropriately high (Figure 4) and the patient 'fits the profile' of one likely to manifest this type of stroke. Patients are usually in their 70s or 80s, and (see Section 2 above) may be demented prior to the occurrence of stroke, though often they are not. Hemorrhages may occur serially in different lobes of the cerebral hemispheres over months or even years. Posterior fossa hemorrhage with CAA is rare, and then almost always into the cerebellum rather than brainstem [31]. Primary hemorrhage into the subarachnoid space as a result of CAA is rare but reported; usually subarachnoid bleeding with CAA is secondary to direct extension of blood from underlying brain parenchyma. In one review, which undertook an analysis of all papers reporting clinicopathologic features of CAA-related hemorrhage up to 1987, it was noted that approximately a third of affected patients also had clinical evidence of hypertension [32].

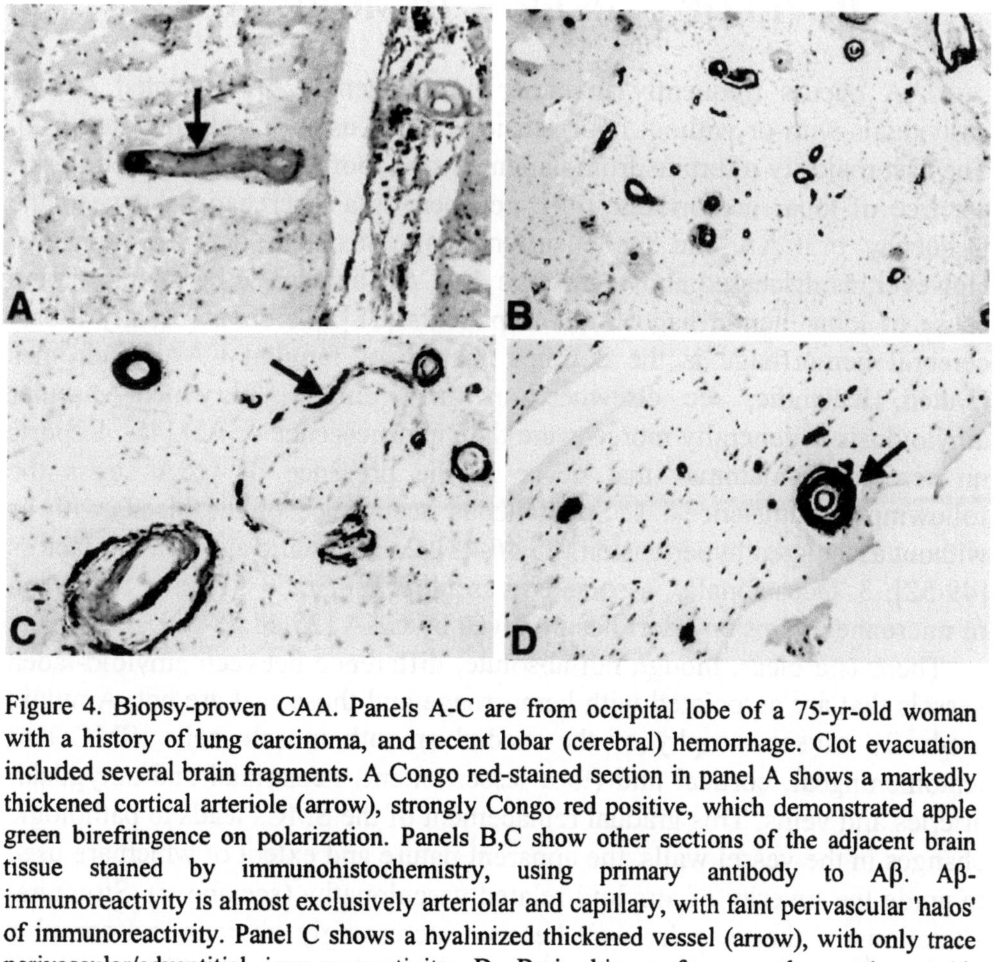

Figure 4. Biopsy-proven CAA. Panels A-C are from occipital lobe of a 75-yr-old woman with a history of lung carcinoma, and recent lobar (cerebral) hemorrhage. Clot evacuation included several brain fragments. A Congo red-stained section in panel A shows a markedly thickened cortical arteriole (arrow), strongly Congo red positive, which demonstrated apple green birefringence on polarization. Panels B,C show other sections of the adjacent brain tissue stained by immunohistochemistry, using primary antibody to Aβ. Aβ-immunoreactivity is almost exclusively arteriolar and capillary, with faint perivascular 'halos' of immunoreactivity. Panel C shows a hyalinized thickened vessel (arrow), with only trace perivascular/adventitial immunoreactivity. D. Brain biopsy from another patient, with markedly thickened (e.g. arrow) Abeta-immunoreactive arterioles and capillaries. (Magnifications: A,B,C X 65; D X 130).

Of greater interest than the phenomenology of CAA-related cerebral hemorrhage is the likely sequence of cellular events in affected arterioles that leads to their rupture, with resultant extravasation of blood into brain parenchyma (see also Sections 5,6 below). In this regard, the Icelandic form of CAA (see Ólafsson and Thorsteinsson, this volume) may share features with Aβ CAA. In both conditions, the pivotal event probably leading to arteriolar rupture is degeneration of the vascular media, with replacement of medial smooth muscle cells by fibrillar amyloid [42,43]. Cellular and molecular events that may contribute to this degeneration are also reviewed by Verbeek *et al* and Kalaria (this volume).

5. WHY DO ONLY SOME PATIENTS WITH CAA DEVELOP CEREBRAL HEMORRHAGE?

CAA occurs frequently in either non-demented or demented elderly individuals with or without hypertension, and is usually well tolerated [32]. The vast majority of brains from patients with (sporadic or age-related) CAA are free of lobar hemorrhage [44], i.e. there is a discrepancy between the frequency of CAA and the occurrence of coexistent brain hemorrhage. However, as discussed above (Section 4), CAA appears to be the proximate cause of lobar hemorrhage in some patients [28]. Cerebral (usually lobar) cerebral hemorrhage is the defining feature of familial CAA syndromes (Dutch, Icelandic, see elsewhere in this volume), in which vascular amyloidosis is generally more severe than in senescence or AD [45]. Reports on cerebral hematomas that occur in the presence of CAA stress the following circumstances: 1. Spontaneous occurrence of the bleed, with or without associated hypertension [25,46-48]; 2. Traumatic aggravating factors [49-52]; 3. Occasionally, associated vasculitis [28,36,53-57]; and formation of microaneurysms on arterioles involved by CAA [25,28,58-61].

There is a clear, though not absolute, difference between amyloid-laden vessels that are associated with hemorrhage and those that are not. Amyloid gradually causes atrophy of the medial smooth muscle cells (SMCs) of leptomeningeal, cortical and (to a lesser extent) neostriatal and amygdalar arteries and veins. This gradual replacement of the SMCs leads to pathologic changes in the vessel walls, the apparent nature and extent of which are used to grade the severity of amyloid-related vasculopathy (see above). Structural alterations of the vascular wall become increasingly obvious and complex with amyloid accumulation [25,62,63]. These changes probably impair the function of affected vessels by reducing their flexibility, and range of contraction or dilatation. In contrast to normal vessels that are usually collapsed at microscopic examination of the CNS, the lumina of amyloid-laden arterioles are often patent, which likely reflects their loss of contractility, and presence of significant rigidity in their walls [44].

Amyloid disrupts vascular architecture and weakens vessel walls, causing cracks, focal fragmentation, 'vessel-within-a-vessel' appearance ('double-barreling'), and microaneurysm formation with or without fibrinoid necrosis [63]. These changes, notably present in Grade 3 CAA, play a major role in the causation of lobar hemorrhage. Amyloid deposits within cerebral blood vessel walls appear well-tolerated up to a threshold, e.g. CAA of Grades 1 and 2, beyond which the structural changes defined as Grade 3 occur, increasing dramatically the risk of lobar hemorrhage. In this setting, extrinsic, otherwise banal factors such as minor head trauma might trigger parenchymal hemorrhages.

Deposition of amyloid in the arteriolar wall itself may sufficiently weaken it to the point of rupture [64]. However, microaneurysms and fibrinoid necrosis affecting amyloid-laden arterioles appear to play a critical role, and may be a precipitating factor, in the genesis of lobar hemorrhage [25,51,58,63,65-68]. In necropsy specimens with severe CAA and cerebral hemorrhages, serial sections have disclosed a direct relationship between fibrinoid necrosis and the site of vascular rupture [28]. Fresh hemorrhage has been found in continuity with rupture of a focally dilated cortical microvessel. Furthermore, serial sections have shown an intracortical aneurysm with fibrinoid necrosis in the neck portions at both ends of an amyloid-laden vessel in a brain with multiple hemorrhages [28]. The absence of blood leakage at sites of both wall fragmentation and fibrin clots suggests that, in some instances, rupture of the wall might be sufficiently gradual to allow the clotting cascade to prevent a bleed. Microaneurysms, possibly an indirect result of antecedent fibrinoid necrosis of amyloid-laden arterioles, may play an important part in CAA-related hemorrhage, based upon studies in HCHWA-D [59]. Additional factors that may increase the likelihood of hemorrhage include granulomatous angiitis or simply the presence of foreign body giant cells around vascular amyloid (see below). Systemic risk factors for CAA-related cerebral hemorrhage, administration of anticoagulants and thrombolytic agents, are discussed by Greenberg (this volume).

In conclusion, the coexistence of CAA and cerebral hemorrhage does not necessarily mean that hemorrhage was the result of CAA. However, when CAA is severe (Grade 3), and accompanied by microaneurysm formation, fibrinoid necrosis or both, its association with cerebral hemorrhage is compelling.

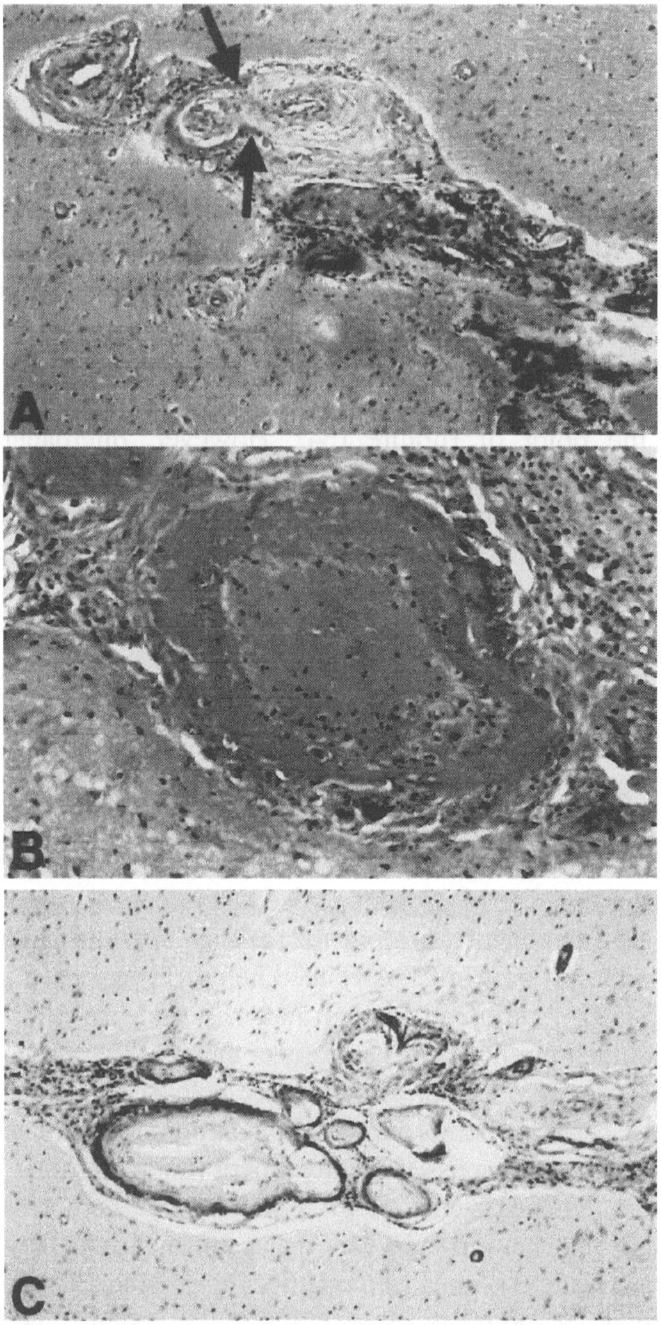

Figure 5. CAA-associated microangiopathies. Panels A-C are all micrographs from sections that originated in brain of a 81-yr-old patient with fatal CAA-related cerebral hemorrhage. A. Microaneurysm formation on a meningeal artery. Arrows indicate 'neck' of the aneurysm, which (distal to the neck) shows a thickened, hyalinized 'dome'. B. Fibrinoid necrosis of an artery. Note lack of any discernible cellular elements in an affected artery. C. Aβ-immunostained section. Scattered parenchymal arteries show immunoreactivity, and adventitia of meningeal vessels is focally immunolabelled. A modest chronic inflammatory infiltrate surrounds the leptomeningeal arteries. (Magnifications: A,C X 65; B X 130).

6. CAA-ASSOCIATED MICROANGIOPATHIES & THEIR RELATIONSHIP TO NEUROLOGIC SYNDROMES & STROKE

The concept of CAA-associated microangiopathies (CAA-AM) probably originated in a seminal paper by Mandybur [63]. He specified these as being vasculopathies that are, in undefined ways, associated with CAA and possibly represent the immediate cause(s) of CAA-associated 'stroke', especially vascular rupture leading to hemorrhage. The morphologic features of CAA-AM, as he conceptualized this, include a 'potpourri' of morphologically heterogeneous anomalies: (a) glomerular formations, (b) microaneurysms, (c) obliterative intimal changes, (d) 'double-barreling', (e) chronic inflammatory or transmural infiltrates, (f) hyaline arteriolar degeneration (sometimes with the formation of microaneurysms), and (g) fibrinoid necrosis of involved vessels (Figure 5). There is nothing in the clinical presentation of a given patient, with the possible exception of hemorrhage itself, that predicts who will have severe CAA-AM though (as discussed above), fibrinoid necrosis of amyloid-infiltrated microvessels and microaneurysm formation are strongly associated *with* brain parenchymal hemorrhage. The role of obliterative intimal change in causing microinfarcts is suspected though not proven, while inflammation may aggravate a tendency to microaneurysm formation; both, by analogy to the putative role of Charcot-Bouchard aneurysms in hypertensive microvascular disease, may precipitate cerebral parenchymal bleeds [69,70]. Currently, granulomatous angiitis/vasculitis associated with CAA is usually differentiated from simple chronic inflammation around amyloid-infiltrated arterioles [36,71]. Secondary vessel wall calcification with CAA, though extraordinarily rare, should be added to the list of CAA-AM [72].

Not surprisingly, CAA-AMs occur with high frequency in a genetically determined form of Aβ CAA, HCHWA-D [73]. Since a major problem in research on sporadic or AD-associated human CAA is the heterogeneity of the patient population under investigation, examining CAA-AM (and its possible role in CAA-related stroke) is simplified by studying patients in whom CAA evolves with a somewhat predictable time course, tempo and set of complications. In autopsy brain tissue from 29 patients with HCHWA-D, the severity and extent of HCHWA-D was evaluated to yield a CAA-AM 'score' for each brain. This score was subsequently correlated, in a retrospective fashion, with clinical features of the patients, including the numbers of 'cerebrovascular lesions' (hemorrhages/infarcts) they had been known to experience during life, duration of illness, and presence or severity of hypertension and systemic atherosclerosis [59]. An association was found between CAA-AM and the number of cerebrovascular lesions.

Microaneurysmal dilatation of CAA-affected arterioles appeared to be the most significant 'predictor' of CAA-related stroke. A moderate association was noted between atherosclerosis and the CAA-AM score; however, hypertension did not show a significant association with CAA-AM in this patient group [59].

Of interest, given the prominence of macrophages in progression of atherosclerotic lesions, these cells were easily demonstrable in the cerebral arterial walls of patients with HCHWA-D and significant CAA-AM [73], while immunoreactive Aβ was often minimal and adventitial, especially in cases of prominent arteriolar fibrosis and microaneurysm formation. Histiocytes/macrophages may thus have a role in the pathogenesis of CAA-AM.

An especially intriguing CAA-AM is CAA-associated angiitis/vasculitis. This often takes the form of giant cell/granulomatous angiitis (GCA) [71,74] (Figure 6). Patients with this combination of pathologic abnormalities present with various forms of stroke and neurobehavioral manifestations, and over a broad range of ages. The presence of macrophages around arterioles involved by CAA raises the question of whether these cells are reacting to amyloid in the vessel wall, or contributing to its deposition. Both immunohistochemical and ultrastructural observations suggest the former, i.e. that GCA most likely represents a foreign body response to amyloid proteins, possibly causing worsening secondary destruction of the affected vessel wall [71]. Molecular studies have failed to show mutations in either exon 17 of the APP gene or exon 2 of the cystatin C gene in patients with CAA/GCA [71]. Conceivably, CAA as a cause of primary CNS angiitis, especially in the elderly, may have been *under*diagnosed in the past.

ACKNOWLEDGEMENTS

Work in H.Vinters' laboratory has been generously supported by NIH grants NS26312, P30AG10123, P50AG16570 and P01AG12435. Alex Brooks, Joyce Wong, Diana Lenard Secor, Justine Garakian and Yi Ding have provided expert technical assistance. Ms. Carol Appleton prepared photographs and Ms. Annetta Pierro assisted with manuscript preparation.

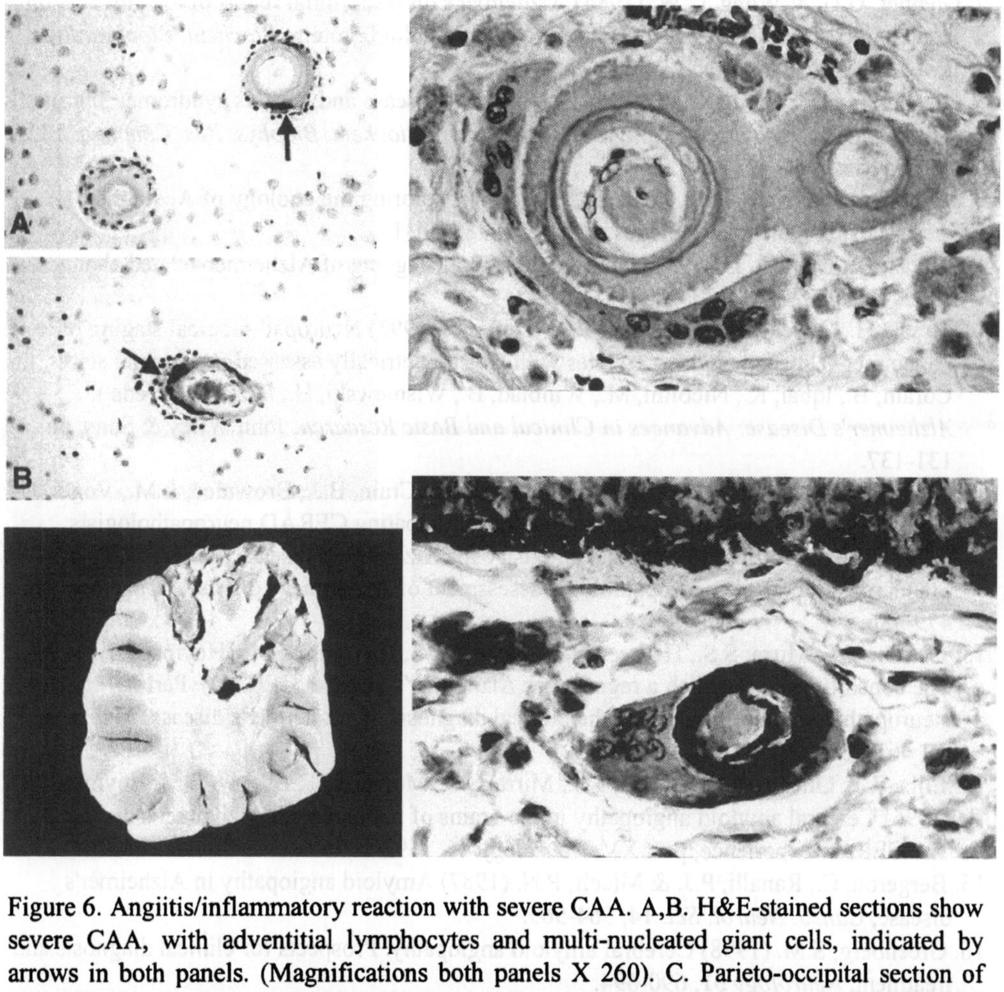

Figure 6. Angiitis/inflammatory reaction with severe CAA. A,B. H&E-stained sections show severe CAA, with adventitial lymphocytes and multi-nucleated giant cells, indicated by arrows in both panels. (Magnifications both panels X 260). C. Parieto-occipital section of brain from a patient with severe CAA and granulomatous angiitis shows extensive encephalomalacia. D,E. Micrographs from sections of brain illustrated in (C) show severe CAA with surrounding foreign body giant cells (D), and focal vessel wall fragmentation with surrounding giant cell (E). (D: H & E-stained section; E: Aβ immunohistochemistry).

REFERENCES

1. Kosik, K.S. (1992) Alzheimer's disease: A cell biological perspective, *Science* **256**, 780-783.
2. Vinters, H.V., Wang, Z.Z., & Secor, D.L. (1996) Brain parenchymal and microvascular amyloid in Alzheimer's disease, *Brain Pathol.* **6**, 179-195.
3. Cummings, J.L., Vinters, H.V., Cole, G.M., & Khachaturian, Z.S. (1998) Alzheimer's disease. Etiologies, pathophysiology, cognitive reserve, and treatment opportunities, *Neurology* **51** (Suppl 1), S2-S17.
4. Selkoe, D.J. (1994) Alzheimer's disease: A central role for amyloid. *J. Neuropathol. Exp. Neurol.* **53**, 438-447.
5. Vinters, H.V. (1998) Alzheimer's disease: a neuropathologic perspective, *Current Diagnostic Pathology* **5**, 109-117.
6. Mandybur, T.I. (1975) The incidence of cerebral amyloid angiopathy in Alzheimer's disease, *Neurology* **25**, 120-126.
7. Glenner, G.G. & Wong, C.W. (1984) Alzheimer's disease: Initial report of the purification and characterization of a novel cerebrovascular amyloid protein, *Biochem. Biophys. Res. Commun.* **120**, 885-890.
8. Glenner, G.G. & Wong, C.W. (1984) Alzheimer's disease and Down's syndrome: Sharing of a unique cerebrovascular amyloid fibril protein, *Biochem. Biophys. Res. Commun.* **122**, 1131-1135.
9. Lendon, C.L., Ashall, F. & Goate, A.M. (1997) Exploring the etiology of Alzheimer disease using molecular genetics, *JAMA* **277**, 825-831.
10. Braak, H. & Braak, E. (1991) Neuropathological stageing of Alzheimer-related changes, *Acta Neuropathol.* **82**, 239-259.
11. Braak, H., Duyckaerts, C., Braak, E. & Piette, F. (1993) Neuropathological staging of Alzheimer-related changes correlates with psychometrically assessed intellectual status, in Corain, B., Iqbal, K., Nicolini, M., Winblad, B., Wisniewski, H., & Zatta, P. (eds.), *Alzheimer's Disease: Advances in Clinical and Basic Research*, John Wiley & Sons, pp. 131-137.
12. Mirra, S.S., Heyman, A., McKeel, D., Sumi, S.M., Crain, B.J., Brownlee, L.M., Vogel, F.S., Hughes, J.P., van Belle, G., Berg, L. & Participating CERAD neuropathologists (1991) The consortium to establish a registry for Alzheimer's disease (CERAD). Part II. Standardization of the neuropathologic assessment of Alzheimer's disease, *Neurology* **41**, 479-486.
13. Gearing, M., Mirra, S.S., Hedreen, J.C., Sumi, S.M., Hansen, L.A. & Heyman, A. (1995) The consortium to establish a registry for Alzheimer's disease (CERAD). Part X. Neuropathology confirmation of the clinical diagnosis of Alzheimer's disease, *Neurology* **45**, 461-466.
14. Ellis, R.J., Olichney, J.M., Thal, L.J., Mirra, S.S., Morris, J.C., Beekly, D. & Heyman, A. (1996) Cerebral amyloid angiopathy in the brains of patients with Alzheimer's disease: The CERAD experience, part XV, *Neurology* **46**, 1592-1596.
15. Bergeron, C., Ranalli, P.J. & Miceli, P.N. (1987) Amyloid angiopathy in Alzheimer's disease, *Can. J. Neurol. Sci.* **14**, 564-569.
16. Greenberg, S.M. (1998) Cerebral amyloid angiopathy. Prospects for clinical diagnosis and treatment, *Neurology* **51**, 690-694.
17. Greenberg, S.M. & Vonsattel, J-P.G. (1997) Diagnosis of cerebral amyloid angiopathy. Sensitivity and specificity of cortical biopsy, *Stroke* **28**, 1418-1422.

18. Vinters, H.V. & Gilbert, J.J. (1983) Cerebral amyloid angiopathy: Incidence and complications in the aging brain II. The distribution of amyloid vascular changes, *Stroke* **14**, 924-928.
19. Masuda, J., Tanaka, K., Ueda, K. & Omae, T. (1988) Autopsy study of incidence and distribution of cerebral amyloid angiopathy in Hisayama, Japan, *Stroke* **19**, 205-210.
20. Vinters, H.V. (1992) Cerebral amyloid angiopathy and Alzheimer's disease: two entities or one? *J. Neurol. Sci.* **112**, 1-3.
21. Blumenthal, H.T. & Premachandra, B.N. (1990) The aging-disease dichotomy. Cerebral amyloid angiopathy--an independent entity associated with dementia, *JAGS* **38**, 475-482.
22. The National Institute on Aging, and Reagan Institute Working Group on Diagnostic Criteria for the Neuropathological Assessment of Alzheimer's Disease (1997) Consensus recommendations for the postmortem diagnosis of Alzheimer's disease. *Neurobiol. Aging* **18**, S1-S2.
23. Verbeek, M.M., Eikelenboom, P. & de Waal, R.M.W. (1997) Differences between the pathogenesis of senile plaques and congophilic angiopathy in Alzheimer disease, *J. Neuropathol. Exp. Neurol.* **56**, 751-761.
24. Gudmundsson, G., Hallgrímsson, J., Jónasson, T.A., & Bjarnason, O. (1972) Hereditary cerebral hemorrhage with amyloidosis, *Brain* **95**, 387-404.
25. Okazaki, H., Reagan, T.J., & Campbell, R.J. (1979) Clinicopathologic studies of primary cerebral amyloid angiopathy, *Mayo Clin. Proc.* **54**, 22-31.
26. Gilles, C., Brucher, J.M., Khoubesserian, P., & Vanderheagen, J.J. (1984) Cerebral amyloid angiopathy as a cause of multiple intracerebral hemorrhages, *Neurology* **34**, 730-735.
27. Torack, R.M. (1975) Congophilic angiopathy complicated by surgery and massive hemorrhage. Light and electron microscopic study, *Am. J. Pathol.* **81**, 349-366.
28. Vonsattel, J.P., Myers, R.H., Hedley-Whyte, E.T., Ropper, A.H., Bird, E.D., & Richardson, E.P. Jr. (1991) Cerebral amyloid angiopathy without and with cerebral hemorrhages: a comparative histological study, *Ann. Neurol.* **30**, 637-649.
29. Russell, D.S. (1954) Discussion: the pathology of spontaneous intracranial hemorrhage, *Proc. Royal Soc. Medicine* **47**, 689-693.
30. Cole, F.M., & Yates, P.O. (1967) The occurrence and significance of intracerebral microaneurysms, *J. Pathol. Bacteriol.* **93**, 393-411.
31. Vinters, H.V. (1998) Cerebral amyloid angiopathy, in Barnett, H.J.M., Mohr, J.P., Stein, B.M., and Yatsu, F.M. (eds.), *Stroke. Pathophysiology, Diagnosis, and Management*, 3/Ed., Churchill Livingstone, New York, pp. 945-962.
32. Vinters, H.V. (1987) Cerebral amyloid angiopathy. A critical review, *Stroke* **18**, 311-324.
33. Jellinger, K. (1977) Cerebrovascular amyloidosis with cerebral hemorrhage, *J. Neurol.* **214**, 195-206.
34. Vinters, H.V., & Duckwiler, G.R. (1992) Intracranial hemorrhage in the normotensive elderly patient, *Neuroimaging Clin. N. America* **2**, 153-169.
35. Coria, F., & Rubio, I. (1996) Cerebral amyloid angiopathies, *Neuropathol. Appl. Neurobiol.* **22**, 216-227.
36. Fountain, N.B., & Eberhard, D.A. (1996) Primary angiitis of the central nervous system associated with cerebral amyloid angiopathy: Report of two cases and review of the literature, *Neurology* **46**, 190-197.
37. Ohshima, T., Endo, T., Nukui, H., Ikeda, S-i., Allsop, D., & Onaya, T. (1990) Cerebral amyloid angiopathy as a cause of subarachnoid hemorrhage, *Stroke* **21**, 480-483.
38. Gray, F., Dubas, F., Roullet, E., & Escourolle, R. (1985) Leukoencephalopathy in diffuse hemorrhagic cerebral amyloid angiopathy, *Ann. Neurol.* **18**, 54-59.

39. Greenberg, S.M., Vonsattel, J.P.G., Stakes, J.W., Gruber, M., & Finklestein, S.P. (1993) The clinical spectrum of cerebral amyloid angiopathy: Presentations without lobar hemorrhage, *Neurology* **43**, 2073-2079.
40. Fisher, M. (1994) *Clinical Atlas of Cerebrovascular Disorders*, Mosby Wolfe, London.
41. Yong, W.H., Robert, M.E., Secor, D.L., Kleikamp, T.J., & Vinters, H.V. (1992) Cerebral hemorrhage with biopsy-proved amyloid angiopathy, *Arch. Neurol.* **49**, 51-58.
42. Vinters, H.V., Secor, D.L., Read, S.L., Frazee, J.G., Tomiyasu, U., Stanley, T.M., Ferreiro, J.A., & Akers, M-A. Microvasculature in brain biopsy specimens from patients with Alzheimer's disease: An immunohistochemical and ultrastructural study, *Ultrastruct. Pathol.* **18**, 333-348.
43. Wang, Z.Z., Jensson, O, Thorsteinsson, L., & Vinters, H.V. (1997) Microvascular degeneration in hereditary cystatin C amyloid angiopathy of the brain, *APMIS* **105**, 41-47.
44. Scholz, W. (1938) Studien zur Pathologie der Hirngefässe II. Die drusige Entartung der Hirnarterien und -capillaren. (Eine From seniler Gefässerkrankung), *Zeitschrift für die gesamte Neurologie und Psychiatrie (Berlin)* **162**, 694-715.
45. Luyendijk, W., Bots, G.T.A.M., Vegter-van der Vlis, M., Went, L.N., & Frangione, B. (1988) Hereditary cerebral hemorrhage caused by cortical amyloid angiopathy, *J Neurol Sci* **85**, 267-280.
46. Neumann, M.A. (1960) Combined amyloid vascular changes and argyrophilic plaques in the central nervous system, *J. Neuropathol. Exp. Neurol.* **19**, 370-382.
47. Ojemann, R.G., & Heros, R.C. (1983) Spontaneous brain hemorrhage, *Stroke* **14**, 468-475.
48. Cosgrove, G.R., Leblanc, R., Meagher-Villemure, K., & Ethier, R. (1985) Cerebral amyloid angiopathy, *Neurology* **35**, 625-631.
49. Ulrich, G., Taghavy, A., & Schmidt, H. (1973) Zur Nosologie und Ätiologie der kongophilen Angiopathie (Gefässform der cerebralen Amyloidosis), *Zeitschrift für die gesamte Neurologie und Psychiatrie (Berlin)* **206**, 39-59.
50. Greene, G.M., Godersky, J.C., Biller, J., Hart, M.N., & Adams, H.P. Jr. (1990) Surgical experience with cerebral amyloid angiopathy, *Stroke* **21**, 1545-1549.
51. Regli, F., Vonsattel, J.P., Perentes, E., & Assal, G. (1981) L'angiopathie amyloïde cérébrale. Une maladie cérébro-vasculaire peu connue. Étude d'une observation anatomo-clinique, *Revue Neurologique* **137**, 181-194.
52. Brandenburg, W., & Hallervorden, J. (1954) Dementia pugilistica mit anatomischem Befund, *Virchows Archiv für Pathologische Anatomie und Physiologie und fur Klinische Medizin* **325**, 680-709.
53. Murphy, M.N., & Sima, A.A.F. (1985) Cerebral amyloid angiopathy associated with giant cell arteritis: a case report, *Stroke* **16**, 514-517.
54. Probst, A., & Ulrich, J. (1985) Amyloid angiopathy combined with granulomatous angiitis of the central nervous system: report on two patients, *Clin. Neuropathol.* **4**, 250-259.
55. Le Coz, P., Mikol, J., Ferrand, J., Woimant, F., Masters, C., Beyreuther, K., Haguenau, M., Cophignon, J., & Pepin, B. (1991) Granulomatous angiitis and cerebral amyloid angiopathy presenting as a mass lesion, *Neuropathol. Appl. Neurobiol.* **17**, 149-155.
56. Shintaku, M., Osawa, K., Toki, J., Maeda, R., & Nishiyama, T. (1986) A case of granulomatous angiitis of the central nervous system associated with amyloid angiopathy, *Acta Neuropathol.* **70**, 340-342.
57. Fountain, N.B., & Lopes, B.S. (1999) Control of primary angiitis of the CNS associated with cerebral amyloid angiopathy by cyclophosphamide alone, *Neurology* **52**, 660-662.
58. Ferreiro, J.A., Ansbacher, L.E., & Vinters, H.V. (1989) Stroke related to cerebral amyloid angiopathy: the significance of systemic vascular disease, *J. Neurol.* **236**, 267-272.

59. Natté, R., Vinters, H.V., Maat-Schieman, M.L.C., Bornebroek, M., Haan, J., Roos, R.A.C., & van Duinen, S.G. (1998) Microvasculopathy is associated with the number of cerebrovascular lesions in hereditary cerebral hemorrhage with amyloidosis, Dutch type, *Stroke* **29**, 1588-1594.
60. Bruni, J., Bilbao, J.M., & Pritzker, K.P.H. (1977) Vascular amyloid in the aging central nervous system. Clinico-pathological study and literature review, *Can. J. Neurol. Sci.* **4**, 239-244.
61. Masuda, J., Tanaka, K., Ueda, K., & Omae, T. (1988) Autopsy study of incidence and distribution of cerebral amyloid angiopathy in Hisayama, Japan, *Stroke* **19**, 205-210.
62. Mandybur, T.I., & Bates, S.R.D. (1978) Fatal massive intracerebral hemorrhage complicating cerebral amyloid angiopathy, *Arch. Neurol.* **35**, 246-248.
63. Mandybur, T.I. (1986) Cerebral amyloid angiopathy: the vascular pathology and complications, *J. Neuropathol. Exp. Neurol.* **45**, 79-90.
64. Rengachary, S.S., Racela, L.S., Watanabe, I., & Abdou, N. (1980) Neurosurgical and immunological implications of primary cerebral amyloid (Congophilic) angiopathy, *Neurosurgery* **7**, 1-9.
65. Corsellis, J.A.N., & Brierley, J.B. (1954) An unusual type of pre-senile dementia (atypical Alzheimer's disease with amyloid vascular changes), *Brain* **77**, 571-586.
66. Dahme, E., & Schröder, B. (1979) Kongophile Angiopathie, cerebrovasculäre Mikroaneurysmen und cerebrale Blutungen beim alten Hund, *Zentralbl. Veterinar. Med. [A]* **26**, 601-613.
67. Dubas, F., Gray, F., Roullet, E., & Escourolle, R. (1985) Leucoencéphalopathies artériopathiques (17 cas anatomo-cliniques), *Revue Neurologique* **141**, 93-108.
68. DeWitt, L.D., Davis, K.R., Hedley-Whyte, E.T., Fenton, M.L., McGrail, K., & Louis, D.N. (1991) Case Records of the Massachusetts General Hospital. Case 27-1991. A 75-year-old man with dementia, myoclonic jerks, and tonic-clonic seizures, *N. Engl. J. Med.* **325**, 42-54.
69. Ellison, D., Love, S., Chimelli, L., Harding, B., Lowe, J., Roberts, G.W., & Vinters, H.V. (1998) *Neuropathology. A Reference Text of CNS Pathology*, Mosby, London, pp. 10.1-10.25.
70. Vinters, H.V., Farrell, M.A., Mischel, P.S., & Anders, K.H. (1998) *Diagnostic Neuropathology*, Marcel Dekker, New York, pp. 110-146.
71. Anders, K.H., Wang, Z.Z., Kornfeld, M., Fray, F., Soontornniyomkij, V., Reed, L.A., Hart, M.N., Menchine, M., Secor, D.L., & Vinters, H.V. (1997) Giant cell arteritis in association with cerebral amyloid angiopathy: Immunohistochemical and molecular studies, *Hum. Pathol.* **28**, 1237-1246.
72. Mackenzie, I.R.A. (1997) Cerebral amyloid angiopathy with extensive mineralization, *Clin. Neuropathol.* **16**, 209-213.
73. Vinters, H.V., Natté, R., Maat-Schieman, M.L.C., van Duinen S.G., Hegeman-Kleinn, I., Welling-Graafland, C., Haan, J., & Roos, R.A.C. (1998) Secondary microvascular degeneration in amyloid angiopathy of patients with hereditary cerebral hemorrhage with amyloidosis, Dutch type (HCHWA-D), *Acta Neuropathol.* **95**, 235-244.
74. Gray, F., Vinters, H.V., Le Noan, H., Salama, J., Delaporte, P., & Poirier, J. (1990) Cerebral amyloid angiopathy and granulomatous angiitis: Immunohistochemical study using antibodies to the Alzheimer A4 peptide, *Hum. Pathol.* **21**, 1290-1293.

Chapter 9

CHEMICAL ANALYSIS OF AMYLOID β PROTEIN IN CAA

ALEX E. ROHER*, YU-MIN KUO*, ALEXANDER A. ROHER*, MARK R. EMMERLING† and WARREN J. GOUX‡

Haldeman Laboratory for Alzheimer Disease Research, Sun Health Research Institute, Sun City, AZ, USA; †Department of Neuroscience and Therapeutics, Parke-Davis Pharmaceutical Research, Division of Warner-Lambert Company, Ann Arbor, MI, USA; ‡Department of Chemistry, University of Texas at Dallas, Richardson, TX, USA

Substantial evidence demonstrates that amyloid deposition in the cerebral vasculature plays a major role in the pathophysiology of Alzheimer's disease. Chemical and immunohistochemical analyses have demonstrated that $A\beta_{40}$ and $A\beta_{42}$ peptides are present in vascular amyloid deposits with a preponderance of the former peptide. The accumulation of cerebrovascular amyloid leads to obliteration of capillary lumen and destruction arterial myocytes resulting in hypoperfusion and loss of control of cerebral blood flow. Post-translational modifications such as racemization, isomerization, cyclization, oxidation and N-terminal degradation largely contribute to the insolubility and proteolytic resistance exhibited by vascular Aβ filaments. The strong association of vascular Aβ with other proteins, carbohydrates and lipids results in additional amyloid stability. There appears to be a strong positive correlation between the magnitude of $A\beta_{40}$ deposition and the dosage of apolipoprotein E4.

1. INTRODUCTION

One of the most significant pathological hallmarks of Alzheimer disease (AD) is the deposition of insoluble 8-10 nm diameter amyloid fibrils in the extracellular space of the gray matter and in the walls of cortical and leptomeningeal vessels [1-3]. Gradual accumulation of the amyloid in the vascular wall causes degeneration of myocytes and endothelial cells, eventually leading to compromised cerebral blood flow and capillary obliteration [3,4].

M.M. Verbeek et al. (eds.),
Cerebral Amyloid Angiopathy in Alzheimer's Disease and Related Disorders, 157–177.

Historically, Scholz was the first to report that a material similar to the amyloid found in senile plaques also occurred in the cerebral blood vessels of people dying over age 70 [5]. The Congo red staining exhibition of green birefringence under polarized light (termed congophilic angiopathy) indicated that the molecules in the cerebral vascular amyloid were in the β-pleated sheet conformation. In the general population, the occurrence of cerebral amyloid angiopathy (CAA), stained by Congo red, was found to be about 30% for people over 60 years of age [6]. However, in AD the frequency of CAA is close to 100% [7]. The chemical composition of the amyloid deposited in the cerebral vessels was not known until 1984 when Glenner and Wong successfully isolated the AD cerebrovascular amyloid and established its amino acid sequence [8]. It was subsequently found that the cerebrovascular amyloid of AD was identical to the cerebrovascular amyloid found in Down's syndrome [9]. Isolation of amyloid from neuritic plaques also revealed that the amyloid-β (Aβ) peptide was the major molecular component [10]. Refined separation techniques and molecular characterizations have demonstrated that Aβ from AD brains is made of a heterogeneous mixture of peptides with various N-terminal truncations and C-terminal endings at residue 40 or 42 [4,11-13].

Both chemical analysis and immunohistochemical techniques also revealed that the majority of Aβ species found in the neuritic plaques is the form terminating at 42Ala, whereas the Aβ peptide terminating at position 40Val predominates in the vascular amyloid [4,13-19]. The $Aβ_{42}$ peptide is considered the initial culprit in gray matter and vascular amyloid deposition due to its higher rate of aggregation when compared to $Aβ_{40}$ [20,21]. However, the quantity of the latter peptide is directly related to Apolipoprotein E (ApoE) ε4 allele, the most important known risk factor for AD [22,23]. In this chapter, we describe the chemistry of the Aβ peptides deposited in cerebral vessels and discuss their significance in relation to CAA and AD.

2. CHEMICAL ANALYSIS OF Aβ FROM LEPTOMENINGEAL AND CORTICAL VESSELS

2.1 Isolation of Aβ from leptomeningeal vessels

The leptomeninges from individuals who died of AD, preferentially those with an ApoE ε4/ε4 or ε3/ε4 genotypes, are gently pulled from the surface of the cerebral hemispheres and thoroughly rinsed with 100 mM Tris-HCl, pH

8.0 (TB) at 4° C. All blood vessels larger than 1 mm in diameter as well as the membranous portions of the leptomeninges are dissected and discarded. Vessels smaller than 1 mm in diameter are cut into 1-2 mm pieces, washed with TB to eliminate all traces of blood and sieved through a 45 μm nylon mesh. The blood vessels are suspended in 20 vol of TB containing 1 mM $CaCl_2$ followed by the addition of 10 μg/ml of DNAse and 0.3 mg/ml of collagenase CLS3 (Worthington), and incubated at 37° C for 16 hrs in a rotatory shaker. A 300 μm nylon mesh is used to eliminate large undigested debris, and the filtrate centrifuged at 5,000 × g for 20 min. The pellet containing the insoluble Aβ is suspended with the aid of a glass homogenizer in 50 vol. of TB, followed by the addition of 50 vol. of 6% sodium dodecyl sulfate (SDS) also prepared in TB. After 2 h of continuous stirring at room temperature the suspension is centrifuged at 8,000 × g, the pellet is once more suspended in 150 ml of TB and the insoluble material recovered by centrifugation. This final pellet is dissolved in 10 vol. of concentrated glass distilled formic acid, the suspension (1 ml aliquots) loaded into 1.5 ml thick wall polyallomer tubes and centrifuged at 435,000 × g (Beckman 100TLA rotor) for 20 min at 20° C. The supernatants containing the solubilized Aβ are reduced to 500 μl each and submitted to size-exclusion fast performance liquid chromatography (see below).

2.2 Isolation of Aβ from cortical vessels

As in the case of the leptomeningeal amyloid, those individuals carrying the ApoE ε4/ε4 and ε3/ε4 do have more abundant deposits of fibrillar insoluble Aβ around their gray matter vessels. The cerebral cortex from one cerebral hemisphere is coronally sectioned, the gray matter carefully separated from white matter and leptomeninges, and cut into cubes of approximately 8 mm per side. The cubes of cortical tissue are vigorously stirred in 1.5 L of lysing buffer: 100 mM Tris-HCl pH 8.0 (TB) to which 5% SDS (w/v), 5 mM EDTA, and 0.03% NaN_3 are added. To ensure the complete removal of the brain parenchyma, the material is recovered by filtration, lysing buffer discarded and replaced with fresh buffer. The processes of filtration and replacement of fresh buffer is repeated 6 times over the course of 72 hrs. After this treatment, the only remaining tissues are the tufts of blood vessels composed of the SDS insoluble extracellular matrix and the insoluble fibrillar amyloid firmly attached to the vascular basal lamina. To eliminate the soluble SDS, EDTA and loose material, the vascular networks are rinsed 6 times each with 2 L of TB and finally recovered using a nylon mesh of 150 μm.

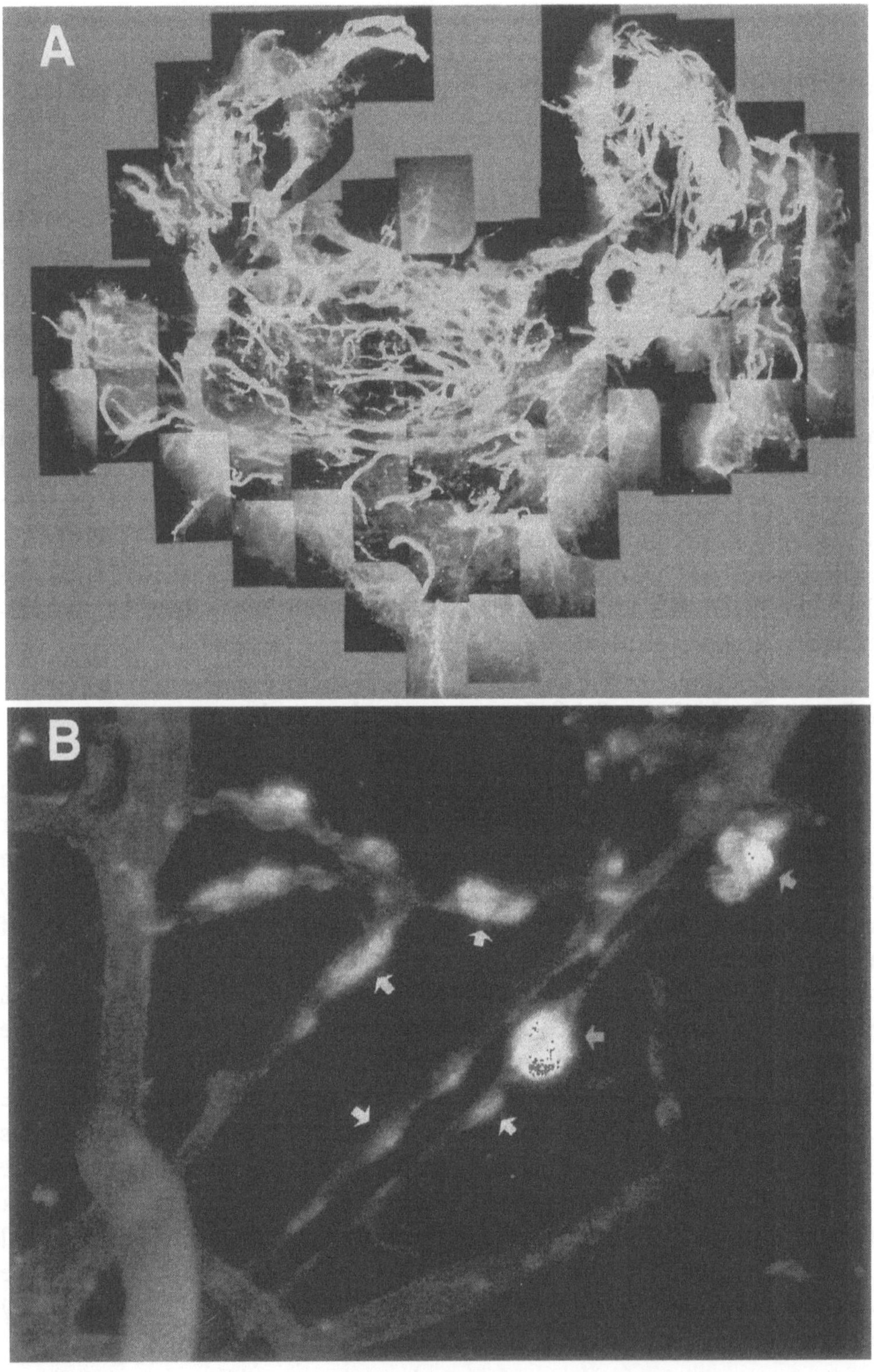
A
B

The resulting tufts of blood vessels loaded with amyloid are shown in Figure 1A and 1B·. Finally, the blood vessels are suspended in 150 ml of TB containing 1mM $CaCl_2$ and submitted to collagenase CLS3 (50 mg) for 16 h at 37° C. The undigested material is separated through a 150 μm nylon sieve and discarded. The filtrate containing most of the vascular amyloid is centrifuged and the pellet submitted to formic acid treatment as described above for the leptomeningeal vascular amyloid.

2.3 Separation of amyloid proteins by size-exclusion FPLC

Aliquots of 500 μl of the formic acid solubilized vascular amyloid were separated on a Pharmacia Biotech FPLC Superose 12 column (1 cm × 30 cm) equilibrated with 80% glass distilled formic acid. The size exclusion chromatography results in the separation of five fractions classified as S1-S5 (Figure 2A). Fractions S1 and S2, representing about 20-25% of the total mass of amyloid, contain a collection of insoluble glycoproteins and glycolipids mixed with aggregated and highly denatured Aβ.

·Figure 1 (page 160). Amyloid deposits in cortical vessels. A) A photographic whole-montage of cortical vessels from an ApoE ε4/ε4 AD patient burdened with amyloid stained by thioflavine-S. The montage depicts the vascular tufts of vessels after lysis of a small cube of parenchymal brain tissue (6 × 8 × 8 mm) by SDS. The only remaining insoluble material is the extracellular matrix and associated amyloid. The tufts of vessels were deposited on a microscopic slide, dried at 45° C for 5 h, fixed for 1 h with absolute ethanol, rinsed with distilled water and stained for 10 min with 1% aqueous thioflavine-S. As can be appreciated, almost every single artery and arteriole in this sample from the temporal lobe is loaded with Aβ. Individuals with ApoE ε3/ε3 genotype only carry about 5 to 10 percent of the amyloid observed in this picture. B) A confocal microscopic view of the amyloid cores intimately associated with the basal lamina of capillaries from the same individual illustrated above. A myriad of cores at all stages of development, from initial deposits indicated by blue arrows to transitional deposits (green arrows), surround the vessels in individuals with the ApoE ε4/ε4 genotype. The larger star-like cores (magenta arrows) totally obliterate the lumen of the capillaries leaving no traces of the endothelial cells or pericytes. At the electron microscopic level, the soluble non-fibrillar Aβ appears to be carried within coated vesicles which fuse with coated pits to generate cytoplasmic recesses at the surface of astroglial end-feet. It is inside of these cytoplasmic recesses where fibrillogenesis apparently occurs [4,24]. This cyto-architectural construct creates the brush-like appearance of the star-like cores of amyloid. In addition, it also suggests that many of the cores of amyloid in the mature and burnt-out neuritic plaques had in their initial stages a capillary at the center of the developing amyloid core [25].

The chromatographic retention times of these fractions indicate Mr ranging from 300 kd to 20 kd. On SDS-PAGE, fractions S1 and S2 do not exhibit discrete bands, but smears instead, suggesting a heterogeneous mixture of insoluble material. Interestingly, fraction S3 (~15 kd) also contains a small amount of glycolipids in addition to trimeric/tetrameric Aβ and a fragment of ApoE that still remains to be characterized. Fractions S4 and S5 (~8 and 4 kd, respectively) represent dimeric and monomeric Aβ with no apparent contamination. However, the latter two fractions are composed of a heterogeneous mixture of $A\beta_{40}$ and $A\beta_{42}$. The ratios of these two Aβ peptides are variable depending upon the vascular amyloid load which is related to the ApoE genotype. Following removal of the formic acid by dialysis against TB and examination by negative-staining electron microscopy, fraction S5 demonstrated a homogeneous assembly of 10-12 nm filaments. Fractions S4 and S3, on the other hand, showed a heterogeneous array of dense clusters and loose aggregates with vaguely defined filaments. Fractions S1 and S2 regenerated into randomly dispersed amorphous aggregates.

2.4 Analysis of lipid, carbohydrates and other compounds associated to vascular deposits of Aβ

As in the case of the paired helical filaments [26-28], analysis of the fibrillar amyloid present in the vascular deposits of AD revealed lipid and carbohydrate components. We have estimated that these non-protein fractions account for approximately 12-15% of the total amyloid mass. The fatty acids present in vascular amyloid were submitted to methanolysis and analyzed as methyl esters by gas chromatography-mass spectrometry (GC-MS). The results of these analyses are shown in Table 1. Monosaccharides released from the acid hydrolysis of the intact samples were analyzed by GC-MS of their trimethylsilyl-methyl glycosides. The molar quantities are shown in Table 2.

Lipids and carbohydrates were also extracted from fractions S1, S2 and S3 by the method of Folch [29]. In order to assess the composition of the organic and aqueous layer, these fractions were analyzed by thin layer chromatography (TLC), using chloroform:methanol:water (55:40:7) as the developing system and compared to lipid and phospholipid standards. Visualization with iodine- and phosphate-sensitive ammonium molybdate reagents revealed that all fractions have polar non-migrating phosphate-containing components as well as apolar phosphate-containing components which migrated with the solvent front.

Table 1. Fatty acid composition found in the hydrolysate of vascular amyloid deposits.

	C14:0	C15:0	C16:0	C18:0	C18:1	C20:0	C22:0	C24:0
S1	3.54		43.8	18.8	12.4			21.46
	(1.23)		(15.2)	(6.54)	(4.31)			(7.44)
S2	4.34		38.2	13.6	32.3			11.56
	(1.76)		(15.5)	(5.53)	(1.31)			(4.69)
S3	3.88	3.81	52.7	39.1		0.31	0.15	
	(3.39)	(3.32)	(46.0)	(34.2)		(0.26)	(0.17)	

Values are reported as percent of the total fatty acid found. Numbers in parenthesis are absolute quantities (nmol), determined by comparing peak areas to that of the internal standard (methyl nonadecanoate).

Table 2. Monosaccharide composition found in the hydrolysate of vascular amyloid deposits.

	Glucose	Galactose	Mannose	Xylose
S1	52.8 (13.34)	10.3 (2.60)	21.1 (5.33)	15.7 (3.97)
S2	49.6 (3.76)	8.10 (6.12)	26.1 (1.98)	16.2 (1.23)
S3	58.6 (104.3)	3.60 (6.40)	34.5 (61.4)	3.31 (5.90)

Values are reported as percent composition of the total monosaccharide found. Numbers in parenthesis are absolute quantities, in nanomol, determined by comparing peak areas to that of the internal standard (myoinositol).

The components of the S1 and S2 were separated by preparative TLC and submitted to nuclear magnetic resonance (NMR). The 500 MHz NMR spectra in both cases showed a broad singlet at about 1.3 ppm arising from near equivalent methylene protons of a long aliphatic chain and a group of resonances near 0.80 ppm most probably originating from terminal aliphatic chain methyls. In some of the spectra derived from fraction S1, there were aromatic proton resonances between 7.2 and 7.8 ppm as well as singlets at 8.58 and 8.20 ppm, a doublet at 6.1 ppm and prominent multiplets at 4.08, 2.42, 2.30 and 2.08 ppm. Resonances at 8.58 and 8.20 and those between 4.08 and 6.1 ppm can be readily assigned to the adenine and ribose ring protons of adenosine monophosphate (AMP). The large resonance at 4.08 ppm arises from the ribose H5' and from another signal coupled to the multiplets at 2.08 and 2.42. These latter resonances are consistent with methylenes or methines α– and β- to an amine. The nuclear magnetic resonance spectrum of the Folch aqueous layer was qualitatively similar to that of the Folch organic layer. Matrix assisted laser desorption ionization-time of flight (MALDI-TOF) mass spectrometry of the Folch aqueous layer demonstrated the presence of peaks at mass/charge of 136, 413.5, 550.8 and 687. There appears to be a common structural unit which accounts for the mass of 136.

In conclusion, our preliminary results indicate that the non-protein components of the amyloid fibrils are carbohydrate and lipid moieties and AMP, as well as other molecules that remain to be characterized. Other studies have demonstrated that the Aβ peptides have a high affinity for the lipid components present in membranes [30-32]. The presence of lipids, carbohydrates and other compounds in the Aβ fibrillar structures of the AD brain may in part be responsible for their resistance to proteolytic degradation. Likewise, they may influence the secondary structure of Aβ resulting in the anomalous folding of this molecule [33,34].

2.5 Biochemical differences between vascular and neuritic plaque core amyloid: the Aβ N-terminal region

To further characterize the chemical structure of the insoluble amyloid purified from the AD brain, the Aβ peptides from the neuritic plaques and vascular deposits were hydrolyzed by trypsin at a 1:50 enzyme to substrate ratio at 37° C for 14 hrs. The resulting enzymatic peptides were freeze dried and subsequently separated by reverse-phase HPLC. The chromatographic profiles of the Aβ tryptic and CNBr peptides derived from the leptomeningeal and cortical vessels are shown in Figs. 2B, 2C, 2D and 2E, respectively.

Important post-translational modifications were found in the Aβ peptides at positions 1 and 7 where the Asp residues isomerized (Iso-Asp) resulting in the shift of the peptide bond from the α- to the β-carbon. This drastic change in the progression of the α-carbon backbone increases the insolubility of the Aβ molecule, augments its resistance to proteolytic degradation and accelerates the generation of irreversibly denatured dimers and trimers/tetramers [21]. Tryptic digestion of chromatographically purified Aβ obtained from neuritic plaque cores, cortical vessels and leptomeningeal vessels demonstrated different degrees of Iso-Asp at position 7. In the neuritic plaque core Aβ about 73% of Asp7 is isomerized while only 25% of the amyloid in cortical vessels carries this modification [13]. In contrast, the Iso-Asp7 in leptomeningeal vessels accounted for less than 10% of the Aβ molecules [4]. A similar situation occurs at residue Asp1 where IsoAsp1 is found in 20% of the neuritic plaque core Aβ and in only 6% of the vascular Aβ [13]. Another important post-translational modification happens at residue Glu3. As the result of the loss of residues Asp1 and Ala2, the N-terminally exposed Glu3 is transformed into pyroglutamyl [35]. The Aβ with this modification is abbreviated as 3pE. The chromatographic profile depicting the separation of the tryptic peptide residues 3pE-5 is shown in

Figure 2F. Table 3 shows the amino acid composition and mass spectrometry of these peptides. The consequences of this change are important for Aβ catabolism since it blocks the degradation of this molecule by aminopeptidases and speeds its aggregation [21]. In the neuritic plaque amyloid, on the average (n=9), 51% of the Aβ molecules start at pyroglutamyl 3 whereas only 11% start at this residue in the vascular amyloid [36].

Figure 2 (page 166). Chromatographic profiles of vascular Aβ. **A)** Elution pattern of leptomeningeal vascular amyloid (solid trace) and gray matter vascular amyloid (hyphened trace) separated on a Superose 12 column. (1 cm × 30 cm) equilibrated with 80% glass distilled formic acid. The chromatography is developed with the same solvent at a flow rate of 15 ml/h at room temperature and monitored at 280 nm. Size-exclusion chromatography resolved the vascular amyloid into five fractions (S1-S5). **B)** Tryptic peptides derived from leptomeningeal vascular amyloid. Purified fraction S5 was submitted to tryptic digestion and the resulting peptides separated on a LKB-RP-Spherisorb ODS2 C18 column (250 x 4 mm, 5 μm bead size) using a water/0.05% trifluoroacetic acid (TFA) as solvent A and acetonitrile/0.05% TFA as solvent B. A linear gradient is developed from 0% → 20% of solvent B in 60 min. This is followed by a second gradient from 20 to 60% solvent B for an additional 60 min. The chromatography is developed at a flow rate of 0.7 ml/min at room temperature and monitored at 214 nm. The tryptic peptide residues 29-42 precipitates as an insoluble core. **C)** Chromatographic profile of the CNBr peptides derived from the leptomeningeal amyloid insoluble tryptic core. This core peptide is dissolved in 500 μl of 80% formic acid followed by the addition of 50 μg of CNBr, incubated at room temperature for 4 h, the volume is reduced by vacuum centrifugation. The resulting CNBr peptides residues 29-35 and 36-42 are separated by HPLC under the same conditions described above. **D)** Elution pattern of the amyloid from the cortical vessels digested by trypsin. **E)** CNBr digested cortical vessel amyloid tryptic insoluble peptides. The chromatographic conditions in C and D were the same as those described above for Figs. A and B. **F)** Elution pattern of the leptomeningeal Aβ digested by trypsin. This HPLC separation was specifically designed to separate the pyroglutamyl containing peptide. The column in this case is a LKB-RP-ODS C18 column (125 mm × 4 mm, 3 μm bead size) and the chromatography developed using a water/0.05% trifluoroacetic acid (TFA) as solvent A and acetonitrile/0.05% TFA as solvent B. A linear gradient is developed from 5% → 8.5% of solvent B in 70 min..The chromatography is developed at a flow rate of 0.7 ml/min at room temperature and monitored at 214 nm. The fraction labeled as CC contains the tryptic peptides 17-28 and 29-40 eluted at 100% acetonitrile concentration. The amino acid sequences and mass spectra of the peptides shown in Figure 2B through Figure 2F are given in Table 3. The numbers on the X-axes represent the chromatographic retention times in minutes.

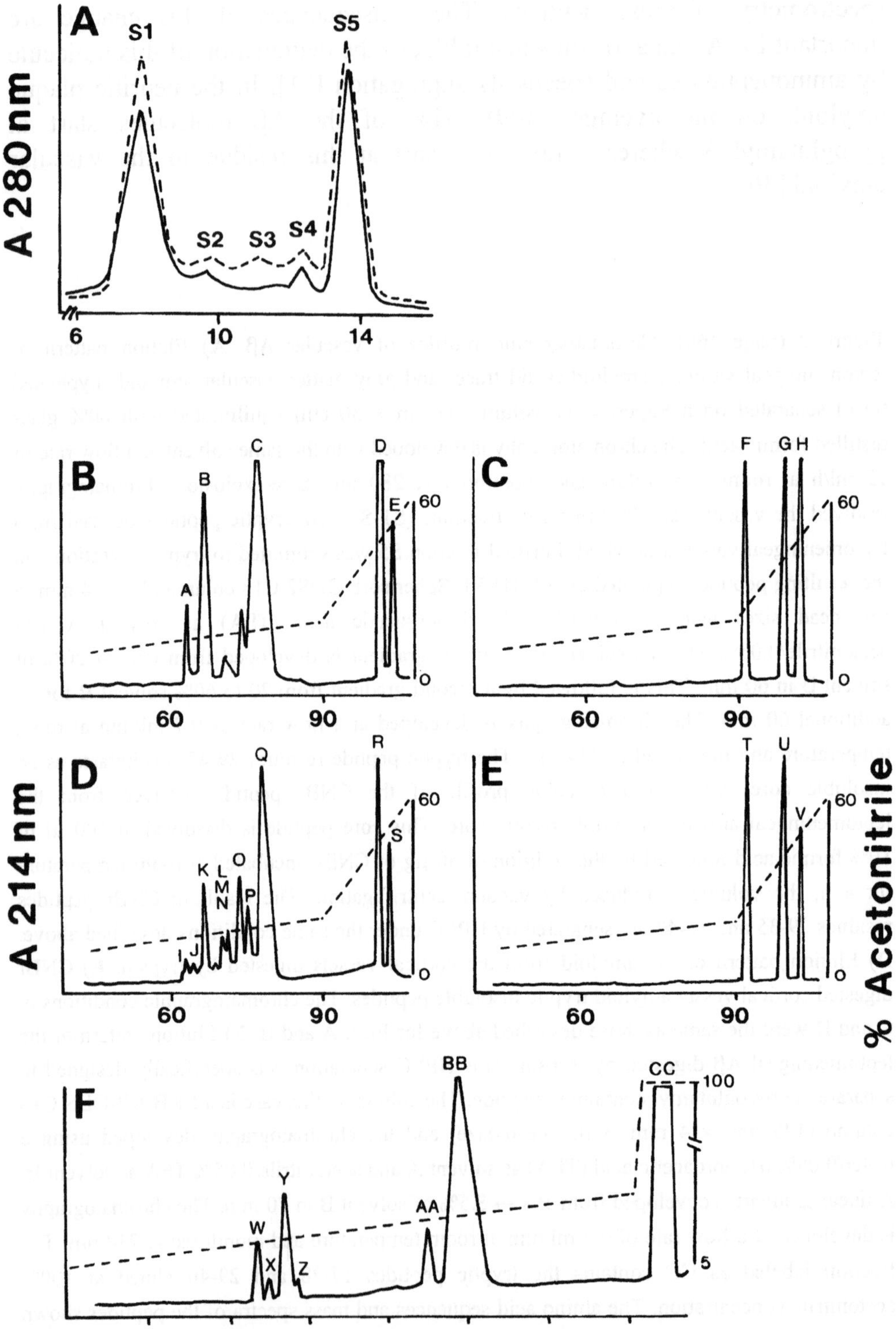
A
S1
S2
S3
S4
S5
A 280 nm
6
10
14
B
A
B
C
D
E
60
O
60
90
C
F
G
H
60
O
60
90
D
I
J
K
L
M
N
O
P
Q
R
S
60
O
A 214 nm
60
90
E
T
U
V
60
O
60
90
% Acetonitrile
F
W
X
Y
Z
AA
BB
CC
100
5
20
40
60

TABLE 3. Mass spectra of tryptic and CNBr peptides derived from vascular Aβ. The corresponding HPLC peaks are shown in Fig. 2.

Peak	Source	Amino Acid Sequence	Residues	Mr	MALDI-MS
A	LVA-Tp	AEFR	2-5	521.6	521.8
B	LVA-Tp	DAEFR	1-5	636.7	637.3
C	LVA-Tp	HDSGYEVHHQK	6-16	1336.5	1337.1
D	LVA-Tp	LVFFAEDVGSNK	17-28	1325.7	1326.1
E	LVA-Tp	GAIIGLMVGGVV	29-40	1085.5	1086.1
F	LVA-Tp-CNBr	VGGVV	36-42	613.8	614.2
G	LVA-Tp-CNBr	GAIIGLM	29-35	625.8	626.0
H	LVA-Tp-CNBr	LVFFAEDVGSNK	17-28	1325.7	1325.9
I	CVA-Tp	DAEFR *	1-5	636.7	637.3
J	CVA-Tp	AEFR	2-5	521.6	522.4
K	CVA-Tp	DAEFR †	1-5	636.7	637.8
L	CVA-Tp	DAEFR ‡	1-5	636.7	637.6
M	CVA-Tp	GYEVHHQK	9-16	997.2	997.5
N	CVA-Tp	SGYEVHHQK	8-16	1084.3	1084.6
O	CVA-Tp	HDSGYEVHHQK §	6-16	1336.5	1336.4
P	CVA-Tp	HDSGYEVHHQK ¶	6-16	1336.5	1336.5
Q	CVA-Tp	HDSGYEVHHQK #	6-16	1336.5	1336.5
R	CVA-Tp	LVFFAEDVGSNK	17-28	1325.7	1326.7
S	CVA-Tp	GAIIGLMVGGVV	29-40	1085.5	1085.0
T	CVA-Tp-CNBr	VGGVV	36-42	613.8	613.7
U	CVA-Tp-CNBr	GAIIGLM ¥	29-35	625.8	627.5
V	CVA-Tp-CNBr	GAIIGLM £	29-35	625.8	627.5
W	LVA-Tp	AEFR	2-5	521.6	522.2
X	LVA-Tp	DAEFR *	1-5	636.7	635.5
Y	LVA-Tp	DAEFR ‡	1-5	636.7	637.3
Z	LVA-Tp	pEFR	3-5	433.5	433.5
AA	LVA-Tp	HDSGYEVHHQK §	6-16	1336.5	1335.9
BB	LVA-Tp	HDSGYEVHHQK #	6-16	1336.5	1336.9

LVA: leptomeningeal vascular amyloid; CVA: cortical vascular amyloid; Tp: tryptic peptide; Tp-CNBr: CNBr hydrolyzed peptide derived from insoluble tryptic core; * Residue 1: L-IsoAspartyl; † Residue 1: D-IsoAspartyl; ‡ Residue 1: L-Aspartyl; § Residue 7: L-IsoAspartyl; ¶ Residue 7: D-IsoAspartyl; # Residue 7: L-Aspartyl; ¥ Residue 35: Homoserine; £ Residue 35: Homoserine lactone.

The characteristic of the AD brain Aβ is a substantial degree of N-terminal degradation which varies from individual to individual. However, this degradation is minimal in the leptomeningeal vascular amyloid where about 20% of Aβ peptides apparently lacks residues Asp1 and 11% lacks residues Asp1-Ala2. In the parenchymo-vascular and neuritic plaque core Aβ, a different situation is encountered since the N-terminal degradation is extensive. We have found that the Aβ in these deposits can start at residues 1, 2, 3, 8, 9, 10 and 11 [13,36]. Other laboratories have reported Aβ molecules with N-termini truncations at every residue from 1 to 11 [11,12,37,38].

2.6 Biochemical differences between vascular and neuritic plaque core amyloid: the Aβ C-terminal region and ApoE genotype

In the neuritic plaques and vascular walls of AD patients the amyloid deposits are a mixture of $A\beta_{40}$ and $A\beta_{42}$. The amino acid sequences of these peptides are identical with the exception that the longer peptide contains 2 extra amino acids (Ile41-Ala42) at the C-terminus. The addition of these two hydrophobic residues renders the $A\beta_{42}$ more insoluble and prone to aggregation than the $A\beta_{40}$ [20]. Oxidation of Met35 to Met sulfone and Met sulfoxide, observed in the AD amyloid, also increases the stability of the fibrils, probably by generating additional hydrogen bonds [21]. Moreover, oxidation of Met35 increases the resistance of Aβ to enzymatic degradation [21]. The amyloid deposited in the neuritic plaques is mostly composed of $A\beta_{42}$ (~90%); the remaining 10% is the more soluble $A\beta_{40}$ [11,13]. The ratio between the two peptides is quite variable in the cerebrovascular amyloid [4,39].

However, $A\beta_{40}$ is the predominant type in those individuals with ApoE ε4/ε4 (Figure 3). The interaction between Aβ and ApoE is very stable since these two molecules remain intimately associated throughout the purification steps involved in the purification of Aβ from the human brain. Furthermore, after separating the different components by FPLC and removal of the formic acid, the regenerated Aβ filaments still carry epitopes for ApoE (Figure 4). Because of these differences in composition, the vascular amyloid is more soluble than the plaque core amyloid. The AD vascular and plaque core $A\beta_{40}$ can be separated from the $A\beta_{42}$ by 5M guanidine hydrochloride, since the former is soluble in this denaturing agent while the latter precipitates into a jelly-like pellet [13]. However, both $A\beta_{40}$ and $A\beta_{42}$ dissociate into monomeric, dimeric and trimeric/tetrameric structures in the presence of 80% formic acid or 5 M guanidine thiocyanate [21,40]. The Aβ species terminating at residue 42, i.e. $A\beta_{1-42}$, $A\beta_7$IsoAsp-42 and $A\beta_3$pE-42 generate as a function of time increasing quantities of dimeric and trimeric/tetrameric oligomers that can not be dissociated into monomeric Aβ [21,40]. These irreversibly denatured oligomers reach equilibrium within the Aβ filamentous structures where they represent 25% (dimeric) and 15% (trimeric/tetrameric) of the total Aβ. The remaining 60% can be dissociated into monomeric units by the action of strong denaturing agents [21,40].

3. DISCUSSION AND CONCLUSION

The origin of the vascular amyloid in AD has been a hotly debated issue. It has been postulated that the Aβ peptide is locally produced by vascular myocytes which express the APP molecule and have been shown to generate Aβ peptides in cell cultures [18,41-43]. A hematogenous source for the Aβ peptides has also been proposed [44]. This second hypothesis has been recently reinforced by the finding of relatively high levels of circulating Aβ [45] and by the ability of this peptide to cross the blood-brain barrier (BBB) either by itself or bound to carrier molecules [46-48]. However, the early deposition of Aβ in both the leptomeningeal and parenchymal arteries starts at the most peripheral layers of the tunica media close to the periarterial spaces [1,49]. These spaces are of great importance since they apparently drain the interstitial fluid of the brain into the lymphatic vessels of the head and neck and finally into the venous circulation [50-52]. In addition, a myriad of capillaries and arterioles contain heavy deposits of amyloid intimately associated to their basal membranes, which suggests an aborted attempt to eliminate Aβ from the brain into the circulation. Furthermore, the deposition of Aβ along the vascular arbor follows a reverse gradient wherein the smaller vessels are more heavily involved and the larger vessels the least implicated [1]. Hence, a third hypothesis postulates the brain cells i.e. neurons and glia as the major source of amyloid and the microvasculature and periarterial spaces as putative pathways for the removal of soluble Aβ into the circulation. Interestingly, the main component of the vascular Aβ is the $A\beta_{40}$ which is by far more soluble than the $A\beta_{42}$, although this longer peptide is considered to be the initial seeding molecule for amyloid deposits [15,53]. However, both peptides appear to have a high affinity for components of the vascular extracellular matrix such as collagen IV and heparan sulfate proteoglycan [54,55]. In the final analysis, the origins of the vascular amyloid in AD may be multiple since the evidence suggests the participation of myocytes, neurons and glia. In addition, peripheral tissues such as platelets [56,57] and leukocytes [58], and skeletal muscle [59-61], which in AD has been found to have elevated Aβ, may also participate in the pathogenesis of cerebrovascular amyloidosis.

The pathogenesis and role of cerebrovascular amyloidosis in AD remains shrouded in mystery. However, the answers may lay in understanding why Aβ begins to accumulate in the cerebral vasculature by age 50 [62]. Inflammatory episodes triggered by ionizing radiation, degenerative vasculitis or cerebrovascular malformations cause amyloid deposition [63]. It has also been suggested that changes in molecular conformation induced by chaperone molecules may precipitate the aggregation of Aβ [49]. The interactions of Aβ with molecules like ApoE or α_1-antichymotrypsin or elements of the extracellular matrix still remain to be deciphered. The presence of lipids in these deposits may be significant since several lipids have been shown to accelerate Aβ aggregation [64].

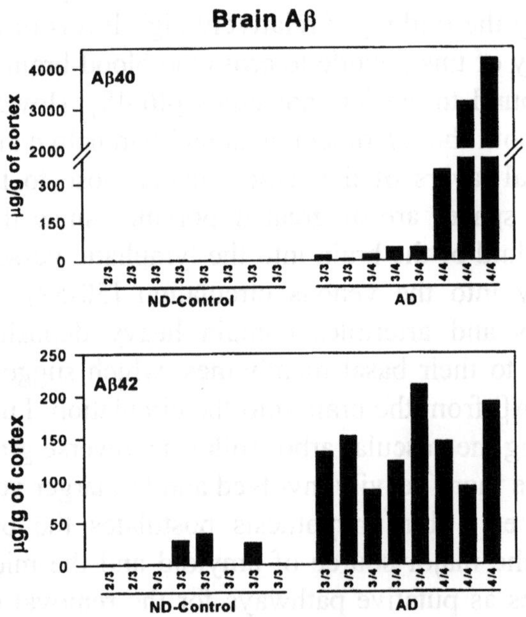

Figure 3. A histogram showing the abundance of $A\beta_{40}$ and $A\beta_{42}$ in the brains of non-demented (ND) and AD individuals in relation to their ApoE genotype. These measurements were done in samples of cerebral cortex, and therefore, they represent the sum of vascular and non-vascular amyloid. Notice that the presence of $A\beta_{42}$ is necessary to manifest AD, but the amount of $A\beta_{40}$ is extremely high, 15 to 150 times more in those individuals carrying ApoE ε4/ε4 than in those carrying ApoE ε3/ε4 or no ε4. The majority of the Aβ in these ApoE4 individuals is deposited in the vascular walls.

The brain nuclei responsible for innervating the cerebral blood vessels such as the cholinergic nucleus basalis of Meynert (nbM), the noradrenergic locus ceruleus and the serotonergic dorsal raphe nucleus are gravely damaged in AD [65,66]. Experimental elimination of these innervations in animal models produces dramatic changes in cerebral blood flow [67]. We have recently induced the deposition of Aβ around cortical vessels in rabbits in which the nucleus basalis magnocellularis (equivalent to the nbM in humans) was selectively lesioned. Thus, it appears that cortical cholinergic deafferentiation with its impact on brain perfusion results in cerebrovascular amyloidosis.

In conclusion, the pathophysiological consequences of the vascular amyloidosis in AD are severe since the deposition of amyloid in arteries of brain and leptomeninges completely destroyed the myocytes causing a loss of control in cerebral blood flow. At the level of the capillary network, especially in those individuals carrying the ApoE ε4, the continuous deposition of Aβ will eventually obliterate the capillary lumen, creating areas of ischemia with devastating effects on nearby neurons. Thus, in individuals with ApoE ε4/ε4 allele, their high risk for suffering AD at an earlier age appears to be related to their severe vascular amyloidosis in which $A\beta_{40}$ is mostly represented. Therapeutic intervention directed against amyloid accumulation needs to be implemented at the very early stages of Aβ deposition. Denaturation of Aβ into irreversible dimers and oligomers as well as the accumulation of post-translational modifications, and the association of Aβ filaments with lipids and carbohydrates results in very stable and insoluble molecular polymers which are difficult to degrade and cumbersome to remove. The accumulation of amyloid in the neuritic plaques of the cerebral cortex is considered by some researchers as an epiphenomenon of lesser importance with limited impact on AD. However, a large body of evidence clearly indicates that amyloid deposition in the cerebral vasculature severely and extensively damages precious nutrient and oxygen supply to the brain and is beyond doubt a major player in the dementing process.

ACKNOWLEDGEMENTS.

We are grateful to Dr. Dean C. Luehrs for constructive suggestions during the preparation of this manuscript. This work was in part supported by the State of Arizona Center for Alzheimer's Disease Research.

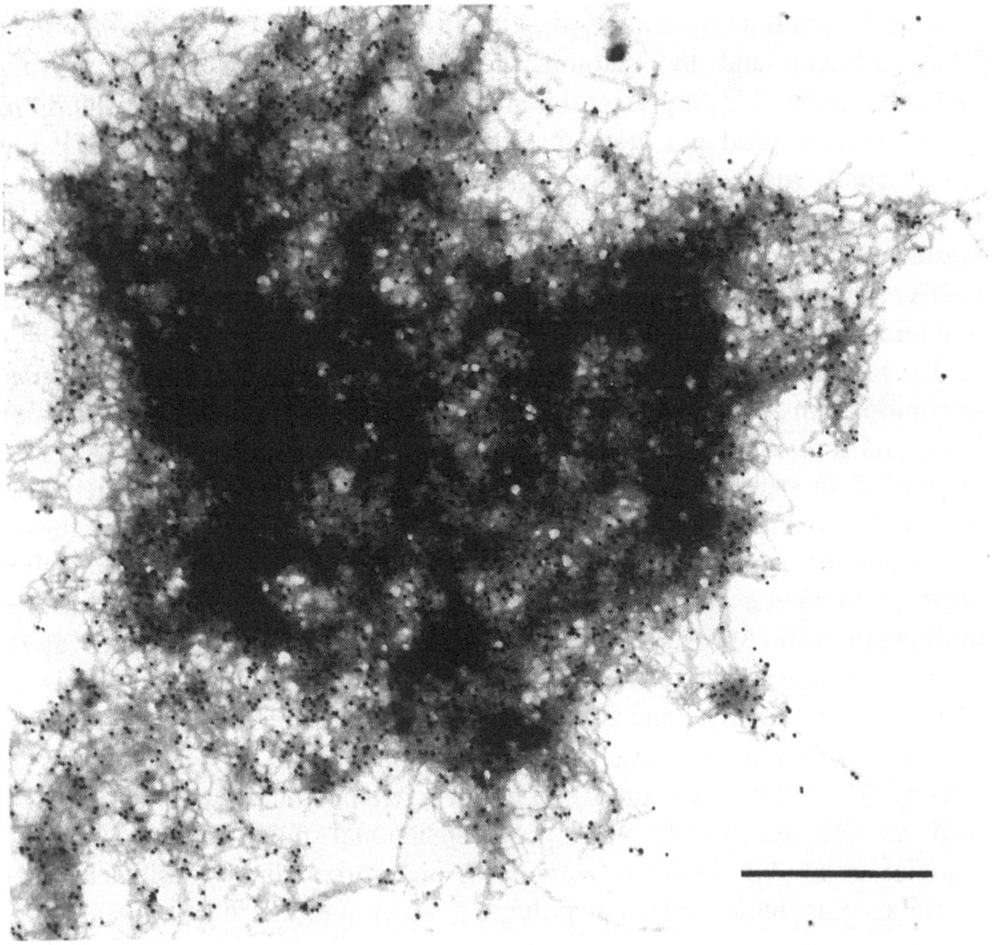

Figure 4. Electron micrograph demonstrating the presence of ApoE associated to Aβ filaments. From the FPLC fractions S3-S5 in formic acid solution, the filaments were regenerated by dialysis against Tris-buffer. Prior to negative staining, the filaments were reacted with a polyclonal antibody against ApoE (kindly donated by Dr. Charles Bisgaier, Esperion Therapeutics, Ann Arbor, MI). A secondary antibody attached to colloidal gold revealed that the ApoE epitopes are relatively abundant and randomly distributed along the surface of the filaments. The size-exclusion chromatography suggests that the ApoE epitopes in our preparations represent either the N- or the C-terminal regions of the molecule, since the ApoE elution time corresponds to 12-18 kd rather than the 34 kd representing the whole molecule. Bar = 500 nm.

REFERENCES

1. Weller, R.O., Massey, A., Newman, T.A., Hutchings, M., Kuo, Y.M., and Roher, A.E. (1998) Cerebral amyloid angiopathy: amyloid β accumulates in putative interstitial fluid drainage pathways in Alzheimer's disease, *Am.J.Pathol.* **153**, 725-733.

2. Selkoe, D.J. (1994) Normal and abnormal biology of the β-amyloid precursor protein, *Annu. Rev. Neurosci.* **17**, 489-517.
3. Wisniewski, H.M., Wegiel, J., Wang, K.C., and Lach, B. (1992) Ultrastructural studies of the cells forming amyloid in the cortical vessel wall in Alzheimer's disease, *Acta Neuropathol.(Berl.)* **84**, 117-127.
4. Roher, A.E., Lowenson, J.D., Clarke, S., Woods, A.S., Cotter, R.J., Gowing, E., and Ball, M.J. (1993a) β-Amyloid-(1-42) is a major component of cerebrovascular amyloid deposits: implications for the pathology of Alzheimer disease, *Proc. Natl. Acad. Sci. U.S.A.* **90**, 10836-10840.
5. Scholz, W. (1938) Studien zur Pathologie der Hirngefässe. II. Die drusige Entartung der Hirnarterien und Capillaren, *Zeitschrift für die gesamte Neurologie und Psychiatrie* **162**, 694-715.
6. Esiri, M.M. (1987) Cerebral congophilic angiopathy, in R.A. Griffiths and S.T. McCarthy (eds.), *Degenerative Neurological Disease in the Elderly*, Bristol, Wright, 79-87.
7. Premkumar, D.R., Cohen, D.L., Hedera, P., Friedland, R.P., and Kalaria, R.N. (1996) Apolipoprotein E-epsilon4 alleles in cerebral amyloid angiopathy and cerebrovascular pathology associated with Alzheimer's disease, *Am. J. Pathol.* **148**, 2083-2095.
8. Glenner, G.G. and Wong, C.W. (1984b) Alzheimer's disease: initial report of the purification and characterization of a novel cerebrovascular amyloid protein, *Biochem. Biophys. Res. Commun.* **120**, 885-890.
9. Glenner, G.G. and Wong, C.W. (1984a) Alzheimer's disease and Down's syndrome: sharing of a unique cerebrovascular amyloid fibril protein, *Biochem. Biophys. Res. Commun.* **122**, 1131-1135.
10. Masters, C.L., Simms, G., Weinman, N.A., Multhaup, G., McDonald, B.L., and Beyreuther, K. (1985) Amyloid plaque core protein in Alzheimer disease and Down syndrome, *Proc. Natl. Acad. Sci.U.S.A.* **82**, 4245-4249.
11. Miller, D.L., Papayannopoulos, I.A., Styles, J., Bobin, S.A., Lin, Y.Y., Biemann, K., and Iqbal, K. (1993) Peptide compositions of the cerebrovascular and senile plaque core amyloid deposits of Alzheimer's disease, *Arch. Biochem. Biophys.* **301**, 41-52.
12. Mori, H., Takio, K., Ogawara, M., and Selkoe, D.J. (1992) Mass spectrometry of purified amyloid beta protein in Alzheimer's disease, *J. Biol. Chem.* **267**, 17082-17086.
13. Roher, A.E., Lowenson, J.D., Clarke, S., Wolkow, C., Wang, R., Cotter, R.J., Reardon, I.M., Zurcher-Neely, H.A., Heinrikson, R.L., and Ball, M.J. (1993b) Structural alterations in the peptide backbone of β-amyloid core protein may account for its deposition and stability in Alzheimer's disease, *J. Biol. Chem.* **268**, 3072-3083.
14. Fukumoto, H., Asami-Odaka, A., Suzuki, N., Shimada, H., Ihara, Y., and Iwatsubo, T. (1996) Amyloid β protein deposition in normal aging has the same characteristics as that in Alzheimer's disease. Predominance of Aβ 42(43) and association of Aβ 40 with cored plaques, *Am. J. Pathol.* **148**, 259-265.
15. Mann, D.M., Iwatsubo, T., Ihara, Y., Cairns, N.J., Lantos, P.L., Bogdanovic, N., Lannfelt, L., Winblad, B., Maat-Schieman, M.L., and Rossor, M.N. (1996) Predominant deposition of amyloid-β 42(43) in plaques in cases of Alzheimer's disease and hereditary cerebral hemorrhage associated with mutations in the amyloid precursor protein gene, *Am. J. Pathol.* **148**, 1257-1266.
16. Mak, K., Yang, F., Vinters, H.V., Frautschy, S.A., and Cole, G.M. (1994) Polyclonals to β-amyloid(1-42) identify most plaque and vascular deposits in Alzheimer cortex, but not striatum, *Brain Res.* **667**, 138-142.

17. Iwatsubo, T., Odaka, A., Suzuki, N., Mizusawa, H., Nukina, N., and Ihara, Y. (1994) Visualization of Aβ 42(43) and Aβ 40 in senile plaques with end- specific Aβ monoclonals: evidence that an initially deposited species is Aβ 42(43), *Neuron* **13**, 45-53.
18. Wisniewski, H.M., Frackowiak, J., and Mazur-Kolecka, B. (1995) *In vitro* production of β-amyloid in smooth muscle cells isolated from amyloid angiopathy-affected vessels, *Neurosci. Lett.* **183**, 120-123.
19. Alonzo, N.C., Hyman, B.T., Rebeck, G.W., and Greenberg, S.M. (1998) Progression of cerebral amyloid angiopathy: accumulation of amyloid-β_{40} in affected vessels, *J. Neuropathol. Exp. Neurol.* **57**, 353-359.
20. Jarrett, J.T., Berger, E.P., and Lansbury, P.T.J. (1993) The C-terminus of the β protein is critical in amyloidogenesis, *Ann. N.Y. Acad. Sci.* **695**, 144-148.
21. Kuo, Y.M., Webster, S., Emmerling, M.R., De Lima, N., and Roher, A.E. (1998) Irreversible dimerization/tetramerization and post-translational modifications inhibit proteolytic degradation of Aβ peptides of Alzheimer's disease, *Biochim. Biophys. Acta* **1406**, 291-298.
22. Ishii, K., Tamaoka, A., Mizusawa, H., Shoji, S., Ohtake, T., Fraser, P.E., Takahashi, H., Tsuji, S., Gearing, M., Mizutani, T., Yamada, S., Kato, M., St.George-Hyslop, P.H., Mirra, S.S., and Mori, H. (1997) $A\beta_{1\text{-}40}$ but not Aβ1-42 levels in cortex correlate with apolipoprotein E epsilon4 allele dosage in sporadic Alzheimer's disease, *Brain Res.* **748**, 250-252.
23. Gearing, M., Mori, H., and Mirra, S.S. (1996) Aβ-peptide length and apolipoprotein E genotype in Alzheimer's disease, *Ann. Neurol.* **39**, 395-399.
24. Roher, A., Gray, E.G., and Paula-Barbosa, M. (1988) Alzheimer's disease: coated vesicles, coated pits and the amyloid- related cell, *Proc. R. Soc. Lond. B. Biol. Sci.* **232**, 367-373.
25. Miyakawa, T., Shimoji, A., Kuramoto, R., and Higuchi, Y. (1982) The relationship between senile plaques and cerebral blood vessels in Alzheimer's disease and senile dementia. Morphological mechanism of senile plaque production, *Virchows Arch. B. Cell Pathol. Incl. Mol. Pathol.* **40**, 121-129.
26. Goux, W.J., Rodriguez, S., and Sparkman, D.R. (1995) Analysis of the core components of Alzheimer paired helical filaments. A gas chromatography/mass spectrometry characterization of fatty acids, carbohydrates and long-chain bases, *FEBS Lett.* **366**, 81-85.
27. Goux, W.J., Rodriguez, S., and Sparkman, D.R. (1996) Characterization of the glycolipid associated with Alzheimer paired helical filaments, *J. Neurochem.* **67**, 723-733.
28. Sparkman, D.R., Goux, W.J., Jones, C.M., White, C.L., and Hill, S.J. (1991) Alzheimer disease paired helical filament core structures contain glycolipid, *Biochem. Biophys. Res. Commun.* **181**, 771-779.
29. Folch, J., Lees, M., and Stanley, G.H.S. (1956) A simple method for the isolation and purification of total lipids from animal tissues, *J. Biol. Chem.* **226**, 497-509.
30. Choo-Smith, L.P., Garzon-Rodriguez, W., Glabe, C.G., and Surewicz, W.K. (1997) Acceleration of amyloid fibril formation by specific binding of Aβ-$(_{1\text{-}40})$ peptide to ganglioside-containing membrane vesicles, *J. Biol. Chem.* **272**, 22987-22990.
31. McLaurin, J. and Chakrabartty, A. (1997) Characterization of the interactions of Alzheimer β-amyloid peptides with phospholipid membranes, *Eur. J. Biochem.* **245**, 355-363.
32. Yanagisawa, K. and Ihara, Y. (1998) GM1 ganglioside-bound amyloid β-protein in Alzheimer's disease brain, *Neurobiol. Aging* **19**, S65-S67
33. Choo-Smith, L.P. and Surewicz, W.K. (1997) The interaction between Alzheimer amyloid β $(_{1\text{-}40})$ peptide and ganglioside GM1-containing membranes, *FEBS Lett.* **402**, 95-98.

34. McLaurin, J., Franklin, T., Fraser, P.E., and Chakrabartty, A. (1998) Structural transitions associated with the interaction of Alzheimer β-amyloid peptides with gangliosides, *J. Biol. Chem.* **273**, 4506-4515.
35. Saido, T.C., Iwatsubo, T., Mann, D.M., Shimada, H., Ihara, Y., and Kawashima, S. (1995) Dominant and differential deposition of distinct β-amyloid peptide species, Aβ N3(pE), in senile plaques, *Neuron* **14**, 457-466.
36. Kuo, Y.M., Emmerling, M.R., Woods, A.S., Cotter, R.J., and Roher, A.E. (1997) Isolation, chemical characterization, and quantitation of Aβ 3- pyroglutamyl peptide from neuritic plaques and vascular amyloid deposits, *Biochem. Biophys. Res. Commun.* **237**, 188-191.
37. Gouras, G.K., Xu, H., Jovanovic, J.N., Buxbaum, J.D., Wang, R., Greengard, P., Relkin, N.R., and Gandy, S. (1998) Generation and regulation of β-amyloid peptide variants by neurons, *J. Neurochem.* **71**, 1920-1925.
38. Wang, R., Sweeney, D., Gandy, S.E., and Sisodia, S.S. (1996) The profile of soluble amyloid β protein in cultured cell media. Detection and quantification of amyloid β protein and variants by immunoprecipitation-mass spectrometry, *J. Biol. Chem.* **271**, 31894-31902.
39. Shinkai, Y., Yoshimura, M., Ito, Y., Odaka, A., Suzuki, N., Yanagisawa, K., and Ihara, Y. (1995) Amyloid β-proteins 1-40 and 1-42(43) in the soluble fraction of extra- and intracranial blood vessels, *Ann. Neurol.* **38**, 421-428.
40. Roher, A.E., Chaney, M.O., Kuo, Y.M., Webster, S.D., Stine, W.B., Haverkamp, L.J., Woods, A.S., Cotter, R.J., Tuohy, J.M., Krafft, G.A., Bonnell, B.S., and Emmerling, M.R. (1996) Morphology and toxicity of Aβ-(1-42) dimer derived from neuritic and vascular amyloid deposits of Alzheimer's disease, *J. Biol. Chem.* **271**, 20631-20635.
41. Davis-Salinas, J., Saporito-Irwin, S.M., Cotman, C.W., and Van Nostrand, W.E. (1995) Amyloid β-protein induces its own production in cultured degenerating cerebrovascular smooth muscle cells, *J. Neurochem.* **65**, 931-934.
42. Wisniewski, H.M. and Wegiel, J. (1994) Beta-amyloid formation by myocytes of leptomeningeal vessels, *Acta Neuropathol.(Berl.)* **87**, 233-241.
43. Kalaria, R.N., Premkumar, D.R., Pax, A.B., Cohen, D.L., and Lieberburg, I. (1996) Production and increased detection of amyloid β protein and amyloidogenic fragments in brain microvessels, meningeal vessels and choroid plexus in Alzheimer's disease, *Brain Res. Mol. Brain Res.* **35**, 58-68.
44. Selkoe, D.J. (1989) Molecular pathology of amyloidogenic proteins and the role of vascular amyloidosis in Alzheimer's disease, *Neurobiol. Aging* **10**, 387-395.
45. Kuo, Y.M., Emmerling, M.R., Lampert, H.C., Hempelman, S.R., Kokjohn, T.A., Woods, A.S., Cotter, R.J., and Roher, A.E. (1999) High levels of circulating Aβ42 are sequestered by plasma proteins in Alzheimer disease, *Biochem. Biophys. Res. Commun.* **257**, 787-791.
46. Saito, Y., Buciak, J., Yang, J., and Pardridge, W.M. (1995) Vector-mediated delivery of 125I-labeled β-amyloid peptide Aβ $_{1-40}$ through the blood-brain barrier and binding to Alzheimer disease amyloid of the Aβ $_{1-40}$/vector complex, *Proc. Natl. Acad. Sci.U.S.A.* **92**, 10227-10231.
47. Mackic, J.B., Stins, M., McComb, J.G., Calero, M., Ghiso, J., Kim, K.S., Yan, S.D., Stern, D., Schmidt, A.M., Frangione, B., and Zlokovic, B.V. (1998) Human blood-brain barrier receptors for Alzheimer's amyloid-β 1-40. Asymmetrical binding, endocytosis, and transcytosis at the apical side of brain microvascular endothelial cell monolayer, *J. Clin. Invest.* **102**, 734-743.
48. Zlokovic, B.V., Martel, C.L., Matsubara, E., McComb, J.G., Zheng, G., McCluskey, R.T., Frangione, B., and Ghiso, J. (1996) Glycoprotein 330/megalin: probable role in receptor-

mediated transport of apolipoprotein J alone and in a complex with Alzheimer disease amyloid β at the blood-brain and blood-cerebrospinal fluid barriers, *Proc. Natl. Acad. Sci.U.S.A.* **93**, 4229-4234.
49. Wisniewski, T. and Frangione, B. (1992) Apolipoprotein E: a pathological chaperone protein in patients with cerebral and systemic amyloid, *Neurosci. Lett.* **135**, 235-238.
50. Kida, S., Weller, R.O., Zhang, E.T., Phillips, M.J., and Iannotti, F. (1995) Anatomical pathways for lymphatic drainage of the brain and their pathological significance, *Neuropathol. Appl. Neurobiol.* **21**, 181-184.
51. Zhang, E.T., Inman, C.B., and Weller, R.O. (1990) Interrelationships of the pia mater and the perivascular (Virchow- Robin) spaces in the human cerebrum, *J. Anat.* **170:111-23**, 111-123.
52. Ichimura, T., Fraser, P.A., and Cserr, H.F. (1991) Distribution of extracellular tracers in perivascular spaces of the rat brain, *Brain Res.* **545**, 103-113.
53. Iwatsubo, T., Mann, D.M., Odaka, A., Suzuki, N., and Ihara, Y. (1995) Amyloid β protein (Aβ) deposition: Aβ 42(43) precedes Aβ 40 in Down syndrome, *Ann. Neurol.* **37**, 294-299.
54. Castillo, G.M., Ngo, C., Cummings, J., Wight, T.N., and Snow, A.D. (1997) Perlecan binds to the β-amyloid proteins (Aβ) of Alzheimer's disease, accelerates Aβ fibril formation, and maintains Aβ fibril stability. *J. Neurochem.* **69**, 2452-2465.
55. Snow, A.D., Kinsella, M.G., Parks, E., Sekiguchi, R.T., Miller, J.D., Kimata, K., and Wight, T.N. (1995) Differential binding of vascular cell-derived proteoglycans (perlecan, biglycan, decorin, and versican) to the β-amyloid protein of Alzheimer's disease, *Arch. Biochem. Biophys.* **320**, 84-95.
56. Li, Q.X., Whyte, S., Tanner, J.E., Evin, G., Beyreuther, K., and Masters, C.L. (1998) Secretion of Alzheimer's disease Aβ amyloid peptide by activated human platelets, *Lab. Invest.* **78**, 461-469.
57. Chen, M., Inestrosa, N.C., Ross, G.S., and Fernandez, H.L. (1995) Platelets are the primary source of amyloid β-peptide in human blood, *Biochem. Biophys. Res. Commun.* **213**, 96-103.
58. Nordstedt, C., Naslund, J., Thyberg, J., Messamore, E., Gandy, S.E., and Terenius, L. (1994) Human neutrophil phagocytic granules contain a truncated soluble form of the Alzheimer β/A4 amyloid precursor protein (APP), *J. Biol. Chem.* **269**, 9805-9810.
59. Sarkozi, E., Askanas, V., Johnson, S.A., Engel, W.K., and Alvarez, R.B. (1993) β-Amyloid precursor protein mRNA is increased in inclusion-body myositis muscle, *Neuroreport.* **4** , 815-818.
60. Askanas, V., Alvarez, R.B., and Engel, W.K. (1993) β-Amyloid precursor epitopes in muscle fibers of inclusion body myositis, *Ann. Neurol.* **34**, 551-560.
61. Golde, T.E., Estus, S., Usiak, M., Younkin, L.H., and Younkin, S.G. (1990) Expression of β amyloid protein precursor mRNAs: recognition of a novel alternatively spliced form and quantitation in Alzheimer's disease using PCR, *Neuron* **4**, 253-267.
62. Funato, H., Yoshimura, M., Kusui, K., Tamaoka, A., Ishikawa, K., Ohkoshi, N., Namekata, K., Okeda, R., and Ihara, Y. (1998) Quantitation of amyloid β-protein (Aβ) in the cortex during aging and in Alzheimer's disease, *Am. J. Pathol.* **152**, 1633-1640.
63. Yamada, M., Itoh, Y., Shintaku, M., Kawamura, J., Jensson, O., Thorsteinsson, L., Suematsu, N., Matsushita, M., and Otomo, E. (1996) Immune reactions associated with cerebral amyloid angiopathy, *Stroke* **27**, 1155-1162.
64. Wilson, D.M. and Binder, L.I. (1997) Free fatty acids stimulate the polymerization of tau and amyloid β peptides. *In vitro* evidence for a common effector of pathogenesis in Alzheimer's disease, *Am. J. Pathol.* **150**, 2181-2195.

65. Zweig, R.M., Ross, C.A., Hedreen, J.C., Steele, C., Cardillo, J.E., Whitehouse, P.J., Folstein, M.F., and Price, D.L. (1988) The neuropathology of aminergic nuclei in Alzheimer's disease, *Ann. Neurol.* **24**, 233-242.
66. Wilcock, G.K., Esiri, M.M., Bowen, D.M., and Hughes, A.O. (1988) The differential involvement of subcortical nuclei in senile dementia of Alzheimer's type, *J. Neurol. Neurosurg. Psychiatry* **51**, 842-849.
67. Sato, A. and Sato, Y. (1992) Regulation of regional cerebral blood flow by cholinergic fibers originating in the basal forebrain, *Neurosci. Res.* **14**, 242-274.

Chapter 10

IMMUNOHISTOCHEMICAL ANALYSIS OF AMYLOID β-PROTEIN ISOFORMS IN CAA

HARUYASU YAMAGUCHI* and MARION L.C. MAAT-SCHIEMAN†
*Gunma University School of Health Sciences, Japan; †Department of Neurology of Leiden University Medical Center Leiden, The Netherlands

Cerebral amyloid angiopathy (CAA) is an amyloid deposition in the wall of arteries and veins in the leptomeninges, and also in the arterioles and capillaries in the cortex. Major isoforms were $A\beta_{x-40}$ ($A\beta_{40}$) and $A\beta_{x-42(43)}$ ($A\beta_{42}$). Immunohistochemical analysis revealed focal deposits of $A\beta_{42}$ in the early stages of CAA in both meningeal and cortical vessels. In advanced stages of CAA, the entire vessel wall is labeled for both $A\beta_{40}$ and $A\beta_{42}$, and $A\beta_{40}$ labeling is predominant. In advanced CAA, $A\beta_{40}$ becomes partly soluble and diffuses out from the vessel walls after dipping the tissue sections in formic acid. The N-terminal of deposited Aβ varies due to proteolytic cleavage and processing. These findings are consistent in normal aged people and in disorders having CAA; i.e.sporadic AD, early-onset familial AD with both APP and presenilin mutations, Down syndrome and HCHWA-D. In an early stage, deposition of amyloid-associated proteins in the basement membrane precedes focal $A\beta_{42}$ deposition. Initial $A\beta_{42}$ deposits may be a seed for further massive deposition of $A\beta_{40}$ (leading to advanced CAA) in both meningeal and cortical vessels.

1. INTRODUCTION

Amyloid β-protein (Aβ) is derived from a larger precursor protein, amyloid β-protein precursor (APP). Aβ is produced and secreted as a normal product by neurons and glia, and is found in cerebrospinal fluid (CSF), plasma and the soluble fraction of brain homogenates. In the wall of cerebral arteries, smooth muscle cells, and also pericytes, produce and secrete Aβ [1,2]. The major isoforms of Aβ consist of 40 and 42(43) amino acids.

Cerebral amyloid angiopathy (CAA) is an amyloid deposition in the wall of arteries and veins in the leptomeninges, and also in the arterioles and capillaries in the cortex. Aβ isoforms in CAA were first characterized by

M.M. Verbeek et al. (eds.),
Cerebral Amyloid Angiopathy in Alzheimer's Disease and Related Disorders, 179–188.

direct chemical analysis [3,4]. Major isoforms were $A\beta_{x-40}$ ($A\beta_{40}$) and $A\beta_{x-42(43)}$ ($A\beta_{42}$), which had C-terminals at the 40th and 42nd (43rd) amino acid of the Aβ sequence, respectively, and various N-terminals from the 1st to 11th amino acid [5] (see Roher *et al.*, this volume). Chemical analysis is superior to immunohistochemical methods in examining the ratio of Aβ isoforms in CAA, although it has the possibility to lose certain isoforms due to their insolubility and low recovery rate from HPLC columns [4]. Using immunohistochemical analysis, however, it is possible to estimate of each vessel affected by CAA whether it is in an early stage (patchy staining) or in an advanced stage (entire wall labeling) [6], whereas the result of chemical analysis is derived from a mixture of both early and advanced CAA. This means that immunohistochemical study is superior in the determination of the Aβ isoforms in initial CAA.

Using chemical analyses, $A\beta_{40}$ appeared to be relatively abundant in leptomeningeal CAA in Alzheimer's disease (AD), when compared to that in senile plaques [3,4,7] (see Roher *et al.,* this volume). Using immunohistochemical techniques, however, $A\beta_{42}$, and not $A\beta_{40}$, is the predominant isoform in initial CAA [7,8]. Here, the authors describe the immunohistochemical analysis of Aβ isoforms in CAA, and discuss the mechanism of amyloid deposition in blood vessel walls.

2. C-TERMINAL HETEROGENEITY OF Aβ IN ALZHEIMER'S DISEASE AND NORMAL AGING

Iwatsubo *et al.* [9] first applied Aβ C-terminal end-specific antibodies to immunohistochemistry of AD brain sections, and reported the labeling of CAA with both 40 and 42 directed antibodies. Further studies confirmed predominant $A\beta_{40}$ immunolabeling of leptomeningeal CAA in AD [7,8,10-13], although some studies showed $A\beta_{42}$ predominancy [14]. In AD brains, leptomeningeal CAA in an advanced stage has large amounts of Aβ, especially Aβ40, the more soluble isoform (Table 1). In these massive deposits, $A\beta_{40}$ becomes partly soluble and diffuses out from the vessel walls after dipping the tissue sections in formic acid to enhance the immunostaining. As shown in Figure 1b, $A\beta_{40}$ spreads diffusely around CAA or flows downward towards the direction of gravity. $A\beta_{42}$, the isoform that is more amyloidogenic than $A\beta_{40}$, does not show such diffusion after formic acid pretreatment. This relatively soluble $A\beta_{40}$ can be extracted from AD brains as a source of Aβ, and a high tissue content of soluble $A\beta_{40}$ is linked to the amounts of CAA present [15]. Leptomeningeal CAA in AD, which is in the advanced stage, shows immunoreactivity for both $A\beta_{40}$ and

$A\beta_{42}$, although some show either of them [7,11,13]. Capillary CAA tends to show $A\beta_{42}$ predominancy (Table 1).

In non-demented middle-aged patients, we were able to find initial CAA deposits, which occupy only a small part of vessel walls. As shown in Figure 2, $A\beta_{42}$ labeling is predominant in the initial deposits (Table 1). $A\beta_{40}$ labeling is not spread by diffusion after formic acid pretreatment in the initial deposits (Figure 2b). Shinkai *et al.* [7] examined the brains of non-demented and AD subjects with both ELISA and immunohistochemistry, and found that $A\beta_{42}$ was predominant in the CAA of non-demented subjects, while $A\beta_{40}$ was predominant in the CAA of AD subjects, which have large amounts of Aβ in advanced CAA. Focal deposition of $A\beta_{42}$ in the vessel wall, may come first, and become a "seed" for further deposition of $A\beta_{40}$, which is a major isoform of soluble Aβ in CSF and plasma.

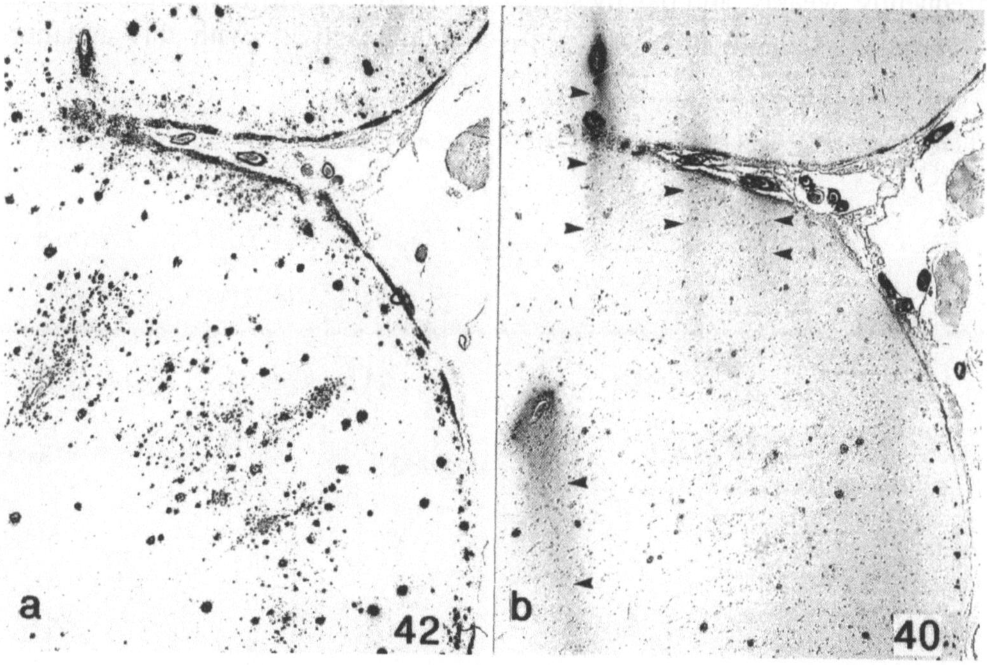

Figure 1. Labeling of CAA with an antibody against the C-terminal of $A\beta_{42}$ (a) and $A\beta_{40}$ (b). Note the diffusion of $A\beta_{40}$ around CAA after formic acid pretreatment (arrowheads); X33.

3. C-TERMINAL HETEROGENEITY OF Aβ AND APP MUTATION/GENE DOSE

Mutations of the APP gene within or around the Aβ coding region cause autosomal-dominant forms of AD. Mutations at codon 717 (near C-terminal of Aβ; Hardy type) cause increased production of $A\beta_{42}$, without changing the total amount of Aβ production [16]. A moderate to severe CAA was present. Affected vessels were most commonly and most intensely immunoreactive for $A\beta_{40}$, and some vessels also showed a weaker and often patchy $A\beta_{42}$ immunoreactivity (See Table 1) [17].

A double mutation at codon 670/671 (near N-terminal of Aβ; Swedish type) causes overproduction of both $A\beta_{40}$ and $A\beta_{42}$ isoforms [18]. A moderate to severe CAA was present in both leptomeningeal and parenchymal vessels, mostly being $A\beta_{40}$-immunoreactive [17]. The staining was strong and uniform throughout the vessel wall. Vessels were less frequently, weakly, and patchy stained for $A\beta_{42}$ [17]. However, Kalaria *et al.* reported a relatively increased $A\beta_{42}$-immunoreactivity with this mutation compared to sporadic AD (See Table 1) [13].

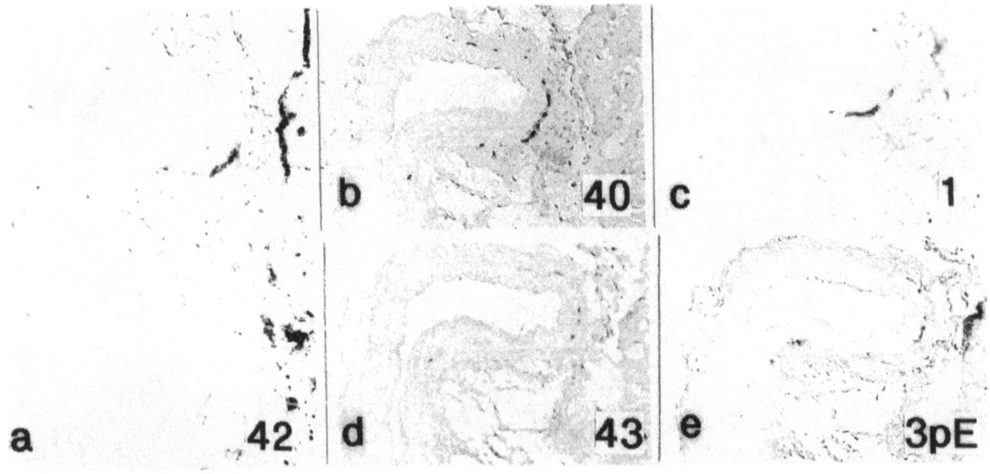

Figure 2. CAA in an early stage in the brain of a non-demented subject labeled with 5 different end-specific Aβ antibodies; C-terminal 42 (a), C-terminal 40 (b), N-terminal 1 (c), C-terminal 43 (d) and pyroglutamate at N-terminal 3 (e). Initial deposition is positive for all antibodies.

Hereditary cerebral hemorrhage with amyloidosis-Dutch type (HCHWA-D) is caused by a point mutation of APP gene at codon 693, which corresponds to the 22nd amino acid of Aβ. The pathology of HCHWA-D is characterized by severe CAA, which causes cerebral hemorrhage and infarction (Figure 4). The CAA in meningeal vessels and also in cortical penetrating arteries and arterioles is in an advanced stage at autopsy. Such vessels were strongly and uniformly stained for $A\beta_{40}$ and the intensity of $A\beta_{42}$ labeling was generally much less, although most vessels were also $A\beta_{42}$-positive (Figure 3a,b) [17]. The authors recently found that capillaries in the cortex showed focal labeling exclusively for $A\beta_{42}$ (Figure 3c,d, Table 1). Such capillaries are Congo red-negative, and seem to be in an initial stage of CAA. In HCHWA-D, initial deposits in capillary CAA consist of the $A\beta_{42}$ isoform (Maat-Schieman, M.L.C. *et al.*, submitted).

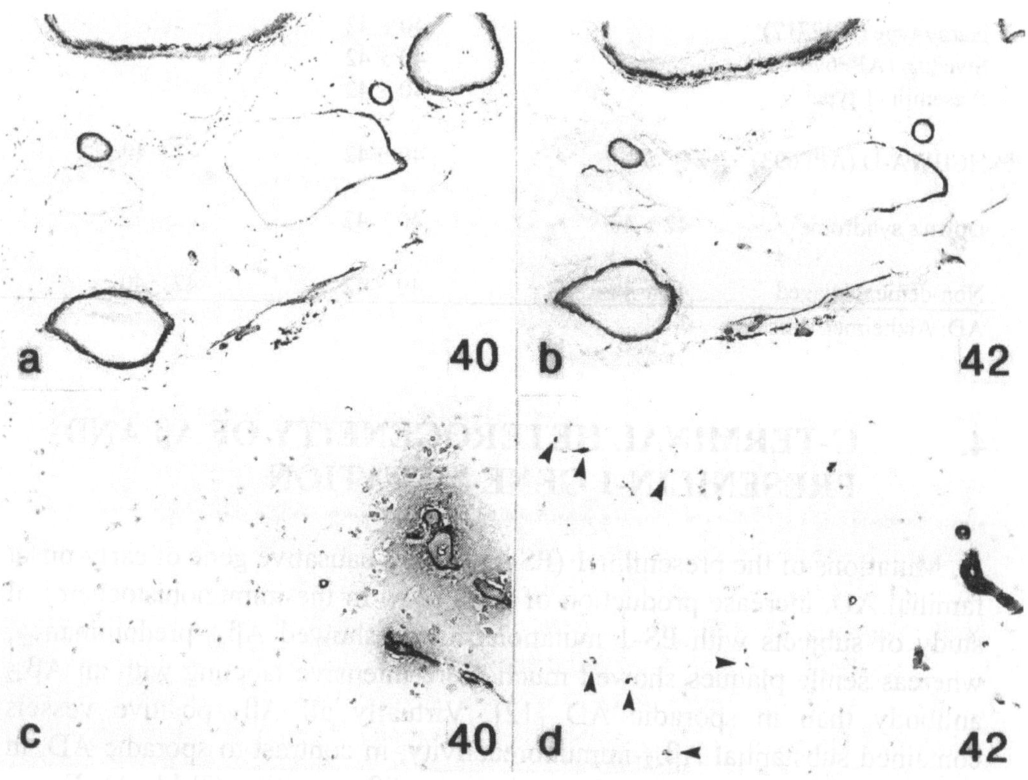

Figure 3. CAA in hereditary cerebral hemorrhage with amyloidosis, Dutch-type. Leptomeningeal CAA shows similar immunoreactivity with both $A\beta_{40}$ (a) and $A\beta_{42}$ (b) antibodies, whereas CAA in cortical capillaries exclusively shows $A\beta_{42}$ labeling (c, arrowheads).

In Down's syndrome patients, APP gene dose is 1.5 times more than in normal subjects, because of its location on chromosome 21. Overproduction of Aβ accelerates Aβ accumulation resulting in the formation of diffuse plaques in the 2nd decade of their life. In Down syndrome, CAA showed $A\beta_{40}$ predominancy, especially in old subjects [19]. Lemere *et al.* [20] reported a similar $A\beta_{40}$ predominancy in both middle aged and older subjects. Kalaria *et al.* [13] found $A\beta_{42}$ predominancy in Down syndrome patients in their 30s, although 1 subject aged 69 also had $A\beta_{42}$ predominancy (Table 1).

Table 1. Summary of the C-terminal heterogeneity of Aβ in cerebral amyloid angiopathy (CAA)

	Leptomeningeal CAA		Capillary CAA
	Initial stage	Advanced stage	
Sporadic AD	42 > 40	40 ≥ 42	42 > 40
Familial AD:			
Hardy type (APP717)		40 ≥ 42	
Swedish (APP670/671)		40 ≥ 42	
Presenilin-1 type		40 ≥ 42	
HCHWA-D (APP693)		40 ≥ 42	42 > 40
Down's syndrome	42 ≥ 40	40 ≥ 42	
Non-demented aged	42 > 40	40 ≥ 42	42 > 40

AD: Alzheimer's disease

4. C-TERMINAL HETEROGENEITY OF Aβ AND PRESENILIN-1 GENE MUTATION

Mutations of the presenilin-1 (PS-1) gene, a causative gene of early-onset familial AD, increase production of $A\beta_{42}$ [21]. In the immunohistochemical study of subjects with PS-1 mutations, CAA showed $A\beta_{40}$ predominancy, whereas senile plaques showed much more intensive labeling with an $A\beta_{42}$ antibody than in sporadic AD [12]. Virtually all $A\beta_{40}$-positive vessels contained substantial $A\beta_{42}$-immunoreactivity, in contrast to sporadic AD, in which only a small subset of vessels were $A\beta_{42}$-positive (Table 1). Fewer $A\beta_{43}$-positive vessels were observed [12].

5. N-TERMINAL HETEROGENEITY OF Aβ

Tekirian *et al.* [22] examined brains of aged dogs, polar bears, as well as AD patients and non-demented subjects, with a variety of Aβ N-terminal directed end-specific antibodies, including $A\beta_{N1}$[D] for L-aspartate at N1, $A\beta_{N1}$[rD] for D-aspartate at N1, $A\beta_{N3}$[pE] for pyroglutamate at N3, and $A\beta_{N17}$[L] for leucine at N17. Each Aβ N-terminal isoform can be present in

CAA in AD and non-demented humans, as well as in aged animals. $A\beta_{N1}$[rD] was usually absent in human CAA, except in AD cases with atherosclerotic and vascular hypertensive changes [22]. As shown in Figure 2c,e, Aβ in CAA has variable N-terminals.

In Down syndrome, N-terminal heterogeneity was examined using the two antibodies mentioned above, $A\beta_{N1}$[D] and $A\beta_{N3}$[pE], but no particular difference was found [12].

Watson *et al.* [23] examined the CAA in HCHWA-D with a panel of N-terminal directed antibodies, and found labeling of CAA with both $A\beta_{N1}$[D] and $A\beta_{N3}$[pE].

6. CELL STAINING WITH Aβ C-TERMINAL ANTIBODIES

In addition to amyloid deposition, which is an extracellular event, Aβ antibodies label cellular components of vessel walls. Immunoreactivity for $A\beta_{40}$ co-localizes with that for presenilin-1 in smooth muscle cells of CAA-bearing meningeal arteries in AD [24]. Lemere *et al.* [20] reported $A\beta_{42}$ labeling of smooth muscle cells in young Down syndrome subjects.

7. MECHANISM OF Aβ DEPOSITION IN CAA

CAA may begin as punctate Aβ deposits within the basement membrane at the junction of media and adventitia in the large meningeal arteries and between smooth muscle cells in the small arteries and arterioles. Such small Aβ deposits have a fibrillar structure under EM [6]. Then, amyloid deposition spreads along the basement membrane, and occupies the entire wall of the vessels, resulting in the loss of smooth muscle cells.

Where does the Aβ come from in the initial stage of CAA? There are 3 possibilities: from CSF through the adventitia, from circulating blood through the endothelium, or from medial smooth muscle cells which secrete Aβ. In large meningeal arteries, Aβ begins to deposit at the outer elastic laminae (media-adventitia junction), and amyloid spreads outside into the adventitia. In middle-size arteries, amyloid continues to accumulate even after the loss of smooth muscle cells. This suggests that CSF Aβ, most of which is produced in the brain parenchyma, may be a major source of Aβ accumulating in the vessel wall.

It is important to think about the reason why arteries outside the central nervous system do not show CAA, although Aβ exists everywhere in the body. Prior to the presence of Aβ immunostaining in the meningeal arteries, labeling for Aβ-associated proteins, such as apolipoprotein E, apolipoprotein J and vitronectin, was found at the media-adventitia junction, the site where Aβ deposition occurs first [25,26]. This means that certain age-related changes in the basement membrane may promote deposition of Aβ42, and that this change may be characteristic of the brain arteries. The initial deposition becomes a seed for further deposition of Aβ, especially of large amounts of Aβ40 in the advanced stage. The N-terminus of Aβ in CAA may be processed by extracellular proteases, resulting in a variety of N-terminal sequences of Aβ in CAA.

In the advanced stages of CAA in meningeal arteries, inflammatory cells remove amyloid, and finally the vessel wall becomes swollen and fragile (aneurysm formation), resulting in bleeding [27].

REFERENCES

1. Davis-Salinas, J., Saporito-Irwin, S.M., Cotman, C.W., van Nostrand, W.E. (1995) Amyloid β-protein induces its own production in cultured degenerating cerebrovascular smooth muscle cells. *J. Neurochem.* **65**, 931-934.
2. Frackowiak, J., Zoltowska, A. and Wisniewski, H.M. (1994) Non-fibrillar β-Amyloid protein is associated with smooth muscle cells of vessel walls in Alzheimer Disease. *J. Neuropathol. Exp. Neurol.* **53,** 637-645.
3. Joachim, C.L., Duffy, L.K., Morris, J.H. and Selkoe, D.J. (1988) Protein chemical and immunocytochemical studies of meningovascular –amyloid protein in Alzheimer's disease and normal aging. *Brain Res.* **474**, 100-111.
4. Roher, A.E., Lowenson, J.D., Clarke, S., Woods, A.S., Cotter, R.J., Gowing, E and Ball, M.J. (1993) β-Amyloid-(1-42) is a major component of cerebrovascular amyloid deposits: implications for the pathology of Alzheimer disease. *Proc. Natl. Acad. Sci. USA* **90**, 10836-40.
5. Roher, A.E., Lowenson, J.D., Clarke, S., Wolkow, C., Wang, R., Cotter R.J., Readon I.M., Zurcher-Neely, H.A., Heinrikson, R.L., Ball, M.J. and Greenberg, B.D. (1993) Structural alterations in the peptide backbone of β-amyloid core protein may account for its deposition and stability in Alzheimer's disease. *J. Biol. Chem.* **268**, 3072-83.
6. Yamaguchi, H., Yamazaki, T., Lemere, C.A., Frosch, M.P. and Selkoe, D.J. (1992) Beta amyloid is focally deposited within the outer basement membrane in the amyloid angiopathy of Alzheimer's disease. An immunoelectron microscopic study. *Am. J. Pathol.* **141**, 249-259.
7. Shinkai, Y., Yoshimura, M., Ito, Y., Odaka, A., Suzuki, N., Yanagisawa, K. and Ihara, Y. (1995) Amyloid β-proteins 1-40 and 1-42(43) in the soluble fraction of extra- and intracranial blood vessels. *Ann. Neurol.* **38**, 421-428.
8. Alonzo, N.C., Hyman, B.T., Rebeck, G.W. and Greenberg, S.M. (1998) Progression of cerebral amyloid angiopathy: accumulation of amyloid-β40 in affected vessels. *J. Neuropathol. Exp. Neurol.* **57**, 353-359.

9. Iwatsubo, T., Okada, A., Suzuki, N., Mizusawa, H., Nukina, N. and Ihara, Y. (1994) Visualization of Aβ42(43) and Aβ40 in senile plaques with end-specific Aβ monoclonals: evidence that an initial deposited species is Aβ42(43). *Neuron* **13**, 45-53.
10. Mak, K., Yang, F., Vinters, H.V., Frautschy, S.A. and Cole, G.M. (1994) Polyclonals to β-amyloid (1-42) identify most plaque and vascular deposits in Alzheimer cortex, but not striatum. *Brain Res.* **667**, 138-142.
11. Yamaguchi, H., Sugihara, S., Ishiguro, K., Takashima, A. and Hirai, S. (1995) Immunohistochemical analysis of COOH-termini of amyloid beta protein (Aβ) using end-specific antisera for Aβ40 and Aβ42 in Alzheimer's disease and normal aging. *Amyloid: Int. J. Clin. Invest.* **2**, 7-16.
12. Lemere, C.A., Lopera, F., Kosik, K.S., Lendon, C.L., Ossa, J., Saido, T.C., Yamaguchi, H., Ruiz, A., Martinez, A., Madrigal, L., Hincapie, L., Arango, J.C., Anthony, D.C., Koo, E.H., Goate, A.M., Selkoe, D.J. and Arango, J.C. (1996) The E280A presenilin 1 Alzheimer mutation produces increased Aβ42 deposition and severe cerebellar pathology. *Nat. Med.* **2**, 1146-1150.
13. Kalaria, R.N., Cohen, D.L., Greenberg, B.D., Savage, M.J., Bogdanovic, NE., Winblad B., Lannfelt, L. and Adem, A. (1996) Abundance of the longer $A\beta_{42}$ in neocortical and cerebrovascular amyloid β-deposits in Swedish familial Alzheimer's disease and Down's syndrome. *Neuroreport* **7**, 1377-1381.
14. Tamaoka, A., Sawamura, N., Odaka, A., Suzuki, N., Mizusawa, H., Shiji, S. and Mori, H. (1995) Amyloid β-protein 1-42/43 (Aβ1-42/3) in cerebellar difuse plaques: enzyme-linked immunosorbent assay and immunohistochemical study. *Brain Res.* **697**, 151-156.
15. Suzuki, N., Iwatsubo, T., Odaka, A., Ishibashi, Y., Kitada, C. and Ihara, Y. (1994) High tissue content of soluble β_{1-40} is linked to cerebral amyloid angiopathy. *Am. J. Pathol.* **145**, 452-460.
16. Suzuki, N., Cheung, T.T., Cai, X.D., Odaka, A., Otvos, L. Jr., Eckman, C., Golde, T.E., Younkin, S.G. (1994) An increased percentage of long amyloid β protein secreted by familial amyloid beta protein precursor (βAPP717) mutants. *Science* **264**, 1336-1340.
17. Mann, D.M.A., Iwatsubo, T., Ihara, Y., Cairns, N.J., Lantos, P.L., Bogdanovic, N., Lannfelt, L., Winblad, B., Maat-Schieman, M.L.C. and Rossor, M.N. (1996) Predominant deposition of amyloid-$\beta_{42(43)}$ in plaques in cases of Alzheimer's disease and hereditary cerebral hemorrhage associated with mutations in the amyloid precursor protein gene. *Am. J. Pathol.* **148**, 1257-1266.
18. Citron, M., Oltersdorf, T., Haass, C., McConlogue, L., Hung, A.Y., Seubert, P., Vigo-Pelfrey, C., Lieberburg, I., Selkoe, D.J. (1992) Mutation of the β-amyloid precursor protein in familial Alzheimer's disease increases β-protein production. *Nature* **360**, 672-674
19. Iwatsubo, T., Mann, D.M.A., Odaka, A., Suzuki, N. and Ihara, Y. (1995) Amyloid β protein (Aβ) deposition: Aβ42(43) precedes Aβ40 in Down syndrome. *Ann. Neurol.* **37**, 294-299.
20. Lemere, C.A., Blusztajn, J.K., Yamaguchi, H., Wisniewski, T., Saido, T.C. and Selkoe DJ (1996) Sequence of deposition of heterogeneous amyloid β-peptides and APO E in Down syndrome: implications for initial events in amyloid plaque formation. *Neurobiol. Dis.* **3**, 16-32.
21. Duff, K., Eckman, C., Zehr, C., Yu, X., Prada, C.M., Perez-tur, J., Hutton, M., Buee, L., Harigaya, Y., Yager, D., Morgan, D., Gordon, M.N., Holcomb, L., Refolo, L., Zenk, B., Hardy, J., Younkin, S. (1996) Increased amyloid-β42(43) in brains of mice expressing mutant presenilin 1. *Nature* **383**, 710-713.

22. Tekirian, T.L., Saido, T.C. Markesbery, W.R., Russell, M.J. Wekstein, D.R., Patel, E. and Geddes, J.W. (1998) N-terminal heterogeneity of parenchymal and cerebrovascular Aβ deposits. *J. Neuropathol. Exp. Neurol.* **57**, 76-94.
23. Watson, D., Maat-Schieman, M., Saido, T., Teplow, D., Biere, A.L., Roos, R. and Selkoe, D (1997) N-terminal heterogegeneity of Aβ caused by HCHWA-D mutation. *Soc. Neurosci.* **23,** 821 (Abstract 321.9).
24. Hayashi, Y., Fukatsu, R., Tsuzuki, K., Yoshida, T., Sasaki, N., Kimura, K., Yamaguchi, H., St. George Hyslop, P.H., Fujii, N. and Takahata, N. (1998) Evidence for presenilin-1 involvement in amyloid angiopathy in the Alzheimer's disease-affected brain. *Brain Res.* **789**, 307-314.
25. Yamaguchi, H., Ishiguro, K., Sugihara, S., Nakazato, Y., Kawarabayashi, T., Sun, X. and Hirai, S. (1994) Presence of apolipoprotein E on extracellular neurofibrillary tangles and on meningeal blood vessels precedes the Alzheimer β-amyloid deposition. *Acta Neuropathol.* **88,** 413-419.
26. Yamaguchi H and Sugihara S (1997) Aβ and its associated proteins with aging: comparison between senile plaques and amyloid angiopathy, in K. Iqbal, B. Winblad, T. Nishimura, M. Takeda M and H.M. Wisniewski (eds.), *Alzheimer's Disease: Biology, Diagnosis and Therapeutics*, Jhon Wiley & Sons, Chichester, pp. 305-309
27. Vinters, H.V., Natte, R., Maat-Schieman, M.L.C., van Duinen, S.G., Hegeman-Kleinn, I., Welling-Graafland, C., Haan, J. and Roos, R.A.C. (199) Secondary microvaascular degeneration in amyloid angiopathy of patients with hereditary cerebral hemorrhage with amyloidosis, Dutch type (HCHWA-D) *Acta Neuropathol.* **95**, 235-244.

Chapter 11

BLOOD BRAIN BARRIER DYSFUNCTION AND CEREBROVASCULAR DEGENERATION IN ALZHEIMER'S DISEASE

RAJ N. KALARIA
Institute for Health of the Elderly, Newcastle General Hospital, Westgate Road, and Department of Psychiatry, University of Newcastle, Newcastle upon Tyne, United Kingdom

The neuropathology of Alzheimer's disease (AD) extends beyond amyloid plaques and neurofibrillary tangles. Recent evidence suggests that more than 30% of AD cases exhibit cerebrovascular pathology, which involves the cellular elements that represent the blood-brain barrier. However, certain vascular lesions such as microvascular degeneration affecting the cerebral endothelium, cerebral amyloid angiopathy and periventricular white matter lesions are evident in virtually all cases of AD. Furthermore, clinical studies have clearly demonstrated blood-brain barrier dysfunction in AD patients who exhibit peripheral vascular abnormalities such as hypertension, cardiovascular disease and diabetes. Whether these vascular lesions along with perivascular denervation are coincidental or causal in the pathogenetic processes of AD remains to be defined. In this chapter, I review accumulated biochemical and morphological evidence in context with the variable but distinct cerebrovascular pathology described to be associated with AD. I also consider genetic influences such as apolipoprotein E in relation to cerebrovascular lesions that may shed light on the pathophysiology of the cerebral vasculature. The compelling vascular pathology associated with AD suggests that transient and focal breach of the blood-brain barrier occurs in late onset AD and may involve an interaction of several factors, which include perivascular mediators as well as peripheral circulation derived factors that perturb the endothelium. These vascular abnormalities are likely to worsen clinical presentation of AD and progression of dementia.

1. INTRODUCTION

Alzheimer's disease (AD) is the most common cause of dementia in the elderly. The underlying processes that lead to dementia in this disorder are

M.M. Verbeek et al. (eds.),
Cerebral Amyloid Angiopathy in Alzheimer's Disease and Related Disorders, 189–206.

not understood but late onset AD is very likely acquired by the interaction of both hereditary and environmental factors. AD is pathologically defined by the presence of senile plaques and neurofibrillary tangles. The presence of other pathologies including vascular lesions is often ignored or regarded as insignificant. Brains of subjects with AD often bear cerebrovascular pathology consisting of degenerative microangiopathy, cerebral amyloid angiopathy (CAA), cerebral infarcts and intracerebral hemorrhages. 'Pure AD' is typically considered as plaque and tangle pathology but does such clear-cut pathology exist in the elderly? Interestingly, even Alzheimer in his original report describing the pathology in the brain of Auguste D. [1] had written that besides "one or several fibrils in otherwise normal cells, and "numerous small miliary foci ...and...storage of peculiar material in the cortex, one sees endothelial proliferation and also occasionally neovascularisation." Is it likely that Alzheimer described degeneration of the cortical microvessels as the seat of the blood-brain barrier (BBB) evident by modern methods rather than angiogenesis or was it that Auguste D. had suffered cerebral infarcts and the "endothelial proliferation" was a consequence of the infarction? Both of these possibilities and the fact that there was moderate atherosclerosis in the basal brain arteries of Auguste D provide a basis for thinking that vascular pathology was also evident in the original case of Alzheimer, which we use today to define AD. Nevertheless, it remains to be known whether the brain vascular pathology found in AD is coincident with the disease process or whether peripheral vascular abnormalities including those that alter cerebral perfusion are causal factors in AD. Indeed, BBB abnormalities may also be caused by vascular factors linked to cardiovascular disease. These include hypertension, atrial fibrillation, and aortic and carotid atherosclerosis that may decrease cerebral perfusion and increase risk of stroke or transient ischemic attacks in AD.

2. CEREBROVASCULAR PATHOLOGY AND THE BLOOD-BRAIN BARRIER IN AD

At least a third of the patients with AD may exhibit a variety of brain vascular lesions (Table 1). These are often ignored as coincidental findings at autopsy. The cerebrovascular pathology of AD encompasses a variety of lesions including changes in endothelial and vascular smooth muscle cells, macroscopic and micro- infarction, hemorrhage and white matter changes related to small vessel disease [2,3]. In addition, amyloid β protein is involved in the degeneration of both the larger perforating arterial vessels as well as the cerebral capillaries that represent the BBB. While these microvascular changes imply that the integrity of the cerebral vasculature is impaired in AD as a result of the progressive cortical pathology, they also entail the long-term influence of peripheral vascular factors. These include long-standing hypertension, atrial fibrillation, coronary or carotid artery disease and diabetes; conditions that may promote cerebral hypoperfusion during aging. Whether, vascular and neurodegenerative pathologies are additive in the way in which they influence clinical presentation or progression of dementia [4,5] is the subject of some debate. It also remains to be known whether each of these lesions are simple manifestations of brain aging or intrinsic to the pathogenesis of AD and the cause of dementia [6,7].

Table 1. The variety and percentage of distribution of cerebrovascular lesions in AD cases

Vascular lesions	Specific feature(s) or markers involved
Degeneration of cerebral microvessels (100%)	Loss of endothelial markers, CD34, GLUT1 Thickening of basement membrane, collagen IV
Localisation of serum proteins (80%)	P component, complement, ApoE in AD lesions
Presence of cerebral amyloid angiopathy (CAA) (98%)	Aβ peptides, Cystatin C protein, inflammatory markers
Presence of lobar and intracerebral hemmorhages (10%)	CAA-related intracerebral hemorrhages
Large cerebral infarcts and cortical microinfarcts (36%)	Variably distributed and sized
Diffuse white matter disease (35%)	Periventricular and deep white matter lesions

Data (%) derived from series published previously [5] and unpublished results (Kalaria *et al*) on 300 cases.

2.1 BBB FUNCTIONAL PROTEINS IN AD

Selective biochemical changes, not necessarily related to ageing, are evident in the cerebral microvasculature of AD and Down's syndrome subjects (Table 2). We previously reported impairment of the BBB associated glucose transporter (GLUT1) in isolated brain microvessels and in cortical membrane fractions obtained from subjects with AD and age-matched controls [8]. The rationale for this study was based on evidence from positron emission tomography studies showing that in AD, the brain has a low metabolic rate for glucose, especially in those regions which are most affected pathologically [9,10]. This may implicate impaired GLUT1 in AD although under normal conditions, glucose transport across the BBB is not rate limiting. However, under pathophysiological conditions such as in seizures and perhaps ageing, transport of glucose can become a limiting factor for brain oxidative metabolism. Using the binding properties of [^{3}H]cytochalasin B and immunochemical methods [8], we also confirmed that many collagen IV positive capillaries were absent in GLUT1 immunoreactivity [11,12]. These findings, along with unchanged binding in the cerebellum and putamen indicated somewhat selective abnormality of the GLUT1 in AD, particularly at the BBB. The reduction of the GLUT1 protein in AD subjects could not be attributed to either ageing or postmortem factors [8] and suggested abnormalities in the post-translational modification of the protein. Whether amyloid β deposition in cerebral microvessels in AD impairs synthesis or increases degradation of the GLUT1 protein remains to be shown. Our recent preliminary studies in microvessels suggest that GLUT1 mRNA is also decreased in AD. It is more likely that the decreased GLUT1 in AD is related to the morphological alterations in the cerebral endothelium as described above. Since the expression of GLUT1 is limited to the endothelium bearing tight junctions [13] these results imply increased permeability of the BBB with consequent down regulation of the GLUT1. Our observations were corroborated by positron emission tomography studies showing that the transport of glucose (kinetic parameter k_1) into brains of AD subjects is diminished [14]. It is also of interest that Marcus *et al* [15] observed decreased uptake of 2-deoxy-D-glucose uptake implying reduced hexokinase activity in isolated cerebral microvessels of AD subjects. This decreased activity directly relates to our reduced GLUT1 findings and suggests that microvessels deprived of glucose may be using other sources of metabolic fuel.

Further studies were pursued to determine if different proteins associated with BBB transport functions were affected in AD [13]. We assessed gamma-glutamyl transferase along with other enzymes known to be associated with the endothelium in isolated brain microvessels (Table 2). There were no statistically significant changes in the activities of plasma membrane associated enzymes including angiotensin converting enzyme and alkaline phosphatase between AD and age-matched controls but the activities were consistently lower in vascular fractions from AD subjects. Previous morphological evidence has shown that the cerebral endothelium is enriched in mitochondria [16] and that cerebral microvessels of a number of species are enriched in monoamine oxidase localised to the outer membrane of mitochondria. We measured monoamine oxidase activities in cerebral microvessels from AD subjects and age-matched controls. We found no significant changes in total monoamine oxidase in microvessels although carnitine acetyltransferase activity, an enzyme also localised to the mitochondria but not necessarily restricted to them, was significantly reduced in microvessels from AD subjects. Cholinesterases previously known to be localised in the cerebral microvasculature were also assessed. We found that activities of both acetylcholinesterase and butyrylcholinesterase were significantly reduced in cerebral microvessels from AD subjects compared to age-matched controls (Table 2). These differential alterations cannot be readily explained but may relate to the deposition of amyloid β in the vasculature since cholinesterases interact with amyloid and related proteins. However, the findings support the notion that selective or focal BBB changes occur in the neocortex in AD.

Table 2. Features of the BBB and markers in AD

Cellular Feature	Specific markers found to be affected
Cerebral endothelium	Loss of glucose transporter, Na+/K+ ATPase
Endothelial membranes/ microvascular endfeet	Moderate loss of AlkP, GGT, AChE, BChE
Basement membrane	Increase in collagen proteins and perlecans
Endothelial mitochondria	Loss of carnitine acetyltransferase
Cerebral endothelium (oxidative stress)	Increased glucose-6-phosphatase
Vascular smooth muscle cells	Loss of alpha actin and accumulation of amyloid β

Data derived from series published previously [13] and unpublished results (Kalaria *et al*) on 300 cases. Abbreviations: AChE, acetylcholinesterase; AlkP, alkaline phosphatase; BuChE, butyrylcholinesterase; GGT, gamma glutamyl transpeptidase.

Previous studies have suggested that the cerebral microvasculature is immunologically activated in AD. Cell adhesion molecules such as the intercellular adhesion molecule-1 (ICAM-1) induced during inflammatory processes are upregulated in response with amyloidotic pathology [17-20] and are increased in brain capillaries and perivascular cells in AD subjects [21]. We reported that ICAM-1 immunoreactivity within capillary profiles associated with amyloid β plaques and soluble ICAM-1 determined by immunoassay were significantly increased in frontal and temporal cortex of AD subjects compared to age-matched controls [22]. Similarly, the vascular cellular adhesion molecule was increased in both capillaries and neocortical extracts from brains of AD subjects (Kalaria *et al*, unpublished observations). These findings support the activation of the cerebral endothelium that may lead to focal or transient increases in permeability of the BBB.

2.2 SERUM PROTEINS AND THE BBB

The localisation of proteins originating in or extravasated from the circulation is considered an index of BBB breakdown and provide support for microvascular abnormalities in AD. This is particularly important if the factors or proteins derived from the circulation are not produced by the brain cellular elements. Both amyloid β deposits and neurofibrillary pathology in AD acquire several specific proteins whose role is unclear in these lesions. One of the first studies [23], reported positive albumin and IgG immunoreactivity in amyloid plaques and tangles contained in brains from AD patients. More recent studies showed that these diffuse deposits of serum proteins were also evident in ageing controls and concluded that such localisation was not proof of BBB breakdown in AD [24]. However, antemortem factors may explain the lack of clear differences between ageing controls and AD subjects [25-27]. To elucidate this issue of BBB permeability in AD, we showed that circulating proteins that particularly localise with amyloidotic lesions may accumulate over a protracted period during focal and transient leaks in the BBB. The abundance of pentraxins such as P component and C reactive protein immunoreactivities in cerebral lesions in AD but lack of their mRNA in brain suggested that these relatively large circulation-derived proteins and possibly others, as yet uncharacterised, originate from the liver during the pathogenetic process. The specific binding of P component and amyloid β could explain the localisation of some but not the more common serum proteins such as albumin and immunoglobulins in cerebral amyloid β deposits [28]. It is feasible that proteins, which do not characteristically interact with amyloid β, would either be sequestered or readily removed via the brain drainage systems. Nonetheless, these observations provide indirect evidence for an immunological link between the brain parenchyma and the circulation.

While the BBB is accepted to be unconditionally impaired in vascular dementia, it has been proposed that BBB dysfunction is involved in the aetiology and pathogenesis of AD [13,29,30] The cerebrospinal fluid (CSF)/ serum albumin ratio is a generally accepted method of assessing BBB function in living subjects. Increased CSF/serum ratios have been reported in AD patients, particularly in those exhibiting peripheral vascular disease [13,31,32], but not apparent in others [33,34]. However, Skoog *et al* [35] reported that 85-year-olds with AD had a higher CSF/serum albumin ratio than non-demented individuals, and that there were indications of a disturbed BBB function even before onset of the disease in a population-based study. It is noteworthy that chronic hypertension, considered a strong risk factor for AD, is another factor which could cause increased vascular permeability with protein extravasation [36,37]. A relative BBB dysfunction may increase the possibility that substances from serum penetrate the BBB and reach the brain, where they may interact with neurons, perhaps initiating a cascade with amyloid accumulation and Alzheimer encephalopathy.

2.3 ENDOTHELIAL DEGENERATION AND BASEMENT MEMBRANE PATHOPHYSIOLOGY

In addition to CAA and associated intracerebral hemorrhage, AD subjects exhibit profound changes in cerebral microvessels often independent of amyloid deposition [2]. Several elegant studies using morphological and biochemical methods have demonstrated abnormalities in various cellular elements of cerebral microvessels or capillaries that relate to BBB function. These include degeneration of vascular smooth muscle cells [38,39], focal constrictions and degenerative changes in smooth muscle cells [40], degeneration and focal necrotic changes of the endothelium [11,41], vascular basement membrane alterations accompanied by accumulation of collagen [30,42], loss of perivascular nerve plexus [43], decreased mitochondrial content and increased pinocytotic vesicles [44], and loss of tight junctions [16]. Using differential immunocytochemical methods, we [11,45] have further defined the convolutional abnormalities and "collapsed" or attenuated capillaries in cortical lobes of AD subjects [46]. The differential labelling was characterised by selective degeneration of the endothelium in capillary profiles, which yet retained their basement lamina, as evident by markers such as collagen IV. This phenomenon was observed in virtually all amyloid β laden cortical lobes of more than 95% of the AD as well as Down's syndrome subjects [11]. Both the length and number of degenerated microvessel profiles were significantly correlated with neocortical amyloid β deposits but there was no apparent relationship between the degenerated microvessels and neurofibrillary tangles or existing pyramidal neurons. This vascular phenomenon along with a profound microangiopathy is concomitant with amyloid β deposition and implies abnormalities in the patency of the brain microvasculature in AD [41]. The observations support previous conclusions on disturbances in local perfusion and oxygen tension as a consequence such that neurons furthest away from the capillaries are divested [43]. However, it should be realised that the brain is a dynamic organ and that reactive or compensatory responses are presumably not limited to neurons or glia [47] but that the cerebral endothelium must also be constantly changing even in ageing [2]. It is conceivable that the BBB may be able to sustain subtle damage within certain regions and that the deposition of extracellular amyloid predisposes the endothelium to further degeneration. Although, these vascular abnormalities combined with the vascular amyloid deposition imply breach of the BBB in AD, clear functional evidence to support this is not apparent from non-invasive imaging and permeability studies [13]. However, it is possible that breakdown of the BBB occurs focally and transiently over a protracted period in association with reactive mechanisms that direct repair and growth [47].

3. CEREBRAL AMYLOID ANGIOPATHY, INTRACEREBRAL HEMORRHAGES AND APOLIPOPROTEIN E

Amyloid β associated CAA has been reported to be present in 62-97% of AD subjects and is consistently present in Down's syndrome [3]. Our recent analysis on isolated cerebral vessels in parallel with brain tissue from a series of 300 cases of AD indicates that CAA is more frequent in AD than previously thought [5]. It involves the leptomeninges, small pial vessels, intracortical arterioles as well as brain capillaries [48]. The lesion was characterised by sporadic focal deposits in surface vessels to complete infiltration of numerous meningeal and intracortical vessels throughout all cortical lobes [49]. There can be little doubt that cerebral vascular amyloid deposition resulting in CAA compounds the ageing-related microvascular abnormalities in AD [2]. However, amyloid β associated CAA may co-exist with other neurodegenerative disorders such as Creutzfeldt-Jakob disease [50] and it may be exclusive to certain disorders such as hereditary cerebral hemorrhage or haemorrhagic stroke with amyloidosis of the Dutch type and the Flemish type. CAA was frequently prominent in the occipital lobes and more profound in the sulci compared to the gyri of the neocortex. Vascular amyloid β deposits were rare in the large cranial arteries or muscular vessels of peripheral organs even in patients with relatively high degree of cerebral amyloid β burden. CAA could result from head injury or sporadic hemorrhage strokes causing vascular amyloid accumulation when cerebral vessels are subjected to trauma, oxidative stress or hemodynamic stress [51]. The characteristic cerebral distribution of CAA also implicates that the process may be largely limited to brain vessels associated with a tight or continuous endothelium and when exposed to molecular triggers which may include soluble amyloid β itself that may even originate in perivascular plaques [2]. An intriguing hypothesis has been proposed to explain the mechanism of CAA and accumulation of amyloid β in brain. Weller *et al* [52] have suggested that the vascular deposition of amyloid is related to the lack of clearance of amyloid β via the interstitial drainage pathways. Irrespective of the mechanism of CAA, it is highly likely that the distinctive vascular deposition in AD and the amyloid angiopathies compromise BBB function and promote chronic hypoperfusion [53].

CAA is considered an important cause of intracerebral and lobar hemorrhages. We have estimated that up to 10% of AD subjects exhibit CAA related intracerebral hemorrhages [5,54]. We have moreover shown that AD subjects with evidence of intracerebral hemorrhage exhibit higher proportions of the more pathogenic amyloid β_{42} compared to amyloid β_{40} in the vasculature (Kalaria *et al*, unpublished observations). Whereas intracerebral bleeds characterise the Dutch and Flemish variants of cerebral hemorrhage with amyloidosis, it can cause premature death in the elderly and AD patients. Consideration of the Dutch disease provides certain clues to a link between CAA and stroke. It is thought that the first stroke-like episode triggers multiple cerebral bleeds which may be accompanied by white matter lesions that in turn lead to rapid decline of cognitive functions [55]. We have previously proposed that certain routinely classified AD cases with predominant microvascular lesions and vascular amyloid β deposition might in fact be CAA variants of AD [48]. This would be consistent with the pathological features of the Dutch hereditary disease where severe CAA is present in the absence of profound cortical pathology.

Previously implicated as a susceptibility factor for cardiovascular disease, the inheritance of the ε4 allele of apolipoprotein E gene (*APOE)* is considered to be the most important genetic factor in non-familial AD. A three to four-fold increased frequency of the *APOE*- ε4 allele is not only linked to late-onset AD, but also to middle-aged individuals with coronary heart disease [56] and atherosclerosis [57]. In accord with the implications that apolipoprotein E might promote pathological alterations in the vascular wall, it is intriguing that the *APOE* ε4 allele is also a strong factor in the development of CAA in AD [5,58]. We examined the frequencies of *APOE*-ε4 alleles in age-matched controls and subgroups of 200 AD subjects exhibiting CAA and other frequently associated vascular lesions. *APOE*- ε4 allele frequency (48%) in AD subjects with moderate to severe CAA was six times higher than those who exhibited mild CAA. In the subjects with severe CAA, the occurrence of an ε4 allele was increased by a factor of 17 (95% CI, 7.56-38.9). This was despite the fact that neocortical amyloid β plaque densities in the advanced and mild CAA groups were similar and that all the subjects had met the accepted neuropathological criteria for AD. More remarkably, the ε4 allele frequency was highly associated with AD subjects exhibiting lobar or intracerebral hemorrhage, all of whom had advanced CAA. These findings suggested the *APOE*- ε4 allele to be a significant factor in the development of CAA in AD and revealed the possibility that *APOE* is a factor in CAA and other vascular abnormalities associated with AD. Our observations on the relationship between *APOE*- ε4 allele and CAA-related intracerebral hemorrhage were confirmed by others [59-62] (see McCarron and Nicoll, this volume), but it was later demonstrated that the ε4 allele does not appear to be an independent risk factor for CAA-related hemorrhage. Surprisingly, recent studies implicate the *APOE*- ε*2* allele as the strong factor for intracerebral hemorrhage in amyloid-laden vessels that may cause rupture of the vessel walls by inducing specific cellular changes [60,63,64].

Recent observations suggest *APOE*- ε4 allele frequency also increases the risk of dementia in stroke-survivors [65] and that the allele frequency is similarly increased in vascular dementia [66]. A report by Bronge *et al* [67] has further suggested that *APOE*- ε4 homozygotes have more extensive white matter lesions seen upon magnetic resonance imaging in the deep white matter than those with the ε3/ε3 genotype. These reports are in accord with our postmortem studies showing that the ε4 allele frequency in 36% of the AD subjects, who exhibited concomitant cerebrovascular pathology resulting from single infarcts, multiple microinfarcts, ischemic white matter lesions, or petechial hemorrhages, was significantly higher than in those without such pathology. That additive effects of peripheral vascular disease and *APOE* may also be important in AD has been implied by some studies [68-70]. We found that 77% of the patients with AD had cardiovascular disease defined by the presence of variable arteriosclerosis of the aortic blood vessels at autopsy (see also Breteler, this volume). We also reported that *APOE* frequencies in patients who exhibited cardiovascular disease were significantly different from those who did not and over 60% of the AD subjects with arteriosclerosis carried at least one *APOE*- ε4 allele. Interestingly, those who carried the *APOE*-ε4 allele were almost three times likely to have had both AD and cardiovascular disease. These findings are consistent with those of Hofman *et al* [68] implicating an interaction between carotid artery thickening and *APOE*-ε4 in AD.

APOE is considered to have more widespread effects than any other genetic factor implicated in AD but the mechanism(s) by which it exerts its effect remains largely unknown. However, several scenarios involving molecular events in brain pathology and physiological actions on the cardiovascular system have been proposed. An increased re-utilization of apolipoprotein E-lipid complexes may explain our recent finding that apolipoprotein E in CSF is decreased both in AD and vascular dementia [71,72] Both hypertension and hyperlipidaemia are major risk factors for atherosclerosis [73]. Since high serum cholesterol level during middle age was associated with an increased risk for AD in old age, it is conceivable that a part of the effect of the *APOE* ε4 allele on the risk for AD is mediated through high serum cholesterol concentrations. Alternatively, it can have direct effects on the endothelium rendering it leaky and leading to compromise of BBB functions [73]. In view of these developments, it remains to be seen whether cholesterol-lowering drugs will be useful to prevent or slowdown progression of AD. Indeed, the outcome of such an approach may help to define the nature of the interaction between apolipoprotein E, atherosclerosis [68], and ischemic white matter lesions [71,74] in the aetiology of AD.

4. MACRO AND MICRO-INFARCTION IN AD

Individuals with AD may be at increased risk for stroke and cerebral infarction, the main diagnostic features in vascular dementia. Current findings suggest that almost 35% of AD subjects bear evidence of cerebral infarction at autopsy [5,54]. Cortical lobar infarcts as well as microinfarcts, which generally tend to be more frequent and invariably localized in the temporal and occipital lobes, occur in AD (Table 1). Both remote and recent infarcts have been evident at autopsy. The microinfarcts may or may not be associated with petechial hemorrhages. Multiple microinfarcts may occur as a result of endothelial damage or luminal blockage or thrombotic events induced in microvessels. CAA may also increase the tendency of cerebral infarction in AD. Neither the strategic location nor the volume of such infarcts has been precisely determined in AD and correlated with measures of cognitive decline. Multiple lacunar infarcts, the type that define Binswanger's disease are usually not seen despite the presence of profound white matter disease in AD. It is not understood whether the isolated infarcts are a consequence of or responsible for triggering the classical pathological lesions in AD. However, certain features or conditions may rapidly precipitate strokes in AD patients. This is corroborated by circumstantial evidence suggesting that rapid development of cerebral infarctions in some AD patients occurs because of a strong interaction between severity of cerebral amyloid angiopathy, i.e. amyloid infiltration of the media and adventitia of cerebral vessels, and hypertension [5,54]. Nevertheless, the molecular and cellular mechanisms, which precipitate strokes and the significance of microvascular disease in the presence or absence of AD pathology needs yet to be elucidated. Although cerebral infarction might proceed independently of the pathogenetic process in AD, recent longitudinal clinical and epidemiological studies suggest that stroke episodes may lead to characteristic degenerative changes of AD demonstrated by the progressive onset and course of dementia [75,76]. Kokmen *et al* [76] suggested that stroke may account for as cause of AD in as many as 50% of demented cases in older age groups. Such studies have also emphasised that AD is three times likely to precipitate in the elderly after a stroke episode or a transient ischemic attack. While their substantial presence is likely to influence the processes that cause dementia in AD, prospective follow up studies are necessary to evaluate their exact role in AD, especially where pre-existing stroke episodes or cardiovascular disease are not considered as an exclusion criteria for the diagnosis of AD.

5. WHITE MATTER LESIONS AND VASCULOPATHY IN AD

Although white matter lesions may influence the course of AD there is no clear consensus to suggest that volume or localization of the lesions is predictive of AD. Ischemic white matter lesions associated with lipohyalinosis and narrowing of the lumen of the small perforating arteries as well as arterioles which nourish the deep white matter, have also been amply described in AD [74,77-80]. Neuropathological correlative studies comparing magnetic resonance imaging (MRI) findings with post-mortem neuropathological examination have determined that the hyperintense deep white matter lesions, identified in more than 60% of AD patients as well as dementia with Lewy bodies [81], consist mainly of de-myelination, reactive gliosis and arteriosclerosis [82]. In late-onset AD subjects white matter lesions are more differentially distributed than in early-onset AD [80] and in vascular dementia patients [83] and they have more severe leukoaraiosis [80]. These findings upon MRI have, however, not been confirmed by post-mortem examination. In a recent study of individuals who were found to be not demented but exhibited extensive neuropathology of AD, de la Monte [84] suggested that the white matter degeneration precedes the cortical atrophy in AD. It is known that long-standing hypertension causes lipohyalinosis and thickening of the vessel walls [85-87]. In this context, it is intriguing that AD-type pathology may be precipitated by hypertensive insults. For example, Sparks *et al* [88] found an increased amount of neurofibrillary tangles and senile plaques in the brains of non-demented individuals with hypertension. However, knowledge about risk factors for the development of ischemic white matter lesions, their progress over time, influence on the clinical course of AD and their relationship with other vascular pathologies remains to be still clarified.

6. CONCLUSIONS

Cerebrovascular abnormalities underscore the role for the blood-brain barrier in the etiopathogenesis of AD. There are profound morphological and biochemical changes such as that in the glucose transporter of the cortical microvasculature in subjects with late onset AD. In addition, amyloid β protein is involved in the degeneration of both the larger perforating arterial vessels as well as the cerebral capillaries that represent the BBB. These vascular changes not only imply that the integrity of the cerebral microvasculature is impaired in AD but may relate to long-term peripheral influences associated with cardiovascular disease or peripheral vascular disease. Genetic factors such as *APOE* and the ε4 allele may modify or attenuate cerebrovascular function during aging and increase the susceptibility to pathogenetic processes in AD. I suggest the microvascular disease causes transient and focal breach of the BBB in late onset AD and it results from an interaction of several factors, which include perivascular mediators and circulation derived damage to the endothelium. These may collectively contribute to the deterioration of cognitive functions in AD.

ACKNOWLEDGEMENTS

Supported by grants from MRC (UK), NIH (NINDS) and a Zenith award from the National Alzheimer's Association (Chicago, USA).

REFERENCES

1. Alzheimer, A. (1907) Uber eine eigenartig Erkrankung der Hirnrinde, *Allg. Z Psychiatrie Psych. Ger. Med.* **64**,146-148.
2. Kalaria, R.N. (1996) Cerebral vessels in ageing and Alzheimer's disease. *Pharm. Therap.* **72,** 193-214.
3. Vinters, H.V. (1987) Cerebral amyloid angiopathy: a critical review. *Stroke* **18,** 311-324.
4. Heyman, A., Fillenbaum, G.G., Welsh-Bohmer, K.A., Gearing, M., Mirra, S.S., Mohs, R.C., Peterson, B.L., and Pieper, C.F. (1998) Cerebral infarcts in patients with autopsy-proven Alzheimer's disease. CERAD Part XVIII. *Neurology* **51,** 159-162.
5. Premkumar, D.R.D., Cohen, D.L., Hedera, P., Friedland, R.P., and Kalaria, R.N. (1996) Apolipoprotein E ε4 alleles in cerebral amyloid angiopathy and cerebrovascular pathology in Alzheimer's disease. *Am. J. Pathol.* **148,** 2083-2095.
6. Hachinski, V., and Munoz, D. (1997) Cerebrovascular pathology in Alzheimer's disease: cause, effect or epiphenomeneon. *Ann. NY Acad. Sci.* **826,**1-6.
7. Pasquier, F., Leys, D., and Scheltens P. (1998) The influence of coincidental vascular pathology on symptomatology and course of Alzheimer's disease. *J. Neural. Transm. Suppl.* **54,** 117-27.
8. Kalaria, R.N., and Harik, S.I. (1989) Reduced glucose transporter at the blood-brain barrier and cerebral cortex in Alzheimer's disease. *J. Neurochem.* **53,** 1083-1088.
9. Frey, K.A., Minoshima, S., and Kuhl, D.E. (1998) Neurochemical imaging of Alzheimer's disease and other degenerative dementias. *Q. J. Nucl. Med.* **42,**166-178.
10. Mielke, R., and Heiss, W.D. (1998) Positron emission tomography for diagnosis of Alzheimer's disease and vascular dementia. *J. Neural. Transm. Suppl.* **53,** 237-250.

11. Kalaria, R.N, and Hedera, P. (1995) Differential degeneration of the endothelium and basement membrane of capillaries in Alzheimer's disease. *Neuroreport* **6,** 477-480.
12. Kawai, M., Kalaria, R.N., Harik, S.I., and Perry G. (1990) The relationship of amyloid plaques to cerebral capillaries in Alzheimer's disease. *Am. J. Pathol.* **37,** 1435-1446.
13. Kalaria, R.N. (1992) The blood-brain barrier and cerebral microcirculation in Alzheimer's disease *Cerebrovascular Brain Metab.* **4,** 226-260.
14. Jagust, W.J., Seab, J.P., Huesman, R.H., Valk, P.E., Mathis, C.A., Reed, B.R., Coxson, P.G., and Budinger, T.F. (1991) Diminished glucose transport in Alzheimer's disease. *Cereb. Blood Flow Metab.* **11,** 323-330.
15. Marcus, D.L., and Freedman, M.L. (1997) Decreased brain glucose metabolism in microvessels from patients with Alzheimer's disease. *Ann. NY Acad. Sci.* **826,** 248-253.
16. Stewart, P.A., Hayakawa, K., Akers, M.A., and Vinters HV. (1992) A mophometric study of the blood brain barrier in Alzheimer's disease. *Lab. Invest.* **67,** 34-742.
17. Akiyama, H., Kawamata, T., Yamada, T., Tooyama, I., Ishii, T., and McGeer, P.L. (1993) Expression of intercellular adhesion moleculae (ICAM)-1 by a subset of astrocytes in Alzheimer disease and some other degenerative neurological disorders, *Acta Neuropathol.* **85,** 628-634.
18. Frohman, E.M., Frohman, T.C., Gupta, S., de Fougerolles. A., and van den Noort, S. (1991) Expression of intercellular adhesion molecule (ICAM-1) in Alzheimer's disease. *J. Neurol. Sci.* **106,** 105-111.
19. Verbeek, M.M., Otte-Holer, I., Westphal, J.R., Wesseling, P., Ruyiter, D.J., and de Waal, R.M. (1994) Accumulation of intercellular adhesion molecule-1 in senile plaques in brain tissue of patients with Alzheimer's disease. *Am. J. Pathol.* **144,** 104-116.
20. Verbeek, M.M., Otte-Holer, I., Wesseling, P., Ruyiter, D.J., and de Waal, R.M. (1996) Differential expression of intercellular adhesion molecule-1 (ICAM-1) in the A beta-containing lesions in brains of patients with dementia of the Alzheimer type. *Acta Neuropathol.* **91,** 608-615.
21. Kalaria, R.N. (1993) The immunopathology of Alzheimer's disease. *Brain Pathol.* **3,** 333-347.
22. Lerner, A., Cohen, D. L,, Pax, A.B., Friedland, R., and Kalaria, R.N. (1993) Cell adhesion molecules in the vascular and neocortical pathology of Alzheimer's disease. *Soc. Neurosci. Abstr.* **19,** 623.
23. Wisniewski, H.M., and Kozlowski, P.B. (1982) Evidence for blood-brain barrier changes in senile dementia of the Alzheimer type (SDAT). *Ann. NY Acad. Sci.* **396,** 119-131.
24. Munoz, D.G., Erkinjuntti, T., Gaytan-Garcia, S., and Hachinski, V. (1997) Serum protein leakage in Alzheimer's disease revisited. *Ann. NY Acad. Sci.* **826,** 173-189.
25. Eikelenboom, P., and Stam, F.C. (1982) Immunoglobulins and complement factors in senile plaques. *Acta Neuropathol. (Berlin)* **57,** 239-242.
26. Licandro, A., Ferla, S., and Tavolato, B. (1983) Alzheimer's disease and senile brains: an immunofluorescence study. *Riv. Patol. Nerv. Ment.* **104,** 75-87.
27. Mann, D.M.A., Davies, J.S., Hawkes, J., and Yates, P.O. (1982) Immunohistochemical staining of senile plaques. *Neuropathol. Appl. Neurobiol.* **8,** 55-61.
28. Kalaria, R.N., Golde, T.E., Cohen, M.L., and Younkin, S.G. (1991) Serum amyloid P in Alzheimer's disease. Implications for dysfunction of the blood-brain barrier. *Ann. NY Acad. Sci.* **640,** 145-148.
29. Hardy, J., Mann, D., Wester, P., and Winblad, B. (1986) An integrative hypothesis concerning the pathogenesis and progression of Alzheimer's disease. *Neurobiol. Aging* **7,** 489-502.

30. Perlmutter, L.S., Myers, M.A., and Barrón, E. (1994) Vascular basement membrane components and the lesions of Alzheimer's disease: Light and electron microscopic analyses. *Microscopy Research and Technique* **28,** 204-215.
31. Blennow, K., Wallin, A., Fredman, P., Karlsson, I., Gottfries, C.G., and Svennerholm, L. (1990) Blood-brain barrier disturbance in patients with Alzheimer's disease is related to vascular factors. *Acta Neurol. Scand.* **81,** 349-351.
32. Hampel, H., Muller-Spahn, F., Berger, C., Haberl, A., Ackenheil, M., and Hock C. (1995) Evidence of blood-cerebrospinal fluid-barrier impairment in a subgroup of patients with dementia of the Alzheimer type and major depression: a possible indicator for immunoactivation. *Dementia* **6,** 348-354.
33. Kay, A., May, C., Papadopoulos, N., Costello, R., Atack,.J.R., Luxenberg, J.S., Cutler, N.R., and Rapoport, S.I. (1987) CSF and serum concentration of albumin and IgG in Alzheimer's disease. *Neurobiol. Aging* **8,** 21-25.
34. Mecocci, P., Parnetti, L., Reboldi, G.P., Santucci, C., Gaiti, A., Ferri, C., Gernini, I., Romagnoli, M., Cadini, D., and Senin U. (1991) Blood.brain barrier in a geriatric population: barrier function in degenerative and vascular dementia. *Acta Neurol. Scand.* **84,** 210-213.
35. Skoog, I., Wallin, A., Fredman, P., Hesse, C., Aevarsson, O., Karlsson, I., Gottfries, C.G., and Blennow, K. (1998a) A population-study on blood-brain barrier function in 85-year-olds. Relation to Alzheimer's disease and vascular dementia. *Neurology* **50,** 966-971.
36. Johansson, B.B. (1994) Pathogenesis of vascular dementia: The possible role of hypertension. *Dementia* **5,** 174-176.
37. Nag, S. (1984) Cerebral changes in chronic hypertension: Combined permeability and immunohistochemical studies. *Acta Neuropathol. (Berlin)* **62,** 178-184.
38. Kawai, M., Kalaria, R.N., Cras, P., Siedlak, S.L., Shelton, E.R., Chan, H.W., Greenberg, B.D., and Perry, G. (1993) Degeneration of amyloid precursor protein-containing smooth muscle cells in cerebral amyloid angiopathy. *Brain Res.* **623,** 142-146.
39. Perry, G., Smith, M.A., McCann, C.E., Siedlak, S.L., Jones, P.K., and Friedland,.R.P. Cerebrovascular muscle atrophy is a feature of Alzheimer's disease. *Brain Res.* **791,** 63-6.
40. Miyakawa, T. (1997) Electron microscopy of amyloid fibrils and microvessels. *Ann. NY Acad. Sci.* **826,** 25-34.
41. Kalaria, R.N., and Hedera, P. (1996) Beta-amyloid vasoactivity in Alzheimer's disease. *Lancet* **347,** 1492-1493.
42. Kalaria, R.N., and Pax, A.B. (1995) Increased collagen content of cerebral microvessels in Alzheimer's disease. *Brain Res.* **705,** 349-352.
43. Scheibel, A.B., Duong, T., and Tomiyasu, U. (1987) Denervation microangiopathy in senile dementia, Alzheimer type. *Alz. Dis. Assoc. Dis.* **1,** 19-37.
44. Claudio, L. (1996) Utrastructural features of the blood-brain barrier in biopsy tissue from Alzheimer's disease patients. *Acta Neuropathol.* **91,** 6-14.
45. Kalaria, R.N., and Kroon, S. (1992) Expression of leukocyte antigen CD34 by brain capillaries in Alzheimer's disease and neurologically normal subjects. *Acta Neuropathol.* **84,** 606-612.
46. Moody, D.M., Brown, W.R., Challa, V.R., Ghazi-Birry, H.S., and Reboussin, D.M. (1997) Cerebral microvascular alterations in aging, leukoaraiosis, and Alzheimer's disease. *Ann. NY Acad. Sci.* **826,** 103-16
47. Kalaria, R.N., Cohen, D.L., Premkumar, D.R.D., LaManna, J.C., and Lust, W.D. (1998) Vascular endothelial growth factor in Alzheimer's disease and experimental cerebral ischemia. *Mol. Brain Res.* **62,** 101-105.

48. Cohen, D. L., Hedera, P., Premkumar, D.R.D., Friedland, R.P., and Kalaria, R.N. (1997) Amyloid-β angiopathies masquerading as Alzheimer's disease. *Ann. N Y Acad. Sci.* **826,** 390-395.
49. Kalaria, R.N., Pax, A.B,, Premkumar, D.R.D., and Lieberburg, I. (1996) Production and increased detection of amyloid β-protein and amyloidogenic fragments in brain microvessels, meningeal vessels and choroid plexus in Alzheimer disease. *Mol. Brain Res.* **35,** 58-68.
50. Gray, F., Chretien, F., Cesaro, P., Chatelain, J., Beaudry, P., Laplanche, J.L., Mikol, J., Bell, J., Gambetti, P., and Degos, J.D. (1994) Creutzfeldt-Jakob disease and cerebral amyloid angiopathy. *Acta Neuropathol.* **88,** 106-111.
51. de la Torre, J.C. (1997) Hemodynamic consequences of deformed microvessels in the brain in Alzheimer's disease. *Ann. NY Acad. Sci.* **826,** 75-91.
52. Weller, R.O., Massey, A., Newman, T.A., Hutchings, M., Kuo, Y.M., and Roher, A.E. (1998) Cerebral amyloid angiopathy: amyloid beta accumulates in putative interstitial fluid drainage pathways in Alzheimer's disease. *Am. J. Pathol.* **153,** 725-733.
53. De Jong, G.I., De Vos, R.A.I,, Jansen Steur, E.N.H., and Luiten, P.G.M. (1997) Cerebrovascular hypoperfusion: a risk factor for Alzheimer's disease? *Ann. NY Acad. Sci.* **826,** 56-74.
54. Olichney, J.M., Hansen, L.A., Hofstetter, C.R., Grundman, M., Katzman, R., and Thal, L.J. (1995) Cerebral infarction in Alzheimer's disease is associated with severe amyloid angiopathy and hypertension. *Arch. Neurol.* **52,** 702-708.
55. Haan, J., Lanser, J.B.K., Zijderveld, I., van der Does, I.G.F., and Roos, R.A.C. (1990) Dementia in hereditary cerebral hemorrhage with amyloidosis-Dutch type. *Arch. Neurol.* **47,** 965-968.
56. Wilson, P.W.F., Myers, R.H., Larson, M.G., Ordovas, J.M., Wolf, P.A., and Schaefer, E.J. (1994) Apolipoprotein E alleles, dyslipidemia, and coronary heart disease: the Framingham offspring study. *JAMA* **272,** 1666-1671.
57. Davignon, J., Gregg, R.E., and Sing, C.F. (1988) Apolipoprotein E polymorphism and atherosclerosis. *Arteriosclerosis* **8,** 1-21.
58. Kalaria, R.N., and Premkumar, D.R.D. (1995) Apolipoprotein E genotype and cerebral amyloid angiopathy In Alzheimer's disease. *Lancet* **346,** 1424.
59. Greenberg, S.M., Vonsattel, J.P., Rebeck, G.W., and Hyman, B.T. (1995) Apolipoprotein E epsilon 4 and cerebral hemorrhage associated wirth amyloid angiopathy. *Ann. Neurol.* **38,** 254-259.
60. Nicoll, J.A., Burnett, C., Love, S., Graham, D.I., Dewar, D., Ironside, J.W., Stewart, J., and Vinters, H.V. (1997) High frequency of apolipoprotein epsilon 2 allele in hemorrhage due to cerebral amyloid angiopathy. *Ann. Neurol.* **41,** 716-721.
61. Olichney, J.M., Hansen, L.A., Hofstetter, C.R., Grundman, M., Katzman, R., and Thal, L.J. (1996) Apolipoprotein epsilon 4 allele is associated with increased neuritic plaques and cerebral amyloid angiopathy in Alzheimer's disease and Lewy body variant. *Neurology* **47,** 190-196.
62. Zarow, C., Zaias, B., Lyness, S.A., and Chui H. (1999) Cerebral amyloid angiopathy in Alzheimer's disease is associated with apolipoprotein E4 and cortical neuron loss. *Alz. Dis. Assoc. Disord.* **13,** 1-8.
63. Greenberg, S.M., Vonsattel, J.P., Segal, A.Z., Chiu, R.I., Clatworthy, A.E., Liao, A., Hyman, B.T., and Rebeck, G.W. (1998) Association of apolipoprotein E epsilon 2 and vasculopathy in cerebral amyloid angiopathy. *Neurology* **50,** 961-965.

64. McCarron, M.O., and Nicoll, J.A. (1998) High frequency of apolipoprotein E epsilon 2 allele is specific for patients with cerebral amyloid angiopathy-related hemorrhage. *Neurosci Lett.* **247,** 45-48.
65. Margaglione, M., Seripa, D., Gravina, C. *et al.* (1998) Prevalence of apolipoprotein E alleles in healthy subjects and survivors of ischemic stroke: an Italian Case-Control Study. *Stroke* **29,** 399-403.
66. Frisoni, G.B., Calabresi, L., Geroldi, C., Bianchetti, A., D'Acquarica, A.L., Govoni, S., Trabucchi, M., and Franceschini, G. (1994) Apolipoprotein E ε 4 allele in Alzheimer's disease and vascular dementia. *Dementia* **5,** 240-242.
67. Bronge, L., Fernaeus, S.E., Blomberg, M., Ingelson, M., Lannfelt, L., Isberg, B., and Wahlund, L.O. (1999) White matter lesions in Alzheimer patients are influenced by apolipoprotein E genotype. *Dement. Geriatr. Cogn. Disord.* **10,** 89-96.
68. Hofman, A., Ott, A., Breteler, M.M.B., Bots, M.L., Slooter, A.J.C., van Harskamp, F., van Duijn, C.N., Van Broeckhoven, C., and Grobbee, D.E. (1997) Atherosclerosis, apolipoprotein E, and the prevalence of dementia and Alzheimer's disease in the Rotterdam Study. *Lancet* **349,** 151-154.
69. Kalaria, R.N. (1997) Apolipoprotein E, arteriosclerosis and Alzheimer's disease. *Lancet* **349,** 1174-1175.
70. Kosunen, O., Talasniemi, S., Lehtovirta, M., Heinonen, O., Helisalmi, S., Mannermaa, A., Paljarvi, L., Ryynanen, M., Riekkinen, P.J. Sr, and Soininen, H. (1995) Relation of coronary atherosclerosis and apolipoprotein E genotypes in Alzheimer patients. *Stroke* **26,** 743-748.
71. Skoog, I., Hesse, C., Fredman, P., Andreasson, L-A., Palmertz, B., and Blennow, K. (1997) Apolipoprotein E in cerebrospinal fluid in 85-year-olds. Relation to dementia, apolipoprotein E polymorphism, cerebral atrophy, and white-matter lesions. *Arch. Neurol.* **54,** 267-272.
72. Skoog, I., Hesse, C., Aevarsson, O., Landahl, S., Wahlström, J., Fredman, P., and Blennow, K. (1998b) A population study of Apo E genotype at the age of 85: relation to dementia, cerebrovascular disease and mortality. *J. Neurol. Neurosurg. Psychiatr.* **64,** 37-43.
73. Notkola, I-L., Sulkava, R., Pekkanen, J., Erkinjuntti, T., Ehnholm, C., Kivinen, P., Tuomilehto, J., and Nissinen, A. (1998) Serum total cholesterol, apolipoprotein E e4 allele, and Alzheimer's disease. *Neuroepidemiology* **17,** 14-20.
74. Wallin, A. (1998) The overlap between Alzheimer's disease and vascular dementia: the role of white matter changes. *Dement. Geriatr. Cogn. Disord.* **Suppl 1,** 30-35.
75. Henon, H., Pasquier, F., Durieu, I., Godefroy, O., Lucas, C., Lebert, F., and Leys D. (1997) Pre-existing stroke in patients: baseline frequency, associated factors and outcome. *Stroke* **28,** 2429-2436.
76. Kokmen, E., Whisnant, J.P., O'Fallon, W.M,, Chu, C.P., and Beard, C.M. (1996) Dementia after ischemic stroke: a population based study in Rochester, Minnesota (1960-1984). *Neurology* **46,** 154-159.
77. Diaz, F,J., Merskey, H., Hachinski, V.C., Lee, D.H., Boniferro, M., Wong, C.J., Mirsen, T.R., and Fox, H. (1991) Improved recognition of leukoaraiosis and cognitive impairment in Alzheimer's disease. *Arch. Neurol.* **48,** 1022-1025.
78. Englund, E. (1998) Neuropathology of white matter changes in Alzheimer's disease and vascular dementia. *Dement. Geriatr. Cogn. Disord.* 9 Suppl **1,** 6-12.
79. Leys, D., Pruvo, J.P,, Parent, M., Vermersch, P., Soetaert, G., Steinling, M., Delacourte, A., Défossez, A., Rapoport, A., Clarisse, J., and Petit, H. (1991) Could Wallerian

degeneration contribute to "leuko-araiosis" in subjects free of any vascular disorder? *J. Neurol. Neurosurg. Psychiatr.* **54,** 46-50.

80. Scheltens, P., Barkhof, F., Valk, J., Algra, P.R., Gerritsen van der Hoop, R., Nauta, J., Wolters, E.C. (1992) White matter lesions on Magnetic Resonance Imaging in clinically diagnosed Alzheimer's disease. Evidence for a heterogeneity. *Brain* **115,** 735-748.
81. Barber, R., Scheltens, P., Gholkar, A., Ballard, C., McKeith, I., Ince, P., Perry, R., and O'Brien, J. (1999) White matter lesions on magnetic resonance imaging in dementia with lewy bodies, Alzheimer's disease, Vascular dementia, and normal aging, *J. Neurol. Neurosurg. Psych.* **67,** 66-72.
82. Van Gijn, G. (1998) Leukoariosis and vascular dementia. *Neurology* **51(suppl 3),** S3-S8.
83. Wahlund, L.O., Basun, H., and Almkvist, O., *et al.* (1994) White matter hyperintensities in dementia: does it matter? *Magn. Reson. Imaging* **12,** 387-394.
84. de la Monte, S.M. (1989) Quantitation of cerebral atrophy in preclinical and end-stage Alzheimer's disease. *Ann. Neurol.* **25,** 450-459.
85. De Reuck, J., Crevits, L., De Coster, W., Sieben, G., and vander Eecken, H. (1980) Pathogenesis of Binswanger chronic progressive subcortical encephalopathy. *Neurology* **30,** 920-928.
86. Román, G.C. (1987) Senile dementia of the Binswanger type. A vascular form of dementia in the elderly. *JAMA* **258,** 1782-1788.
87. Skoog, I. (1997) The relationship between blood pressure and dementia: a review. *Biomed. Pharmacother.* **51,** 367-375.
88. Sparks, D.L,, Hunsaker, III J.C., Scheff, S.W., Kryscio, R.J., Henson, J.L., and Markesbery, W.R. (1990) Cortical senile plaques in coronary artery disease, aging and Alzheimer's disease. *Neurobiol. Aging* **11,** 601-607.

Chapter 12

Aβ-ASSOCIATED PROTEINS IN CEREBRAL AMYLOID ANGIOPATHY

ROBERT M.W. DE WAAL* and MARCEL M. VERBEEK*[†]
*Departments of Pathology and [†]Neurology, University Hospital Nijmegen, Nijmegen, The Netherlands

Like senile plaques, cerebral amyloid angiopathy (CAA) does not result from accumulation of the amyloid β (Aβ) protein alone, but also contains a number of other components, referred to as amyloid β-associated proteins. In this review we will discuss which proteins can be found in the cerebral vessel walls that are affected by CAA, what the possible source of these proteins is, and to what extent they might contribute to the pathogenesis of CAA. The data generated in our laboratory are in agreement with a pathogenesis of CAA that is different from that of the amyloidosis in senile plaques, although part of the amyloid-associated proteins in both these lesions may have the same parenchymal source. Since vascular malfunction is an independent risk factor for developing Alzheimer's disease (AD), and CAA formation may depend on the interactions of Aβ with Aβ-associated proteins, pharmacological interference with these interactions may provide a new approach for treatment of AD and related cerebral vasculopathies.

1. INTRODUCTION

After the initial identification of the amyloid β protein (Aβ) as the major component of cerebrovascular amyloid and, later, of senile plaques, it was discovered that a number of other proteins were deposited in these lesions. Attention mainly focussed on senile plaques, resulting in numerous reports on the presence of a variety of plaque-associated proteins. A few groups of such proteins can be distinguished: inflammatory proteins (including several complement factors and their inhibitors), acute phase proteins (such as α_1-antichymotrypsin), and others like heparan sulfate proteoglycans and apolipoprotein E (ApoE). It is assumed that most of these proteins are deposited as a reaction to amyloid accumulation, but for others a role in the

M.M. Verbeek et al. (eds.),
Cerebral Amyloid Angiopathy in Alzheimer's Disease and Related Disorders, 207–221.

pathogenesis of Alzheimer's disease (AD) has been implied. This was based on various experimental and epidemiological findings, like the ability of these proteins to bind to Aβ *in vitro* and interfering with fibril formation, early appearance of the plaque-associated protein in diffuse plaques, and identification of the protein as a risk factor for AD, e.g. ApoE.

Since there are indications that the pathogenesis of cerebral amyloid angiopathy (CAA) is different from that of senile plaque formation with respect to Aβ isoform composition, the cell types involved in the pathogenesis and the mechanisms of cellular damage [1] it is obvious that differences in expression of Aβ-associated proteins may also occur. In this chapter data on the presence of Aβ-associated proteins in CAA will be summarized and discussed.

2. DISTRIBUTION OF Aβ-ASSOCIATED PROTEINS IN CAA

A large number of Aβ-associated proteins have been identified in senile plaques. These can be categorized as complement factors, inflammatory and acute phase proteins and a number of miscellaneous proteins. In our laboratory we recently systematically studied the expression of these Aβ-associated proteins in cortical vessels affected by CAA that were present in 9 AD brains [2]. In this brain material, leptomeningeal amyloid was present in 7 cases as well. No CAA could be detected in any of the control brains by the Aβ and Congo red stains used. Table 1 provides an overview of the Aβ-associated proteins detected by immunohistochemistry. Congo red staining of amyloid-laden vessels was generally positive, indicating that unfibrillized Aβ was rarely present and that most of the vessels investigated were in an advanced state of CAA formation. In Congo red-negative, Aβ-positive vessels, however, all Aβ-associated proteins that colocalized with CAA in Congo red-positive vessels were present as well.

2.1 Complement factors in CAA.

In line with other reports [3-6] we observed that the complement factors C3, C4d, C5b-9 and clusterin were consistently present in CAA. C1q staining was rather variable, not only among different cases, but also among vessels in an individual case. This is in accordance with previous reports on C1q expression [3-5] and this variability may be explained by the fact that C4d and C3c remain covalently bound to the brain tissue after activation of the complement cascade, whereas this is not the case for C1q. Although C1q binds to aggregated Aβ [7,8], a balance may exist between the production and removal of C1q, resulting in the observed variable expression pattern. Expression of C4-binding protein has been reported in CAA [4]. No staining has been observed for the complement factors C1s, C5 and C1-inhibitor [9].

2.2 Inflammatory and acute phase proteins in CAA.

In contrast to their reported expression in senile plaques [10-13] the acute phase proteins α_1-antichymotrypsin (α_1-ACT) and α_2-macroglobulin and also lactoferrin were not or were barely detectable in vascular amyloid. In the literature conflicting results on the expression of α_1-ACT in CAA have been reported however, varying from no expression to different degrees of expression [12,14]. Our data are in line with those of Rozemuller *et al* [15], indicating that the acute phase proteins α_1-ACT and α_2-macroglobulin are not colocalized with CAA. α_2-Macroglobulin staining was observed decorating the inner and outer basal membranes of CAA-containing vessels. Similarly, staining for the cytokine-induced intercellular adhesion molecule-1 (ICAM-1) was confined to vascular endothelial cells and perivascular pericytes or smooth muscle cells [16] and did not co-localize with Aβ deposits in CAA.

2.3 Miscellaneous proteins

A number of Aβ-associated proteins that do not belong to the above-mentioned categories have been identified in senile plaques. This group comprises, among many others, ApoE, heparan sulfate proteoglycans (HSPGs), amyloid P component, the "non-Aβ component of AD amyloid" (NAC, a proteolytic fragment of α-synuclein), cystatin C (γ-trace protein) and the amyloid precursor protein (APP). In agreement with other reports, we observed strong and consistent expression of ApoE and HSPGs in CAA [17-20]. Since both ApoE and HSPGs may accelerate Aβ fibril formation, these factors may be very important in the development of CAA (see below). Also, amyloid P component was consistently expressed in CAA, confirming previous studies [21,22]. NAC was reported to be expressed in both senile plaques and CAA [23], and since it is amyloidogenic by itself [24], it may play a role in the formation of vascular amyloid. Cystatin C, also referred to as gamma-trace protein, is the amyloidogenic molecule responsible for hereditary cerebral hemorrhage with amyloidosis–Icelandic type (HCHWA-I) (see Ólafsson and Thorsteinsson, this volume). Remarkably, it is also present as an amyloid-associated component in CAA in AD [25,26]. Since it has cysteine protease inhibitory activity, it may have a role in protecting the Aβ deposits from proteolytic degradation. Expression of APP in colocalization with amyloid fibrils has been found in CAA both in AD and in hereditary cerebral hemorrhage with amyloidosis-Dutch type (HCHWA-D), which has led the authors to suggest a vascular source of the Aβ in both types of disease [27].

2.4 Aβ-associated proteins in HCHWA-D.

Vascular amyloidosis is a defining feature of HCHWA-D (see Maat-Schieman *et al*, this olume). Investigations into the occurrence of Aβ-associated proteins in the vascular lesions in HCHWA-D have been directed at extracellular matrix (ECM) components. Collagen I and III, HSPG and fibronectin were associated with amyloid deposition in cerebral vessels [28]. In CAA in AD, arterial vessels heavily laden with amyloid often showed a lack of immunostaining for collagen type I, III, V and VI in the affected vessel parts [29]. Also actin staining was diminished, indicating a loss of muscular cells, which has also been observed in CAA in AD (see Verbeek *et al*, this volume). Although these vessels often show secondary fibrosis, this may explain the susceptibility of these vessels to rupture. The cause of vessel weakening in HCHWA-D seems not related to the absence of basal membrane components, but the extensive loss of smooth muscle cells will obviously have adverse effects on vessel wall compliance.

3. COMPARISON WITH Aβ-ASSOCIATED FACTORS IN SENILE PLAQUES

The fact that in all CAA cases senile plaques were also present allowed us to make a comparison of Aβ-associated proteins present in the two types of lesion. These staining results are summarized in Table 1. All antibodies that gave positive staining for CAA also stained senile plaques, as expected, although there were considerable differences in intensity. Conversely, antibodies directed against α_1-ACT, α_2-macroglobulin, lactoferrin and ICAM-1 gave positive staining of senile plaques, whereas these proteins were not detectable in CAA. The staining results that we obtained in senile plaques were in line with those in the literature [10-12,12-15,20,21,30-33]; an overview is given by Verbeek *et al* [1].

Perivascular Aβ deposits were stained for the entire panel of Aβ-associated factors, including α_1-ACT and ICAM-1, which were not detected in CAA. These deposits are probably more similar to senile plaques than to CAA and therefore similar reactions to Aβ production may occur in both types of lesion. In agreement with this view, microglial cells cluster around approximately 50% of the vessels with perivascular Aβ deposition (Figure 1), in a way similar to that described for neuritic senile plaques [34-36]. Thus, the characteristics of perivascular Aβ deposits resemble very much those of senile plaques and they could be regarded as a "combination lesion" of both CAA and senile plaques.

Table 1: Intensity of staining for Aβ-associated proteins in Aβ-containing lesions of Alzheimer's disease and the approximate percentage of senile plaques stained

Antigen	CAA in cortex[1]	CAA in LM[1]	PA in cortex	Control vessels	Amyloid plaques	Non-amyloid plaques	% SPs stained
Aβ	+++	+++	++	-	+++	++	100
C1q	+ / +++[2]	+/++	+	±	++	+	50-75
C3c	++	++	+	+	+ / ++	+	50-75
C4d	+++	++	+++	±	+++	++ / +++	100
C5b-9	++	++	+	±	+	± / +	50-75
Clusterin[3]	++	++	+	±	+ / ++	+	50-75
α_1-ACT	±	±	+	±	+ / ++	+	50-75
α_2-M	±[4]	±	ND	+	±	±	25-50
HSPG[5]	+++	++	+	++	++	+	75-100
ApoE	+++	+++	++	±	++ / +++	++	75-100
Amyloid P	++	++	+	+	+	± / +	50-75
Lactoferrin	±	±	+	±	+	+	50-75
ICAM-1	-[6]	-	+	±	+	+	75-100

Staining intensities: -, absent; ±, weak; +, moderate; ++, strong; +++, very strong. [1] Indicated is the intensity of staining observed in the amyloid. [2]Staining intensities may vary in different A9-containing vessels. [3]Also referred to as SP40,40 or ApoJ. [4]Accentuation of the inner and outer basement membranes. [5]Identified with an antibody directed against heparitinase-digested HSPG. [6] Endothelial cells and cells in a perivascular position are stained.
Abbreviations: CA, cerebrovascular amyloid; PA, perivascular amyloid β deposits; LM, leptomeninges; α_2-M, α_2-macroglobulin; ND: not determined; SPs, senile plaques. Reproduced with permission from the publisher from: M.M. Verbeek *et al* (1998) *Acta Neuropathol.* **96**: 628-636.

In summary, in comparison with reports on the expression of these factors in the literature, concensus has emerged on the presence of ApoE, HSPGs, several complement factors and their inhibitors and amyloid P component in CAA. However, it seems likely that Aβ-deposition in the cerebral vasculature does not lead to an acute phase reaction involving the excessive production and successive co-deposition of α_1-ACT, α_2-macroglobulin and ICAM-1 as is the case in senile plaques [31]. Since the production of these molecules can be induced by the cytokines interleukin (IL)-6 [37,38] or IL-1 [39,40], it seems likely that production of these cytokines is low or absent in the vascular wall. Certain types of senile plaques contain IL-6 and IL-1-positive microglial cells [11,41,42], suggesting involvement of these interleukins in senile plaque formation. In contrast to the recruitment or activation of microglial cells in classic senile plaques, we did not detect a change in the number of microglial cells around CAA compared to control vessels, suggesting that cells of the monocyte / macrophage lineage cells are not recruited to induce an inflammatory reaction in CAA (Figure 1).

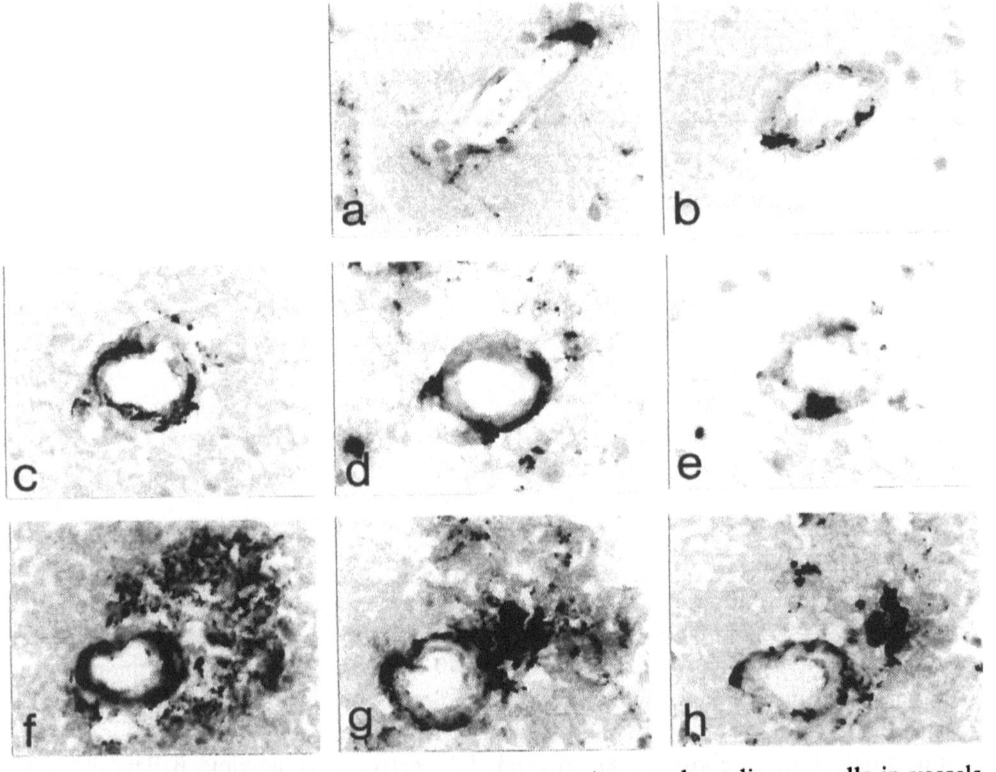

Figure 1. Immunohistochemical staining of monocyte / macrophage lineage cells in vessels with and without amyloid β deposition and in vessels with perivascular Aβ-deposition. Staining of brain tissue of a non-demented control with anti-CD68 antibody KP1 (A) and with the lysosomal marker 25F9 (B) identifies a number of vessel-associated cells. Staining for Aβ (C,F), KP1 (D,G) and 25F9 (E,H) of serial sections of DAT brain tissue reveals that a comparable number of KP1- or 25F9-immunopositive cells are associated with Aβ-containing vessels (C-E). An increase in KP1- or 25F9-immunoreactive cells could be observed around vessels with perivascular Aβ deposits (DA, panel F-H). Magnifications: x 130. Reproduced with permission from the publisher from: M.M. Verbeek *et al* (1998) *Acta Neuropathol.* **96**: 628-636.

4. SOURCE OF THE Aβ-ASSOCIATED PROTEINS

There are a number of possible cellular sources for Aβ-associated proteins that accumulate in CAA, e.g. parenchymal cells (astrocytes, microglia, neurons), vascular cells (endothelial cells, pericytes, smooth muscle cells) and a luminal source (Figure 2). Amyloid deposition in senile plaques is often associated with microglial cell activation [34-36]. These cells may therefore be the source of Aβ-associated proteins that are present in these classic senile plaques. However, Aβ-associated proteins deposited in or around diffuse plaques, where clustering of activated microglial cells is not observed, likely have a different source, e.g. astrocytes or neurons. Aβ-associated proteins in CAA may originate from both these sources, and in addition, from the vascular endothelium and the perivascular smooth muscle cells or pericytes, or even from the circulation. Accumulation of perivascular activated microglial cells around vessel-associated parenchymal amyloid was reported previously [43], but this is not identical to CAA. We found that the intensity and number of cells, belonging to the monocyte/macrophage lineage, in association with amyloid-containing vessels of AD was comparable to the staining observed around unaffected vessels of both control and AD brains [2]. In HCHWA-D, on the other hand, cells of the monocyte/ macrophage lineage colocalized with arterial Aβ. The presence of these cells appeared continuous with perivascular cells and perivascular microglial clustering [44]. Taken together, the available data do not point unanimously to perivascular activated cells of the monocyte lineage as a source of Aβ-associated proteins.

Alternatively, Aβ-associated proteins in CAA may be derived from vascular pericytes or smooth muscle cells. Indeed, ApoE is synthesized by these cells *in vitro*, as are APP, α_1-ACT, HSPG and C1q [45], but it is difficult to extrapolate such data to the situation *in vivo*. In addition, the amounts of Aβ secreted by these cells are minute, indicating that at least Aβ itself is probably not originating only from the vessel wall. On the other hand, we found relatively high amounts of the late complement components C5b-9 (membrane attack complex) in CAA compared to senile plaques, which seems to indicate a local production and activation of complement factors. Very recently, neuronal expression of C1q after ischemia was described in an animal model. This indicates that this cell type may be the source of C1q after Aβ-induced neuronal damage as well [46]. Therefore, most of the above-mentioned Aβ-associated proteins may have multiple sources, probably with an exception for the extracellular matrix (ECM) components of the basal lamina. The proteins of the ECM may contribute to CAA, since early Aβ deposits in CAA have been identified close to the basement membrane [47] and interactions between basal membranes and

APP and Aβ have been reported [48]. ECM components may contribute to Aβ immobilization in the vessel wall, and may be regarded as Aβ-associated proteins, but only excessive production of ECM components can be regarded as a reaction to Aβ deposition.

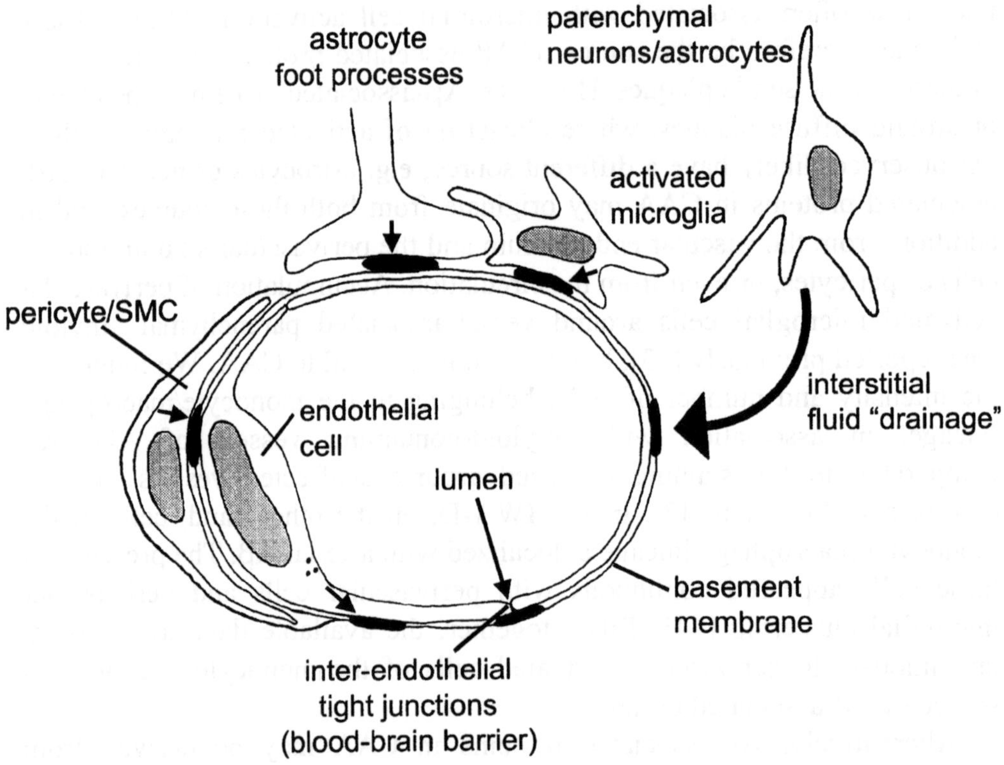

Figure 2. A schematic overview of the possible sources of Aβ-associated proteins (shown in black) in CAA.

A luminal origin of both Aβ and the Aβ-associated proteins seems unlikely due to the presence of the blood-brain barrier and the absence of vascular amyloid in peripheral tissues. Recently, it has been shown that proteins sythesized in the brain parenchyma may be carried by the interstitial fluid drainage towards the periarterial space. This mechanism was suggested for Aβ itself [49], but may also apply to Aβ-associated proteins and thus presents an attractive explanation for the fact that most of the Aβ-associated proteins in senile plaques and CAA are identical.

On the other hand, local factors must play a role as well, in view of the absence of acute phase reactants and activated microglial cells, and of the increased levels of the membrane attack complex. In addition, as discussed elsewhere (Verbeek *et al*, this volume) in response to the cytotoxic effects of Aβ, smooth muscle cells and pericytes produce increasing amounts of APP and Aβ, a mechanism that may also apply to other Aβ-associated proteins. Assuming that transport of factors produced within the central nervous system parenchyma towards the vasculature takes place, it remains puzzling that the acute phase proteins would be an exception.

5. ROLE OF Aβ-ASSOCIATED PROTEINS IN CAA PATHOGENESIS

There are two possible roles for Aβ-associated proteins in the pathogenesis of CAA: they may trigger or facilitate Aβ deposition and/or fibrillization, or, as a secondary effect, increase Aβ-induced cytotoxicity and tissue damage. Theoretically, they may even have no effect. The early appearance of the Aβ-associated factors ApoE and HSPG in vessels in patients affected by CAA, even before Congo red positivity becomes manifest, is a clear indication that these factors play a role in CAA pathogenesis and are not merely an epiphenomenon [17,18]. Data obtained in aged monkeys also support an early role of ApoE in the development of CAA [50]. Both ApoE and HSPG bind to and may accelerate Aβ fibril formation [51,52]. Therefore, it is likely that these factors shift the balance between non-fibrillar and fibrillar Aβ in the cerebral vasculature towards the latter form. As already stated above, a role in the process of Aβ accumulation in the vascular wall may also be assigned to components of the ECM [47,48]. Also, HSPG constituents of the basal membrane may facilitate Aβ immobilisation and subsequent fibrillization. In contrast, it has been found that the basal membrane component laminin inhibits Aβ-peptide fibrillization [53] which, of course, does not preclude a role in binding and targeting Aβ to the extracellular matrix in the vessel wall.

It has been demonstrated that Aβ is toxic for smooth muscle cells and pericytes in the vascular wall (see Verbeek *et al*, this volume). It seems obvious that a certain level of damage must occur before secondary reactions take place that may contribute to cellular damage, because stimuli must be released to attract inflammatory cells or to activate microglia or other cell types. When activated, however, these pathways may increase the damage considerably, for instance, by the activation of the complement pathway, or the generation of free oxygen radicals. For example, it has been shown that Aβ binds and activates C1q [7,8,54], and this process may trigger complement-mediated cell damage. Since many complement factors, like C1q, C3, C4 and C5b-9, are expressed in CAA, the complement pathway especially may be involved in increasing cell death of SMCs in CAA.

6. ROLE OF APOE AS A RISK FACTOR IN CAA

Apart from the early appearance of ApoE in Aβ-containing vessels walls, it has been demonstrated that the ApoE ε4 phenotype is associated with increased deposition of Aβ in CAA [55,56]. Both the ApoE ε4 and ε2 alleles are risk factors for the development of CAA and CAA-related hemorrhages [57-59] (see McCarron and Nicoll, this volume). ApoE may therefore be involved in the pathogenesis of CAA, possibly by enhancing fibril formation [60]. Together with the conditions created by the presence of basal lamina material, this may favor amyloid accumulation. It is not clear what the source of the ApoE deposited in the vessel wall may be. As mentioned above there are several possibilities: delivery from the blood, local production in the cerebral vasculature or drainage from the cerebral compartment to CAA. With regard to the first possibility, it was demonstrated (Zlokovic *et al*, this volume) that ApoE has limited transport across the BBB, indicating that the ApoE found within the brain is produced locally [61]. This cerebral pool of ApoE may be either produced by cells of the cerebral vasculature, such as smooth muscle cells and pericytes [45], or by cells residing in the brain parenchyma, such as astrocytes or neurons [62]. On the other hand, lipids must enter the cerebral compartment, and they are undoubtely accompanied by chaperone molecules such as apolipoproteins and actively transported into the brain.

7. Aβ-ASSOCIATED PROTEINS AS POSSIBLE TARGETS FOR THERAPY

Since Aβ-associated proteins may contribute to Aβ accumulation and fibrillization, they may also be a possible target for therapeutic agents. As the role of CAA in AD becomes increasingly clear, Aβ-associated proteins in the vessel wall should be considered for a possible therapeutic approach. Molecules that target Aβ-associated proteins must fulfill the following criteria: they must be able to traverse the blood-brain barrier formed by endothelial tight junctions, they must inhibit interactions between Aβ and the Aβ-associated proteins, and preferably, be able to dissociate Aβ amyloid fibrils. Identifying specific amino acid sequences on Aβ-associated proteins that mediate the actual interactions with Aβ, as are known (for example) for Aβ-ApoE interactions, will probably be of great importance for the development of fibrillization antagonists.

ACKNOWLEDGMENTS

This work has been financially supported by grants from the Internationale Stichting Alzheimer Onderzoek (ISAO). The authors thank Irene Otte-Höller for immunohistochemistry and preparation of the illustrations.

REFERENCES

1. Verbeek, M.M., Eikelenboom, P., and de Waal, R.M.W. (1997) Differences between the pathogenesis of senile plaques and congophilic angiopathy in Alzheimer's disease, *J. Neuropathol. Exp. Neurol.* **56**, 751-761.
2. Verbeek, M.M., Otte-Höller, I., Veerhuis, R., Ruiter, D.J., and de Waal, R.M.W. (1998) Distribution of Aβ-associated proteins in cerebrovascular amyloid of Alzheimer's disease, *Acta. Neuropathol.* **96**, 628-636.
3. Eikelenboom, P. and Veerhuis, R. (1996) The role of complement and activated microglia in the pathogenesis of Alzheimer's disease, *Neurobiol. Aging* **17**, 673-680.
4. Kalaria, R.N. and Kroon, S.N. (1992) Complement inhibitor C4-binding protein in amyloid deposits containing serum amyloid P in Alzheimer's disease, *Biochem. Biophys. Res. Commun.* **186**, 461-466.
5. Eikelenboom, P. and Stam, F.C. (1984) An immonuhistochemical study on cerebral vascular and senile plaque amyloid in Alzheimer's dementia, *Virchows Arch [Cell Pathol]* **47**, 17-25.
6. Kida, E., Choi Miura, N.H., and Wisniewski, K.E. (1995) Deposition of apolipoproteins E and J in senile plaques is topographically determined in both Alzheimer's disease and Down's syndrome brain, *Brain Res.* **685**, 211-216.
7. Jiang, H., Burdick, D., Glabe, C.G., Cotman, C.W., and Tenner, A.J. (1994) β-Amyloid activates complement by binding to a specific region of the collagen-like domain of the C1qA chain, *J. Immunol.* **152**, 5050-5059.
8. Snyder, S.W., Wang, G.T., Barrett, L., Ladror, U.S., Casuto, D., Lee, C.M., Krafft, G.A., Holzman, R.B., and Holzman, T.F. (1994) Complement C1q does not bind monomeric β-amyloid, *Exp. Neurol.* **128**, 136-142.
9. Eikelenboom, P., Hack, C.E., Rozemuller, J.M., and Stam, F.C. (1989) Complement activation in amyloid plaques in Alzheimer's dementia, *Virchows Arch. B. Cell Pathol. Incl. Mol. Pathol.* **56**, 259-262.
10. Zhan, S.-S., Veerhuis, R., Kamphorst, W., and Eikelenboom, P. (1995) Distribution of beta amyloid associated proteins in plaques in Alzheimer's disease and in the non-demented elderly, *Neurodegeneration* **4**, 291-297.
11. Strauss, S., Bauer, J., Ganter, U., Jonas, U., Berger, M., and Volk, B. (1992) Detection of interleukin-6 and α-2-macroglobulin immunoreactivity in cortex and hippocampus of Alzheimer's disease patients, *Lab. Invest.* **66**, 223-230.
12. Abraham, C.R., Selkoe, D.J., and Potter, H. (1988) Immunochemical identification of the serine protease inhibitor alpha-1-antichymotrypsin in the brain amyloid deposits of Alzheimer's disease, *Cell* **52**, 487-501.
13. Kawamata, T., Tooyama, I., Yamada, T., Walker, D.G., and McGeer, P.L. (1993) Lactotransferrin immunocytochemistry in Alzheimer and normal human brain, *Am. J. Pathol.* **142**, 1574-1585.

14. Shoji, M., Hirai, S., Yamaguchi, H., Harigaya, Y., Ishiguro, K., and Matsubara, E. (1991) Alpha 1-antichymotrypsin is present in diffuse senile plaques, *Am. J. Pathol.* **138**, 247-257.
15. Rozemuller, J.M., Abbink, J.J., Kamp, A.M., Stam, F.C., Hack, C.E., and Eikelenboom, P. (1991) Distribution pattern and functional state of α1-antichymotrypsin in plaques and vascular amyloid in Alzheimer's disease, *Acta. Neuropathol.* **82**, 200-207.
16. Verbeek, M.M., Otte-Höller, I., Wesseling, P., Ruiter, D.J., and de Waal, R.M.W. (1996) Differential expression of intercellular adhesion molecule-1 (ICAM-1) in the Aβ-containing lesions in brains of patients with dementia of the Alzheimer type, *Acta. Neuropathol.* **91**, 608-615.
17. Yamaguchi, H., Ishiguro, K., Sugihara, S., Nakazato, Y., Kawarabayashi, T., Sun, X., and Hirai, S. (1994) Presence of apolipoprotein E on extracellular neurofibrillary tangles and on meningeal blood vessels precedes the Alzheimer β-amyloid deposition, *Acta. Neuropathol.* **88**, 413-419.
18. Schmechel, D.E., Saunders, A.M., Strittmatter, W.J., Crain, B.J., Hulette, C.M., Joo, S.H., Pericak-Vance, M.A., Goldgaber, D., and Roses, A.D. (1993) Increased amyloid β-peptide deposition in cerebral cortex as a consequence of apolipoprotein E genotype in late-onset Alzheimer disease, *Proc. Natl. Acad. Sci. USA* **90**, 9649-9653.
19. Snow, A.D., Mar, H., Nochlin, D., Sekiguchi, R.T., Kimata, K., Koike, Y., and Wight, T.N. (1990) Early accumulation of heparan sulfate in neurons and in the beta-amyloid protein-containing lesions of Alzheimer's disease and Down's syndrome, *Am. J. Pathol.* **137**, 1253-1270.
20. Snow, A.D., Mar, H., Nochlin, D., Kimata, K., Kato, M., Suzuki, S., Hassell, J., and Wight, T.N. (1988) The presence of heparan sulfate proteoglycans in the neuritic plaques and congophilic angiopathy in Alzheimer's disease, *Am. J. Pathol.* **133** , 456-463.
21. Duong, T., Pommier, E.C., and Schiebel, A.B. (1989) Immunodetection of the amyloid P component in Alzheimer's disease, *Acta. Neuropathol.* **78**, 429-437.
22. Westermark, P., Shirahama, T., Skinner, M., Brun, A., Cameron, R., and Cohen, A.S. (1982) Immunohistochemical evidence for the lack of amyloid P component in some intracerebral amyloids, *Lab. Invest.* **46**, 457-460.
23. Uéda, K., Fukushima, H., Masliah, E., Xia, Y., Iwai, A., Yoshimoto, M., Otero, D.A.C., Kondo, J., Ihara, Y., and Saitoh, T. (1993) Molecular cloning of cDNA encoding an unrecognized component of amyloid in Alzheimer disease, *Proc. Natl. Acad. Sci. USA* **90**, 11282-11286.
24. Iwai, A., Yoshimoto, M., Masliah, E., and Saitoh, T. (1995) Non-Aβ component of Alzheimer's disease amyloid (NAC) is amyloidogenic, *Biochemistry* **34**, 10139-10145.
25. Vinters, H.V., Lenard Secor, D., Pardridge, W.M., and Gray, F. (1990) Immunohistochemical study of cerebral amyloid angiopathy. III. Widespread Alzheimer A4 peptide in cerebral microvessel walls colocalizes with gamma trace in patients with leukoencephalopathy, *Ann. Neurol.* **28**, 34-42.
26. Vinters, H.V., Nishimura, G.S., Lenard Secor, D., and Pardridge, W.M. (1990) Immunoreactive A4 and gamma-trace peptide colocalization in amyloidotic arteriolar lesions in brains of patients with Alzheimer's disease, *Am. J. Pathol.* **137**, 233-240.
27. Tagliavini, F., Ghiso, J., Timmers, W.F., Giaccone, G., Bugiani, O., and Frangione, B. (1990) Coexistence of Alzheimer's amyloid precursor protein and amyloid protein in cerebral vessel walls, *Lab. Invest.* **62**, 761-767.
28. van Duinen, S.G., Maat-Schieman, M.L.C., Bruijn, J.A., Haan, J., and Roos, R.A.C. (1995) Cortical tissue of patients with hereditary cerebral hemorrhage with amyloidosis (Dutch) contains various extracellular matrix deposits, *Lab. Invest.* **73**, 183-189.

29. Zhang, W.W., Lempessi, H., and Olsson, Y. (1998) Amyloid angiopathy of the human brain: immunohistochemical studies using markers for components of extracellular matrix, smooth muscle actin and endothelial cells, *Acta Neuropathol. Berl.* **96**, 558-563.
30. Verbeek, M.M., Otte-Höller, I., Westphal, J.R., Wesseling, P., Ruiter, D.J., and de Waal, R.M.W. (1994) Accumulation of intercellular adhesion molecule-1 in senile plaques in brain tissue of patients with Alzheimer's disease, *Am. J. Pathol.* **144**, 104-116.
31. Eikelenboom, P., Zhan, S.-S., van Gool, W.A., and Allsop, D. (1994) Inflammatory mechanisms in Alzheimer's disease, *Trends. Pharmacol. Sci.* **15**, 447-450.
32. Namba, Y., Tomonaga, M., Kawasaki, H., Otomo, E., and Ikeda, K. (1991) Apolipoprotein E immunoreactivity in cerebral amyloid deposits and neurofibrillary tangles in Alzheimer's disease and kuru plaque amyloid in Creutzfeld-Jakob disease, *Brain Res.* **541**, 163-166.
33. Lue, L.-F., Brachova, L., Civin, H., and Rogers, J. (1996) Inflammation, Aβ deposition, and neurofibrillary tangle formation as correlates of Alzheimer's disease neurodegeneration, *J. Neuropathol. Exp. Neurol.* **55**, 1083-1088.
34. Verbeek, M.M., Otte-Höller, I., Wesseling, P., Van Nostrand, W.E., Sorg, C., Ruiter, D.J., and de Waal, R.M.W. (1995) A lysosomal marker for activated microglial cells involved in Alzheimer classic senile plaques, *Acta. Neuropathol.* **90**, 493-503.
35. Rozemuller, J.M., Eikelenboom, P., Stam, F.C., Beyreuther, K., and Masters, C.L. (1989) A4 protein in Alzheimer's disease: primary and secondary cellular events in extracellular amyloid deposition, *J. Neuropathol. Exp. Neurol.* **48**, 674-691.
36. Haga, S., Akai, K., and Ishii, T. (1989) Demonstration of microglial cells in and around senile (neuritic) plaques in the Alzheimer brain, *Acta. Neuropathol.* **77**, 569-575.
37. Ganter, U., Strauss, S., Jonas, U., Weidemann, A., Beyreuther, K., Volk, B., Berger, M., and Bauer, J. (1991) Alpha 2-macroglobulin synthesis in interleukin-6 stimulated human neuronal (SH-SY5Y neuroblastoma) cells. Potential significance for the processing of Alzheimer β-amyloid precursor protein, *FEBS Letters* **282**, 127-131.
38. Castell, J.V., Gómez-Lechón, M.J., David, M., Andus, T., Geier, T., Trullenque, R., Farbra, R., and Heinrich, P.C. (1989) Interleukin-6 is the major regulator of acute phase protein synthesis in adult human hepatocytes, *FEBS Letters* **242**, 237-239.
39. Das, S. and Potter, H. (1995) Expression of the Alzheimer amyloid-promoting factor antichymotrypsin is induced in human astrocytes by IL-1, *Neuron* **14**, 447-456.
40. Dustin, M.L., Rothlein, R., Bhan, A.K., Dinarello, C.A., and Springer, T.A. (1986) Induction by IL-1 and interferon: Tissue distribution, biochemistry, and function of a natural adherence molecule (ICAM-1), *J. Immunol.* **137**, 245-254.
41. Dickson, D.W., Lee, S.C., Mattiace, L.A., Yen, S.-H.C., and Brosnan, C. (1993) Microglia and cytokines in neurological disease, with special reference to AIDS and Alzheimer's disease, *Glia* **7**, 75-83.
42. Griffin, W.S.T., Sheng, J.G., Roberts, G.W., and Mrak, R.E. (1995) Interleukin-1 expression in different plaque types in Alzheimer's disease: Significance in plaque evolution, *J. Neuropathol. Exp. Neurol.* **54**, 276-281.
43. Uchihara, T., Akiyama, H., Kondo, H., and Ikeda, K. (1997) Activated microglial cells are colocalized with perivascular deposits of amyloid-beta protein in Alzheimer's disease brain, *Stroke* **28**, 1948-1950.
44. Maat, S.M., van-Duinen, S.G., Rozemuller, A.J., Haan, J., and Roos, R.A. (1997) Association of vascular amyloid beta and cells of the mononuclear phagocyte system in hereditary cerebral hemorrhage with amyloidosis (Dutch) and Alzheimer disease, *J. Neuropathol. Exp. Neurol.* **56**, 273-284.

45. Verbeek, M.M., Otte-Höller, I., Ruiter, D.J., and de Waal, R.M.W. (1999) Human brain pericytes as a model system to study the pathogenesis of cerebrovascular amyloidosis in Alzheimer's disease, *Cell. Mol. Biol.* **45**, 37-46.
46. Huang, J., Kim, L.J., Mealey, R., Marsh-HC, J., Zhang, Y., Tenner, A.J., Connolly-ES, J., and Pinsky, D.J. (1999) Neuronal protection in stroke by an sLex-glycosylated complement inhibitory protein, *Science* **285**, 595-599.
47. Yamaguchi, H., Yamazaki, T., Lemere, C.A., Frosch, M.P., and Selkoe, D.J. (1992) Beta amyloid is focally deposited within the outer basement membrane in the amyloid angiopathy of Alzheimer's disease, *Am. J. Pathol.* **141**, 249-259.
48. Narindrasorasak, S., Altman, R.A., Gonzalez-DeWhitt, P., Greenberg, B.D., and Kisilevsky, R. (1995) An interaction between basement membrane and Alzheimer amyloid precursor proteins suggests a role in the pathogenesis of Alzheimer's disease, *Lab. Invest.* **72**, 272-282.
49. Weller, R.O., Massey, A., Newman, T.A., Hutchings, M., Kuo, Y.M., and Roher, A.E. (1998) Cerebral amyloid angiopathy: amyloid beta accumulates in putative interstitial fluid drainage pathways in Alzheimer's disease, *Am. J. Pathol.* **153**, 725-733.
50. Summers, J.B., Hill, W.D., Prendergast, M.A., and Buccafusco, J.J. (1998) Co-localization of apolipoprotein E and beta-amyloid in plaques and cerebral blood vessels of aged non-human primates, *Alzheimer's Reports* **1**, 119-128.
51. LaDu, M.J., Falduto, M.T., Manelli, A.M., Reardon, C.A., Getz, G.S., and Frail, D.E. (1994) Isoform-specific binding of Apolipoprotein E to β-amyloid, *J. Biol. Chem.* **269**, 23403-23406.
52. Castillo, G.M., Lukito, W., Wight, T.N., and Snow, A.D. (1999) The sulfate moieties of glycosaminoglycans are critical for the enhancement of beta-amyloid protein fibril formation, *J. Neurochem.* **72**, 1681-1687.
53. Bronfman, F.C., Garrido, J., Alvarez, A., Morgan, C., and Inestrosa, N.C. (1996) Laminin inhibits amyloid-beta-peptide fibrillation, *Neurosci. Lett.* **218**, 201-203.
54. Rogers, J., Cooper, N.R., Webster, S., Schultz, J., McGeer, P.L., Styren, S.D., Civin, W.H., Brachova, L., Bradt, C., Ward, P., and Lieberburg, I. (1992) Complement activation by β-amyloid in Alzheimer disease, *Proc. Natl. Acad. Sci. USA* **89**, 10016-10020.
55. Olichney, J.M., Hansen, L.A., Galasko, D., Saitoh, T., Hofstetter, C.R., Katzman, R., and Thal, L.J. (1996) The apolipoprotein E epsilon 4 allele is associated with increased neuritic plaques and cerebral amyloid angiopathy in Alzheimer's disease and Lewy body variant, *Neurology* **47**, 190-196.
56. Premkumar, D.R.D., Cohen, D.L., Hedera, P., Friedland, R.P., and Kalaria, R.N. (1996) Apolipoprotein E-ε4 alleles in cerebral amyloid angiopathy and cerebrovascular pathology associated with Alzheimer's disease, *Am. J. Pathol.* **148**, 2083-2095.
57. Greenberg, S.M., Rebeck, G.W., Vonsattel, J.-P., Gomez-Isla, T., and Hyman, B.T. (1995) Apolipoprotein E ε4 and cerebral hemorrhage associated with amyloid angiopathy, *Ann. Neurol.* **38**, 254-259.
58. Greenberg, S.M., Vonsattel, J.P., Segal, A.Z., Chiu, R.I., Clatworthy, A.E., Liao, A., Hyman, B.T., and Rebeck, G.W. (1998) Association of apolipoprotein E epsilon2 and vasculopathy in cerebral amyloid angiopathy, *Neurology* **50**, 961-965.
59. Greenberg, S.M., Briggs, M.E., Hyman, B.T., Kokoris, G.J., Takis, C., Kanter, D.S., Kase, C.S., and Pessin, M.S. (1996) Apolipoprotein E ε4 is associated with the presence and earlier onset of hemorrhage in cerebral amyloid angiopathy, *Stroke* **27**, 1333-1337.
60. Wisniewski, T., Castaño, E.M., Golabek, A., Vogel, T., and Frangione, B. (1994) Acceleration of Alzheimer's fibril formation by apolipoprotein E *in vitro*, *Am. J. Pathol.* **145**, 1030-1035.

61. Zlokovic, B.V., Martel, C.L., Mackic, J.B., Matsubara, E., Wisniewski, T., McComb, J.G., Frangione, B., and Ghiso, J. (1994) Brain uptake of circulating apolipoproteins J and E complexed to Alzheimer's amyloid β, *Biochem. Biophys. Res. Commun.* **205**, 1431-1437.

62. Xu, P.T., Gilbert, J.R., Qiu, H.L., Ervin, J., Rothrock, C.T., Hulette, C., and Schmechel, D.E. (1999) Specific regional transcription of apolipoprotein E in human brain neurons, *Am. J. Pathol.* **154**, 601-611.

Chapter 13

NEUROPATHOLOGY OF HEREDITARY CEREBRAL HEMORRHAGE WITH AMYLOIDOSIS-DUTCH TYPE

MARION L.C. MAAT-SCHIEMAN*, SJOERD G. VAN DUINEN[†], REMCO NATTÉ* and RAYMUND A.C. ROOS*

*Departments of *Neurology and [†]Pathology, Leiden University Medical Center, PO Box 9600, 2300 RC Leiden, The Netherlands*

Hereditary cerebral hemorrhage with amyloidosis-Dutch type is caused by an amyloid β (Aβ) precursor protein (APP) gene codon 693 mutation, producing a Gln-for-Glu substitution at residue 22 of Aβ. *In vitro*, the mutation alters the proteolytic processing of APP, increasing the relative quantities of full-length Aβ beginning at Asp^1, and of truncated Aβ peptides beginning at Val^{18} and Phe^{19}, and Aβ peptides with the amino acid substitution exhibit an increased tendency for fibril formation; *in vivo*, the mutation accelerates primarily the deposition of amyloid in cerebral blood vessel walls, resulting in severe amyloid angiopathy with secondary microvascular degeneration and hence cerebral hemorrhages and/or infarcts. Brain parenchymal Aβ deposition is also enhanced but plaque formation is not essentially associated with neurofibrillary degeneration. Evidence suggests similarities between early events in the development of cerebral amyloid angiopathy and senile plaques.

1. INTRODUCTION

Hereditary cerebral hemorrhage with amyloidosis-Dutch type (HCHWA-D) is a rare autosomal dominant cerebral amyloid β (Aβ) disease, affecting members of three families, originating from the coastal villages of Katwijk and Scheveningen in The Netherlands [1]. Patients with this disorder suffer from recurrent strokes with a mid-life onset and most of them develop dementia [2] (see Bornebroek *et al*, this volume). The strokes result from cerebral hemorrhages and/or infarcts due to severe cerebral amyloid angiopathy (CAA) with secondary microvascular degeneration [3,4]. The

M.M. Verbeek et al. (eds.),
***Cerebral Amyloid Angiopathy in Alzheimer's Disease and Related Disorders*, 223–236.**

cause of the disease is a G-to-C transversion in codon 693 of the gene on chromosome 21 encoding the Aβ precursor protein (APP). The point mutation results in a Gln-for-Glu substitution at residue 22 of Aβ [5,6].

2. NEUROPATHOLOGICAL FEATURES

Since the sixties, the brains of almost 40 patients with HCHWA-D have been obtained at autopsy. Presently, material from 35 of these brains - preserved mostly in formalin or as formalin-fixed and paraffin-embedded blocks - is available for investigation. Recently, the study of this material has been greatly facilitated by the availability of highly sensitive Aβ antibodies [7].

2.1 Macroscopy

Brain weight is usually normal, or increased due to edema. On sectioning, one or more lobar hemorrhages are found (except in the few patients dying from intercurrent diseases), sometimes in combination with remnants of older ones. The hemorrhages usually extend to the cortical surface and sometimes to the ventricular wall. They are predominantly found in the temporal, parietal, and occipital regions [8].

2.2 Microscopy

2.2.1 Hemorrhages and infarcts

The cerebral hemorrhages are surrounded by a broad zone of edematous and reactively changed parenchyma. Sometimes, the observed pathology is suggestive of secondary bleeding in infarcted areas. The cerebral cortex and the adjacent subcortical white matter may show scattered small infarcts, which is also sporadically the case in the cerebellum [8]. Local demyelination, axonal loss, and gliosis in the subcortical white matter (leukoencephalopathy) evidence incomplete infarction of these areas.

2.2.2 Amyloid angiopathy

Regional distribution. Aβ deposition is consistently seen in meningocortical blood vessels of the cerebral hemispheres and the cerebellum. The occipital cortex may show particularly severe CAA. Blood vessels in other regions, like the basal ganglia, allocortex, and brain stem can be affected by CAA, but its prevalence in these and other localizations besides the neocortex and cerebellum needs further study. Small foci of Aβ infrequently appear at the junction of the media and adventitia of the arteries of the circle of Willis - which was investigated in a few cases - but Aβ deposits are not found in extracranial arteries.

Local distribution. Aβ deposits are present in arteries and to a lesser extent in veins of the cerebral meninges. In the neocortex, perforating arterioles are preferentially affected. They show Aβ deposition at the level of the upper cortical layers and less frequently also in their course through the deepest layers of the cortex (Figure 1). Capillary involvement is variable, and may be severe, especially in the occipital cortex and in aged patients. Particularly in association with severe cortical amyloid angiopathy, subcortical white matter blood vessels may show Aβ deposition.

In addition to the cerebellar meningeal blood vessels, arterioles and capillaries in the molecular layer of the cerebellum may be affected by CAA [9].

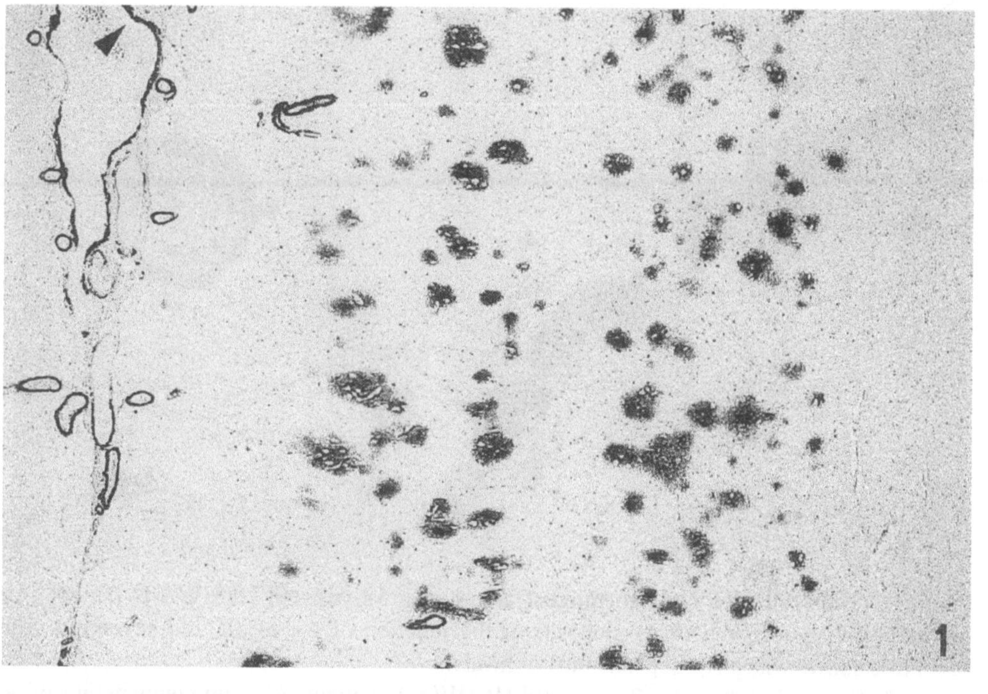

Figure 1. Frontal cortex of a 39-year-old HCHWA-D patient. $A\beta_{42}$ immunolabeling of meningeal arteries and a vein (arrow-head) and of arterioles in the upper cortex. In the larger meningeal arteries, Aβ deposition is focal. Diffuse plaques are distributed throughout the cortex with sparing of layers I and IV.

Morphology. In large meningeal arteries, compact Aβ deposits are localized in the outer media at the junction with the adventitia. Wispy Aβ deposits are found in veins and may be present in the adventitia of large arteries (Figure 2). Smaller meningeal arteries show compact Aβ deposition throughout the thickness of their walls. Cerebrocortical arterioles may harbor only compact Aβ ("l'angiopathie congophile de Pantelakis" [10]). However, they often show one or two layers of radially arranged Aβ around the layer of compact Aβ (Figure 3). Radial arteriolar Aβ develops most prominently in the upper layers of the cerebral cortex and decreases, c.q., disappears towards the pial surface and in the deeper cortical layers [11]. One layer of radial Aβ may protrude from cerebrocortical capillaries.

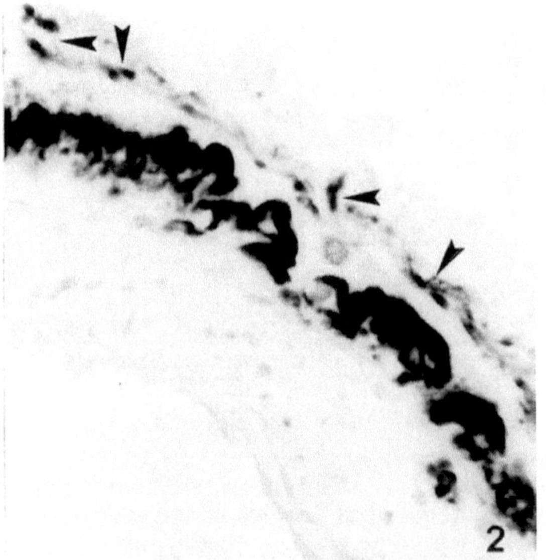

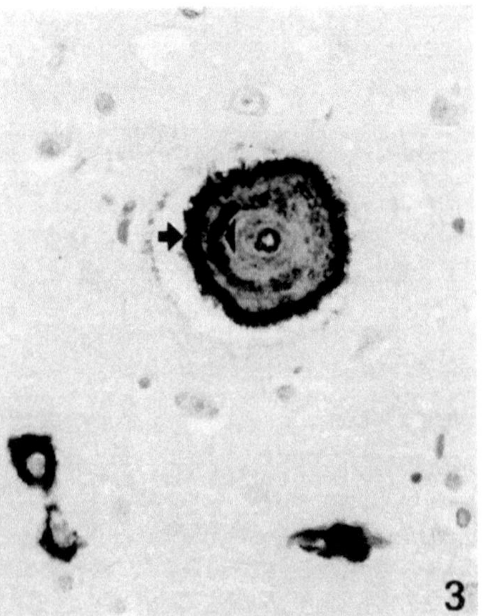

Figure 2. The wall of a large meningeal artery in a 58-year-old HCHWA-D patient. Aβ immunolabeling of compact Aβ deposits at the junction of the media and adventitia. The adventitia shows wispy Aβ deposits (arrow-heads).

Figure 3. Cerebral cortex of a 76-year-old HCHWA-D patient. $A\beta_{42}$ immunolabeling of an arteriole and capillaries. The arteriole shows a thickened wall and narrowing of its lumen. The outer radial layer is indicated by an arrow; remnants of the inner radial layer by an arrowhead

The accumulation of Aβ in the arterial and arteriolar media is associated with degeneration of the medial smooth muscle cells [11,12]. In the course of the process of Aβ deposition, vascular walls may thicken with concomitant narrowing up to virtual occlusion of the lumen (Figure 3). Narrowing of particularly the perforating cortical arterioles is believed to lead to hypoperfusion of the subcortical white matter and hence leukoencephalopathy.

Aβ C- and N-terminal heterogeneity. Biochemical studies have indicated that Aβ from HCHWA-D leptomeningeal blood vessels is predominantly composed of the $A\beta_{1-40}$ isoform and its C-terminal truncated derivatives and of minor amounts of $A\beta_{1-42}$ [13]. The results of immunohistochemistry with antibodies specific for Aβ ending C-terminally at Val_{40} ($A\beta_{40}$) or Ala_{42} ($A\beta_{42}$) were consistent with these findings [13,14]. Using highly sensitive monoclonal end-specific Aβ antibodies, a recent study demonstrated strong labeling for both $A\beta_{40}$ and $A\beta_{42}$ of CAA-affected meningocortical blood vessels in HCHWA-D. Moreover, in a number of blood vessels, particularly capillaries and small arterioles, Aβ deposits were solely composed of $A\beta_{42}$, whereas deposits consisting of $A\beta_{40}$ alone were not observed (see Yamaguchi *et al*, this volume) [9,15]. By immuno-electron microscopy, non-fibrillar $A\beta_{42}$ deposits were recently identified in the basement membranes of such capillaries [16]. On immunohistochemistry with antibodies specific for various Aβ N-termini, HCHWA-D vascular Aβ showed considerable heterogeneity with Aβ species beginning at Asp_1, isomerized Asp_1, racemized Asp_1, $pyroglutamate_3$, Arg_5, and $pyroglutamate_{11}$ [17].

Non-Aβ proteins. HCHWA-D vascular amyloid shows immunostaining for the Aβ-associated proteins amyloid P-component and apolipoprotein E (ApoE) and the extracellular matrix (ECM) components collagen type I and III and fibronectin [18,19,20]. Labeling for heparan sulfate proteoglycans, collagen type IV, and laminin is weak or absent [20]. The distribution of inflammatory proteins in Dutch vascular amyloid has not been studied. Cystatin C immunoreactivity frequently colocalizes with the Aβ deposits in HCHWA-D vessel walls [21].

Perivascular pathology in the cerebral cortex. Congo red-negative Aβ deposits may be present, and are frequent in some patients, in the cerebral parenchyma immediately surrounding CAA-affected cortical blood vessels. Congo red-positive perivascular Aβ-deposits are rare.

Arterioles and capillaries with radial Aβ may be associated with degenerating neurites, as shown by immunohistochemical studies using antibodies to synaptophysin, ubiquitin, or APP [22,23]. Glial markers reveal the association of such vessels with reactive astrocytes and activated microglia. Infiltration of radial Aβ by the latter as well as by perivascular cells may explain why the peripheral part of a subset of CAA-affected arterioles shows strong, irregular immunoreactivities for macrophage / monocyte markers. Cells of the macrophage / monocyte system are also associated with Aβ deposits in meningeal arteries [24]. Clusters of coarse deposits of unknown origin and significance, which are strongly immunoreactive for the ECM components heparan sulfate proteoglycan, laminin, and collagen type III and IV may be found in the parenchyma in the vicinity of or around arterioles with a large amount of amyloid [20].

Secondary microvascular degeneration. CAA-associated microvasculopathies (CAA-AM), including fibrosis or hyalinization, microaneurysm formation, perivascular lymphocytic inflammation, foreign body type granulomatous inflammation, vascular thrombi, vessel wall calcification, and fibrinoid necrosis are a common finding in HCHWA-D brain (see Vinters and Vonsattel, this volume). Vascular fibrosis or hyalinization is the most frequently identified form and it is often associated with microaneurysm formation and macrophage infiltration of the vessel wall [4].

In HCHWA-D patients, the severity and extent of CAA-AM, particularly of microaneurysm formation, correlate with the number of cerebrovascular lesions. This finding supports the notion that CAA-AM contribute to the development of CAA-associated hemorrhages and/or infarcts [25].

2.2.3 Senile plaques

Regional distribution. Parenchymal Aβ deposits in the form of diffuse plaques are distributed throughout the neocortical gray matter with sparing of layer I and less consistently of layer IV (Figure 1). Plaque density, ranging from 10 to 50 plaques/mm^2, varies among patients and does not increase with age. Locally, e.g., in the motor cortex, plaques may be absent. In some patients, the cerebellar molecular layer occasionally harbors diffuse [9]. Further studies are needed to determine the presence of plaques in localizations other than the neocortex and cerebellum.

Neocortical plaque types. Particularly in older patients, a subset of neocortical plaques is associated with tau-negative degenerating neurites and/or reactive glial cells. Such plaques are occasionally cored. Special plaque types are homogeneous plaques and, in elderly patients, plaques composed of coarse, strongly immunoreactive Aβ deposits [9,15,23,26]. The ApoE genotype affects neither the number nor the type of plaques [27].

Aβ C-and N-terminal heterogeneity. Plaques in HCHWA-D brain are invariably $A\beta_{42}$-immunoreactive [9,14,15]. In particular in older patients, a subset of neocortical plaques may show variably intense $A\beta_{40}$

immunolabeling. Plaques associated with degenerating neurites and/or reactive glia are always strongly Aβ_{40}-positive [9,15].

As studied in young HCHWA-D brains, neocortical plaques are reactive for antibodies recognizing Aβ species beginning at Asp$_1$ [28].

Non-Aβ proteins. Both ApoE and amyloid P-component immunolocalize to neocortical plaques [18,19]. In a minority of the plaques, weak immunoreactivity for ECM components is observed [20]. Neocortical plaques are immunolabeled for the complement factors C1q and C3d and the acute phase protein α_1-antichymotrypsin [29]. The distribution of other inflammatory proteins in HCHWA-D plaques and of the various Aβ-associated proteins in different plaque types awaits further investigation.

Figure 4. Transentorhinal cortex of a 76-year-old HCHWA-D patient. Neurofibrillary tangles, neuropil threads, and degenerating neurites surrounding CAA-affected blood vessels (arrow-heads) are positive with an antibody against tau.

2.2.4 Neurofibrillary degeneration

Sporadically, tau-positive neocortical tangles and/or degenerating neurites, the latter associated with plaques and/or CAA-affected blood vessels, may be observed in, as a rule, older HCHWA-D patients and in conjunction with neurofibrillary degeneration affecting the allocortex (Figure 4) [9].

2.2.5 Extracranial Aβ

Evidence of Aβ deposition in extracranial tissues, i.e., in skin, viscera, and choroid of the eye [30] or in extracranial arteries (see above), of patients with HCHWA-D has not been found.

3. *IN VITRO* STUDIES

3.1 Aβ secretion

Whereas the APP mutations causing Alzheimer's disease (AD) increase either total Aβ secretion or the proportion of secreted $A\beta_{42}$, the HCHWA-D mutation has no such effects on APP processing. On the other hand, recent cell transfection experiments indicate that the mutation increases the relative levels of full-length Aβ beginning at Asp_1, and of truncated Aβ peptides beginning at Val_{18} and Phe_{19} [17,31-34].

3.2 Fibrillogenesis of HCHWA-D Aβ

By accelerating the formation of stable β-pleated sheet structures [35-39], the HCHWA-D mutation appears to increase the rates of Aβ fibril nucleation, elongation, and self-association [40]. Both mutant and non-mutant Aβ can be isolated from HCHWA-D cerebrovascular amyloid [41], suggesting that the HCHWA-D mutation influences the assembly of wild-type Aβ peptides by providing HCHWA-D fibril nuclei from which wild-type or mixed fibrils can elongate [17].

3.3 Interactions of HCHWA-D Aβ with other molecules

So far, the interaction of HCHWA-D Aβ with molecules affecting Aβ fibrillogenesis has been studied with regard to ApoE, laminin, and heparin. ApoE produces only a minimal effect on the already accelerated fibrillogenesis of HCHWA-D $A\beta_{1-40}$ [42]. HCHWA-D $A\beta_{1-40}$ binds heparin (a glycosaminoglycan used as a model for tissue heparan sulfates) more readily than wild-type Aβ due to its increased tendency to form fibrils, but not because of a greater affinity for heparin [43]. Laminin inhibits HCHWA-

D $A\beta_{1\text{-}40}$ (as well as wild-type $A\beta_{1\text{-}40}$) fibril formation, probably by acting as an inhibitor of fibril elongation [44].

4. The Dutch, Italian and Flemish APP mutations

In addition to the HCHWA-D (Dutch) mutation, two other pathogenic APP mutations resulting in amino acid substitutions within the Aβ sequence are presently known.

In members of 3 unrelated Italian families, an A-for-G mutation in codon 693 of the APP gene (Italian mutation), predicting a Lys-for-Glu substitution at residue 22 of Aβ, is associated with stroke, cognitive decline, and seizures in some cases. The brain of an approximately 60-year-old patient with this mutation was investigated and showed Aβ deposits in meningocortical blood vessels and in the cerebral parenchyma. The former were primarily $A\beta_{40}$-positive, the latter were predominantly immunolabeled for $A\beta_{42}$. Although extensive, the vascular Aβ deposits appeared amorphous rather than fibrillar as judged by thioflavine S fluorescence. Abnormal tau was absent except in the hippocampus [45,46].

A C-for-G mutation in codon 692 of the APP gene (Flemish mutation), resulting in a Gly-for-Ala substitution at residue 21 of Aβ, was reported in members of a Dutch family, presenting with progressive dementia, or stroke due to cerebral hemorrhage [47]. Brain autopsy, performed in two demented patients, revealed severe CAA, numerous senile plaques, of which many with unusually large amyloid cores, and extensive neurofibrillary degeneration [48]. *In vitro*, this mutation increases total Aβ secretion [34,49], and full-length/partial Flemish peptides appear to aggregate more slowly than the corresponding wild-type Aβ sequences [38,39].

Taken together, the above data indicate that both the site of the amino acid substitution within the Aβ sequence and the substitution itself may be relevant to the phenotype.

5. REMARKS ON THE PATHOGENESIS AND EVOLUTION OF CAA

5.1 Source of Aβ

Aβ, produced by cells of the vessel wall, i.e., smooth muscle cells, pericytes, perivascular cells, and endothelial cells, as well as soluble Aβ derived from blood or cerebrospinal fluid have been designated as potential sources of vascular amyloid [50-54]. Arguing that the normal brain eliminates soluble Aβ along putative periarterial interstitial fluid drainage pathways, a recent study of the pattern of Aβ deposition in AD hypothesized that deposition of brain-derived Aβ in such pathways contributes significantly to CAA [55]. This hypothesis is supported by the very similar pattern of Aβ deposition in HCHWA-D cerebral blood vessels and is also consistent with observations suggestive of the progression of Aβ deposition at the adventitial aspect of CAA-affected HCHWA-D arterioles with a meanwhile degenerated media [11].

5.2 Initial Aβ deposition

Consistent with the hypothesis that $A\beta_{42}$ is deposited first in cerebral vessel walls and then entraps the more soluble $A\beta_{40}$ [55,56], exclusively $A\beta_{42}$-immunoreactive Aβ deposits were identified in HCHWA-D cerebral blood vessels, whereas vascular Aβ deposits consisting of $A\beta_{40}$ alone were absent. Moreover, non-fibrillar deposits of $A\beta_{42}$ were observed in HCHWA-D capillary basement membranes. These observations are suggestive of similarities between early events in the development of CAA and senile plaques.

REFERENCES

1. Van Duinen, S.G., Castaño, E.M., Prelli, F., Bots, G.T.A.M., Luyendijk, W., Frangione, B. (1987) Hereditary cerebral hemorrhage with amyloidosis in patients of Dutch origin is related to Alzheimer disease, *Proc. Natl. Acad. Sci. USA* **84**, 5991-5994
2. Bornebroek, M., Van Buchem, M.A., Haan, J., Brand, R., Lanser, J.B.K., de Bruïne, F.T., Roos, R.A.C. (1996) Hereditary cerebral hemorrhage with amyloidosis-Dutch type: better correlation of cognitive deterioration with advancing age than with the number of focal lesions or white matter hyperintensities, *Alz. Dis. Assoc. Dis.* **10**, 224-231
3. Wattendorff, A.R., Bots, G.Th.A.M., Went, L.N., Endtz, L.J. (1982) Familial cerebral amyloid angiopathy presenting as recurrent cerebral hemorrhage, *J. Neurol. Sci.* **55**, 121-135
4. Vinters, H.V., Natté, R., Maat-Schieman, M.L.C., van Duinen, S.G., Hegeman-Kleinn, I., Welling-Graafland, C., Haan, J., Roos, R.A.C. (1998) Secondary microvascular degeneration in amyloid angiopathy of patients with hereditary cerebral hemorrhage with amyloidosis, Dutch type (HCHWA-D), *Acta Neuropathol.* **95**, 235-244
5. Levy, E., Carman, M.D., Fernandez-Madrid, I.J, Power, M.D., Lieberburg, I., van Duinen, S.G., Bots, G.T.A.M., Luyendijk, W., Frangione, B. (1990) Mutation of Alzheimer's disease amyloid gene in hereditary cerebral hemorrhage, Dutch type, *Science* **248**, 1124-1126
6. Bakker, E., van Broeckhoven, C., Haan, J., Voorhoeve, E., van Hul, W., Levy, E., Lieberburg, I., Carman, M.D., van Ommen, G.J.B., Frangione, B., Roos, R.A.C. (1991)

DNA diagnosis for hereditary cerebral hemorrhage with amyloidosis (Dutch type), *Am. J. Hum. Genet.* **49**, 518-521

7. Yamaguchi, H., Sugihara, S., Ogawa, A., Saido, T.C., Ihara, Y. (1998) Diffuse plaques associated with astroglial amyloid β protein, possibly showing a disappearing stage of senile plaques, *Acta Neuropathol.* **95**, 217-222
8. Luyendijk, W., Bots, G.T.A.M., Vegter-van der Vlis, M., Went, L.N., Frangione, B. (1988) Hereditary cerebral hemorrhage caused by cortical amyloid angiopathy, *J. Neurol. Sci.* **85**, 267-280
9. Maat-Schieman, M.L.C., Yamaguchi, H., van Duinen, S.G., Natté, R., Roos, R.A.C. (1999) C-Terminal heterogeneity of the amyloid β protein (Aβ) in hereditary cerebral hemorrhage with amyloidosis (Dutch) (HCHWA(D)), in R.A. Kyle and M.A. Gertz (eds.), *Amyloid and Amyloidosis 1998*, Proceedings of the VIII International Symposium on Amyloidosis, Parthenon Publishing, Carnforth, UK, 479-481
10. Pantelakis, S. (1954) Un type particulier d'angiopathie sénile du système nerveux central: l'angiopathie congophile. Topographie et frèquence, *Mschr. Psychiat. Neurol.* **1208**, 219-256
11. Maat-Schieman, M.L.C., van Duinen, S.G., Rozemuller, A.J.M., Haan, J., Roos, R.A.C. (1997) Association of vascular amyloid β and cells of the mononuclear phagocyte system in hereditary cerebral hemorrhage with amyloidosis (Dutch) and Alzheimer's disease, *J. Neuropathol. Exp. Neurol.* **56**, 273-284
12. Maat-Schieman, M.L.C., van Duinen, S.G., Bornebroek, M., Haan, J., Roos, R.A.C. (1996) Hereditary cerebral hemorrhage with amyloidosis-Dutch type (HCHWA-D): II - A review of histopathological aspects, *Brain Pathol.* **6**, 115-120
13. Castaño, E.M., Prelli, F., Soto, C., Beavis, R., Matsubara, E., Shoji, M., Frangione, B. (1996) The length of amyloid-β in hereditary cerebral hemorrhage with amyloidosis, Dutch type. Implications for the role of amyloid-β 1-42 in Alzheimer's disease, *J. Biol. Chem.* **271**, 32185-32191
14. Mann, D.M.A., Iwatsubo, T., Ihara, Y., Cairns, N.J., Lantos, P.L., Bogdanovic, N., Lannfelt, L., Winblad, B., Maat-Schieman, M.L.C., Rossor, M.N. (1996) Predominant deposition of amyloid-$\beta_{42(43)}$ in plaques in cases of Alzheimer's disease and hereditary cerebral hemorrhage associated with mutations in the amyloid precursor protein gene, *Am. J. Pathol.* **148**, 1257-1266
15. Maat-Schieman, M.L.C., Yamaguchi, H., van Duinen, S.G., Natté, R., Roos, R.A.C. (1998) C-Terminal heterogeneity of Aβ in hereditary cerebral hemorrhage with amyloidosis (Dutch) (HCHWA-D), *VIII International Symposium on Amyloidosis*, Mayo Press, Rochester, MN, pp. 191 (Abstract 156)
16. Natté, R., Yamaguchi, H., Maat-Schieman, M.L.C., Roos, R.A.C., van Duinen, S.G. (1999) Ultrastructural evidence of non-fibrillar Aβ42 deposits in the capillary basement membrane of patients with hereditary cerebral hemorrhage with amyloidosis, Dutch type, *Alzheimer's reports: vascular factors in Alzheimer's disease* 2(Suppl1), S22
17. Watson, D.J., Selkoe, D.J., Teplow, D.B. (1999) Effects of the amyloid precursor protein Glu^{693}|Gln 'Dutch' mutation on the production of amyloid β-protein, *Biochem. J.* **340**, 703-709
18. Timmers, W.F., Tagliavini, F., Haan, J., Frangione, B. (1990) Parenchymal preamyloid and amyloid deposits in the brains of patients with hereditary cerebral hemorrhage with amyloidosis-Dutch type, *Neurosci. Lett.* **118**, 223-226
19. Haan, J., Van Broeckhoven, C., van Duijn, C.M., Voorhoeve, E., van Harskamp, F., van Swieten, J.C., Maat-Schieman, M.L.C., Roos, R.A.C., Bakker, E. (1994b) The

apolipoprotein E 4 allele does not influence the clinical expression of the amyloid precursor protein gene codon 693 or 692 mutations, *Ann. Neurol.* **36**, 434-437

20. Van Duinen S.G., Maat-Schieman, M.L.C., Bruijn, J.A., Haan, J., Roos, R.A.C. (1995) Cortical tissue of patients with hereditary cerebral hemorrhage with amyloidosis (Dutch) contains various extracellular matrix deposits, *Lab. Invest.* **73**, 183-189
21. Haan, J., Maat-Schieman, M.L.C., van Duinen, S.G., Jensson, O., Thorsteinsson, L., Roos, R.A.C. (1994a) Co-localization of β/A4 and cystatin C in cortical blood vessels in Dutch, but not in Icelandic hereditary cerebral hemorrhage with amyloidosis, *Acta. Neurol. Scand.* **89**, 367-371
22. Tagliavini, F., Giaccone, G., Bugiani, O., Frangione, B. (1993) Ubiquitinated neurites are associated with preamyloid and cerebral amyloid β deposits in patients with hereditary cerebral hemorrhage with amyloidosis Dutch type, *Acta Neuropathol.* **85**, 267-271
23. Maat-Schieman, M.L.C., Radder, C.M., van Duinen, S.G., Haan, J., Roos, R.A.C. (1994a) Hereditary cerebral hemorrhage with amyloidosis (Dutch): a model for congophilic plaque formation without neurofibrillary pathology, *Acta Neuropathol.* **88**, 371-378
24. Maat-Schieman, M.L.C., Rozemuller, A.J., van Duinen, S.G., Haan, J., Eikelenboom, P., Roos, R.A.C. (1994b) Microglia in diffuse plaques in hereditary cerebral hemorrhage with amyloidosis (Dutch). An immunohistochemical study, *J. Neuropathol. Exp. Neurol.* **53**, 483-491
25. Natté, R., Vinters, H., Maat-Schieman, M.L.C., Bornebroek, M., Haan, J., Roos, R.A.C., van Duinen, S.G. (1998) Microvasculopathy is associated with the number of cerebrovascular lesions in hereditary cerebral hemorrhage with amyloidosis, Dutch type, *Stroke* **29**, 1588-1594
26. Maat-Schieman, M.L.C., van Duinen, S.G., Haan, J., Roos, R.A.C. (1992) Morphology of cerebral plaque-like lesions in hereditary cerebral hemorrhage with amyloidosis (Dutch), *Acta Neuropathol.* **84**, 674-679
27. Bornebroek, M., Haan, J., Van Duinen, S.G., Maat-Schieman, M.L.C., Van Buchem, M.A., Bakker, E., Van Broeckhoven, C., Roos, R.A.C. (1997) Dutch hereditary cerebral amyloid angiopathy: structural lesions and apolipoprotein E genotype, *Ann. Neurol.* **41**, 695-698
28. Watson, D.J., Landers, A.D., Selkoe, D.J. (1997) Heparin-binding properties of the amyloidogenic peptides Aβ and amylin. Dependence on aggregation state and inhibition by Congo red, *J. Biol. Chem.* **272**, 31617-31624
29. Rozemuller, J.M., Bots, G.T.A.M., Roos, R.A.C., Eikelenboom, P. (1992) Acute phase proteins but not activated microglial cells are present in parenchymal β/A4 deposits in the brains of patients with hereditary cerebral hemorrhage with amyloidosis-Dutch type, *Neurosci. Lett.* **140**, 137-140
30. Van der Laan J.S., Maat-Schieman, M.L.C., Roos R.A.C., van Duinen, S.G. (1997) No extracranial Aβ in hereditary cerebral hemorrhage with amyloidosis-Dutch, *Brain Pathol.* **7**, 1107 (Abstract P1.I1.10)
31. Felsenstein, K.M., Lewis-Higgins, L. (1993) Processing of the β-amyloid precursor protein carrying the familial, Dutch-type, and a novel recombinant C-terminal mutation, *Neurosci. Lett.* **152**, 185-189
32. Hendriks, L., De Jonghe, C., Cras, P., Martin, J-J., Van Broeckhoven, C. (1996) β-Amyloid precursor protein and early-onset Alzheimer's disease, in Ciba Foundation Symposium 199 ed., *The Nature and Origin of Amyloid Fibrils*, Wiley, Chichester, pp. 170-180
33. Maruyama, K., Tomita, T., Shinozaki, K., Kume, H., Asada, H., Saido, T.C., Ishiura, S., Iwatsubo, T., Obata, K. (1996) Familial Alzheimer's disease-linked mutations at Val^{717} of

amyloid precursor protein are specific for the increased secretion of Aβ42(43), *Biochem. Biophys. Res. Commun.* **227**, 730-735

34. De Jonghe, C., Zehr, C., Yager, D., Prada, C-M., Younkin, S., Hendriks, L., Van Broeckhoven, C., Eckman, C.B. (1998) Flemish and Dutch mutations in amyloid β precursor protein have different effects on amyloid β secretion, *Neurobiol. Dis.* **5**, 281-286
35. Wisniewski, T., Ghiso, J., Frangione, B. (1991) Peptides homologous to the amyloid protein of Alzheimer's disease containing a glutamine for glutamic acid substitution have accelerated amyloid fibril formation, *Biochem. Biophys. Res. Commun.* **179**, 1247-1254
36. Fraser, P.E., Nguyen, J.T., Inouye, H., Surewicz, W.K., Selkoe, D.J., Podlisny, M.B., Kirschner, D.A. (1992) Fibril formation by primate, rodent, and Dutch-hemorrhagic analogues of Alzheimer amyloid β-protein, *Biochemistry* **31**, 10716-10723
37. Fabian, H., Szendrei, G.I., Mantsch, H.H., Otvos, L., Jr. (1993) Comparative analysis of human and Dutch-type Alzheimer β-amyloid peptides by infrared spectroscopy and circular dichroism, *Biochem. Biophys. Res. Commun.* **191**, 232-239
38. Clements, A., Walsh, D.M., Williams, C.H., Allsop, D. (1993) Effects of the mutations Glu22 to Gln and Ala21 to Gly on the aggregation of a synthetic fragment of the Alzheimer's amyloid β/A4 peptide, *Neurosci. Lett.* **161**, 17-20
39. Clements, A., Allsop, D., Walsh, D.M., Williams, C.H. (1996) Aggregation and metal-binding properties of mutant forms of the amyloid Aβ peptide of Alzheimer's disease, *J. Neurochem.* **66**, 740-747
40. Teplow, D.B. (1998) Structural and kinetic features of amyloid β-protein fibrillogenesis, *Amyloid: Int. J. Exp. Clin. Invest.* **5**, 121-142
41. Prelli, F., Levy, E., van Duinen, S.G., Bots, G.T.A.M., Luyendijk, W., Frangione, B. (1990) Expression of a normal and variant Alzheimer's β-protein gene in amyloid of hereditary cerebral hemorrhage, Dutch type: DNA and protein diagnostic assays, *Biochem. Biophys. Res. Commun.* **170**, 301-307
42. Castaño, E.M., Prelli, F., Wisniewski, T., Golabek, A., Kumar, R.A., Soto, C., Frangione, B. (1995) Fibrillogenesis in Alzheimer's disease of amyloid β peptides and apolipoprotein E, *Biochem. J.* **306**, 599-604
43. Watson, D., Maat-Schieman, M., Saido, T., Teplow, D., Biere, A.L., Roos, R., Selkoe, D. (1997) N-Terminal heterogeneity of Aβ caused by the HCHWA-D mutation, *Soc. Neurosci.* **23**, 821 (Abstract 321.9)
44. Bronfman, F.C., Alvarez, A., Morgan, C., Inestrosa, N.C. (1998) Laminin blocks the assembly of wild-type Aβ and the Dutch variant peptide into Alzheimer's fibrils, *Amyloid: Int. J. Exp. Clin. Invest.* **5**, 16-23
45. Bugiani, O., Padovani, A., Magoni, M., Andora, G., Sgarzi, M., Savoiardo, M., Bizzi, A., Giaccone, G., Rossi, G., Tagliavini, F. (1998) An Italian type of HCHWA, *Neurobiol. Aging* **19**, 238 (Abstract 999)
46. Tagliavini, F., Rossi, G., Padovani, A., Magoni, M., Andora, G., Sgarzi, M., Bizzi, A., Savoiardo, M., Carella, F., Morbin, M., Giaccone, G., Bugiani, O. (1999) A new βPP mutation related to hereditary cerebral hemorrhage, *Alzheimer's reports: vascular factors in Alzheimer's disease* 2(Suppl1), S28
47. Hendriks, L., van Duijn C.M., Cras, P., Cruts, M., Van Hul, W., van Harskamp, F., Warren, A., McInnis, M.G., Antonarakis, S.E., Martin, J-J, Hofman, A., Van Broeckhoven, C. (1992) Presenile dementia and cerebral hemorrhage linked to a mutation at codon 692 of the β-amyloid precursor protein gene, *Nat. Genet.* **1**, 218-221
48. Cras, P., van Harskamp, F., Hendriks, L., Ceuterick, C., van Duijn, C.M., Stefanko, S.Z., Hofman, A., Kros, J.M., Van Broeckhoven, C., Martin, J.J. (1998) Presenile Alzheimer

dementia characterized by amyloid angiopathy and large amyloid core type senile plaques in the APP 692Ala|Gly mutation, *Acta Neuropathol.* **96**, 253-260

49. Haass, C., Hung, A.Y., Selkoe, D.J., Teplow, D.B. (1994) Mutations associated with a locus for familial Alzheimer's disease result in alternative processing of amyloid β-protein precursor, *J. Biol. Chem.* **269**, 17741-17748
50. Shoji, M., Hirai, S., Harigaya, Y., Kawarabayashi, T., Yamaguchi, H. (1990) The amyloid β-protein precursor is localized in smooth muscle cells of leptomeningeal vessels, *Brain Res.* **530**, 113-116
51. Ghersi-Egea, J.-F., Gorevic, P.D., Ghiso, J., Frangione, B., Patlak, C.S., Fenstermacher, J.D. (1996) Fate of cerebrospinal fluid-borne amyloid β-peptide: rapid clearance into blood and appreciable accumulation by cerebral arteries, *J. Neurochem.* **67**, 880-883
52. Martel, C.L., Makic, J.B., McComb, J.G., Ghiso, J., Zlokovic, B.V. (1996) Blood-brain barrier uptake of the 40 and 42 amino acid sequences of circulating Alzheimer's amyloid β in guinea pigs, *Neurosci. Lett.* **206**, 157-160
53. Verbeek, M.M., Eikelenboom, P, de Waal, R.M.W. (1997) Differences between the pathogenesis of senile plaques and congophilic angiopathy in Alzheimer disease, *J. Neuropathol. Exp. Neurol.* **56**, 751-761
54. Natté, R., de Boer, W.I., Maat-Schieman, M.L.C., Baelde, H.J., Vinters, H.V., Roos, R.A.C., van Duinen, S.G. (1999) Amyloid β precursor protein-mRNA is expressed throughout cerebral vessel walls, *Brain Res.* **828**, 179-183
55. Weller, R.O., Massey, A., Newman, T.A., Hutchings, M., Kuo, Y-M., Roher, A.E. (1998) Cerebral amyloid angiopathy. Amyloid β accumulates in putative interstitial fluid drainage pathways in Alzheimer's disease, *Am. J. Pathol.* **153**, 725-733
56. Shinkai, Y., Yoshimura, M., Ito, Y., Odaka, A., Suzuki, N., Yanagisawa, K., Ihara, Y. (1995) Amyloid β-proteins 1-40 and 1-42 (43) in the soluble fraction of extra- and intracranial blood vessels, *Ann. Neurol.* **38**, 421-428

Chapter 14

NEUROPATHOLOGY AND GENETICS OF PRION PROTEIN AND BRITISH CEREBRAL AMYLOID ANGIOPATHIES

BERNARDINO GHETTI*, PEDRO PICCARDO*, BLAS FRANGIONE†, RUBÉN VIDAL† and JORGE GHISO†

**Indiana University School of Medicine, Indianapolis, IN, USA, †New York University School of Medicine, New York, NY, USA*

Cerebral amyloid angiopathy (CAA), may be associated with normal aging and Alzheimer disease, and may be a cause of cerebral hemorrhage. Disorders in which amyloid is deposited as CAA include hereditary cerebral hemorrhage with amyloidosis of Icelandic and Dutch types, the Hungarian and Ohio kindreds of meningocerebrovascular amyloidosis and the gelsolin-related spinal and CAA. Recently, two forms of dementia associated with CAA not presenting cerebral hemorrhage have been described. They are the prion protein CAA (PrP-CAA) and the familial British dementia (FBD). PrP-CAA is associated with a stop codon mutation (Y145stop) in the PrP gene. In FBD, a single nucleotide substitution at codon 267 results in an arginine residue in place of the stop codon normally occurring in the precursor molecule and a longer open-reading frame of 277 amino acids instead of 266. The two most characteristic microscopic lesions of PrP-CAA and FBD are amyloid deposits in parenchymal blood vessels and neurofibrillary lesions.

1. INTRODUCTION

Cerebral amyloid angiopathy (CAA), a disorder most commonly associated with normal aging and Alzheimer disease (AD), is often a cause of cerebral hemorrhage. Familial conditions in which amyloid is deposited as CAA include hereditary cerebral hemorrhage with amyloidosis-Icelandic type(HCHWA-I) [1] and Dutch (HCHWA-D) [2] types, the Hungarian and Ohio kindreds of meningocerebrovascular amyloidosis [3,4] and the gelsolin-related spinal and cerebral amyloid angiopathy [5]. In most of the cases, the biochemical identification of the amyloid subunit responsible for

M.M. Verbeek et al. (eds.),
Cerebral Amyloid Angiopathy in Alzheimer's Disease and Related Disorders, 237–247.

the deposits has led to the discovery of a genetic defect associated with the disease. In Table 1 we list the constituent proteins of the various forms of CAA and show their genetic alterations.

The constituent protein of the cerebrovascular amyloid in HCHWA-I is a variant of cystatin C (ACys-L68Q) encoded by a single gene located on chromosome 20. The 110-residue-long ACys-L68Q amyloid subunit is a degradation product of the cysteine protease inhibitor cystatin C, starting at position 11 and bearing an amino acid substitution (glutamine for leucine) [6] as a result of a single nucleotide change, A for T at codon 68 [7] (see Ólafsson and Thorsteinsson, this volume).

In HCHWA-D, amyloid deposits are formed of a mixture of wild-type amyloid β (Aβ) peptide and the Aβ-E22Q variant [8-10]. The Aβ peptide is a fragment of the larger Aβ precursor protein (APP) encoded by a single gene on chromosome 21. The Aβ-E22Q variant is due to a single nucleotide change, G for C, at codon 693 of APP [10] (see Maat-Schieman *et al*, this volume).

Cerebral amyloid deposits consisting of transthyretin (TTR) variants ATTR-D18G [3] and ATTR-V30G [4] have been reported in two families carrying different point mutations in the *TTR* gene mapped to chromosome 18. Familial TTR amyloidosis is usually associated with peripheral neuropathy and involvement of visceral organs whereas signs of central nervous system involvement are an exception. In the Hungarian kindred, a single nucleotide change (A for G) at codon 18 results in the presence of glycine instead of aspartic acid while in the Ohio kindred, a T for G substitution at codon 30 results in the substitution of valine for glycine.

Amyloid deposits in familial amyloidosis, Finnish type (FAF) are composed of a 7 kd internal degradation product (residues 173-243) of human plasma gelsolin, an actin regulatory protein encoded by a gene on chromosome 9. The amyloid subunit (AGel) bears an amino acid substitution at position 187 (aspartic acid for asparagine) [11,12] due to a single G to A transition [13]. The mutation has been detected in Finnish, Dutch, American and Japanese families. A different amino acid substitution at the same codon 187 has been described in patients of Danish and Czech origin suffering from the same disorder. In these cases, a transition of G to T at the same codon results in the presence of a tyrosine instead of the normally occurring aspartic acid [14]. Finnish cases carrying the 187 Asn for Asp mutation have demonstrated a widespread spinal, cerebral and meningeal amyloid angiopathy [5].

In recent years, two novel forms of dementia associated with CAA, but not presenting cerebral hemorrhage have been described. They are the prion protein cerebral amyloid angiopathy (PrP-CAA) and the familial British dementia (FBD) (Table 1, Figures 1 and 2).

Table 1: Familial cerebral amyloid angiopathies

Hereditary Cerebral Hemorrhage with Amyloidosis - Icelandic type	**ACys-Q68 (110 aa) (chromosome 20)**	**11 L→Q 120; 68 CTG → CAG**
Hereditary Cerebral Hemorrhage with Amyloidosis - Dutch type	**Aβ-Q22 (40 aa) (chromosome 21)**	**672 E→Q 711; 693 GAA → CAA**
Cerebral amyloid angiopathy Hungarian and Ohio kindreds	**ATTR-G18 ATTR-G30 (127 aa) (chromosome 18)**	**D→G V→G; 1 127; 18 30; GTG → GGG; GAT → GGT**
Gelsolin-related cerebral amyloid angiopathy	**AGel-N15 AGel-Y15 (71 aa) (chromosome 9)**	**D< N / Y; 173 243; 187 → AAC; GAC → TAC**
Vascular variant of Prion protein cerebral amyloidosis	**APrP-Stop145 (≅70 aa) (chromosome 20)**	**Y→Stop; ≅74 144; 145 TAT→TAG**
Familial British Dementia	**ABri (34 aa) (chromosome 13)**	**Stop→R; 244 277; 267 TGA → AGA**

2. PRION PROTEIN CEREBRAL AMYLOID ANGIOPATHY

Abnormal isoforms of the prion protein (PrP) are found in sporadic and iatrogenic Creutzfeldt-Jakob disease, Kuru and hereditary disorders caused by mutations in the PrP gene *(PRNP)*. Parenchymal deposition of PrP amyloid is the hallmark of Gerstmann-Sträussler-Scheinker disease (GSS), a dominant disorder associated with specific *PRNP* mutations [15]. The composition of PrP amyloid has been determined in GSS with mutations at codons 117, 198, and 217 [16-18]. The major components of amyloid are N- and C-terminally truncated PrP fragments of approximately 7 and 11 kd, spanning residues 81-144/150 and 58-150 respectively [16-18]. These fragments originate from the mutant alleles [17]. In GSS with a mutation at codon 198 or 217, deposition of PrP amyloid in the parenchyma is associated with neurofibrillary lesions composed of paired helical filaments (PHFs) [15].

A rare form of early onset-dementia, clinically similar to AD, associated with a TAT to TAG (Tyrosine to Stop) mutation at codon 145 of *PRNP* [19] and resulting in a stop codon (Y145STOP) in conjunction with Met 129 is associated with PrP amyloid deposition in cerebral vessels [20]. As a result of this genetic defect, the pathologic PrP is synthesized as a truncated isoform. The C-terminus of this PrP is similar to the C-terminus of abnormally degraded PrP that results from mutations at *PRNP* codons 117, 198 and 217. Such a protein is devoid of glycosylation sites and the signal sequence for the glycosyl-phosphatidyl- inositol (GPI) anchor. This molecular defect results in the deposition of vascular PrP amyloid. The disorder represents the first instance of PrP amyloid angiopathy in humans [20].

The two most characteristic microscopic lesions of PrP-CAA are seen in the gray matter and they are: (i) amyloid deposits in parenchymal and leptomeningeal blood vessels as well as in the perivascular neuropil and (ii) neurofibrillary lesions (Figure 1). The amyloid angiopathy primarily involves the walls of small and medium sized vessels of the cerebral and cerebellar gray matter and to a lesser extent the leptomeningeal vessels; amyloid is also prominent in the perivascular space and the surrounding parenchyma. Amyloid is immunoreactive with anti-PrP antibodies but not with anti-Aβ antibodies. Vessel-associated PrP deposits are conspicuous in the gray matter of cerebrum and cerebellum, but are only occasionally observed in the brainstem. While PrP deposits in the capillaries are conspicuously present in the vessel walls, in the arterioles they are abundant in the adventitia, relatively sparing the media. PrP-immunoreactivity is rare in the vessels of the white matter.

Neurofibrillary tangles, neuropil threads and dystrophic neurites are numerous in the cerebral gray matter. Most affected is the hippocampus, where the majority of pyramidal and granule cells are involved. In the pyramidal layer, numerous extracellular tangles are seen. In PrP-CAA, neurofibrillary lesions are recognized by a panel of anti-tau monoclonal antibodies directed to the N-terminus and multiple phosphorylation sites of the microtubule-associated protein tau. Specifically, the intracellular neurofibrillary lesions are immunolabeled by ALZ 50, a monoclonal antibody which recognizes an epitope including residues 2-10 of the tau protein and preferentially recognizes a conformation of assembled protein [21-23]. The anti-tau antibodies immunolabeled a large number of neuropil threads, although with variable intensity. AT8, a phosphorylation-dependent monoclonal antibody which recognizes phosphorylated Ser202/Thr205 [24] labeled the largest number of neuropil threads, whereas ALZ 50 recognized the smallest number. In many instances, abnormal neurites immunolabeled by ALZ 50 and AT8 are closely associated with parenchymal amyloid deposits and amyloid laden blood vessels, resulting in a picture of neuritic immunolabelling that overlapped with PrP-immunopositivity.

Neurofibrillary lesions are composed of PHFs, with a periodicity of 70-80 nm and are decorated by the phosphorylation dependent monoclonal antibody PHF1 which recognizes phosphorylated Ser396/Ser404 of the tau protein [25].

PrP amyloid is immunostained by antibodies raised against synthetic peptides corresponding to residues PrP 90-102 and PrP 127-147, as well as monoclonal antibody 3F4 recognizing residues 109-112 [26-28]. The amyloid is not recognized by antibodies raised against the N-terminus of PrP (PrP 23-40), while it is positive with antibodies recognizing the C-terminus of PrP (PrP 220-231) [20,26], suggesting that wild-type PrP participates in the disease process. By electron microscopy, amyloid is composed of straight 8-10 nm fibrils; bundles of radially oriented amyloid filaments are seen adjacent to the vessel wall, whereas haphazardly oriented filaments are numerous within the wall of vessels internally to the basement membrane. Amyloid fibrils are decorated by PrP 90-102, which also labeled amorphous material intermingled with the fibrils.

Immunoblot analysis of proteins extracted from amyloid showed that the smallest subunit migrated as a broad band at approximately 7.5 kd. This band is immunoreactive with antibodies to epitopes located between residues 90-147 of PrP, and is not reactive with PrP 23-40, and PrP 220-231. Amyloid fractions also contained higher molecular weight peptides migrating as poorly resolved bands of 12-16 and 22-30 kd. These bands exhibited the same pattern of immunoreactivity as the 7.5 kd peptide, suggesting that they consisted primarily of polymers of amyloid protein.

The pattern of immunoreactivity indicates that the major PrP amyloid protein of Y145STOP is truncated at the N- and C-termini. Truncation of the C-terminus of the protein occurs at a similar site (i.e. between aa 144 and 150) in all GSS variants in which the amyloid protein has been analyzed [16-18]. Thus, the formation of a PrP peptide with a C-terminus around residue 150 appears to be important for amyloid formation. The absence of C-terminal fragments of PrP from purified amyloid suggests that the fragments are not incorporated into amyloid filaments, or are removed during the extraction procedure, or contribute to amyloid in an amount not detectable by our procedure. In GSS, the amyloid protein originates from the mutant allele. In Y145STOP, the PrP peptides containing the C-terminus probably derive from wild-type PrP, suggesting that it can also be involved in the pathologic process.

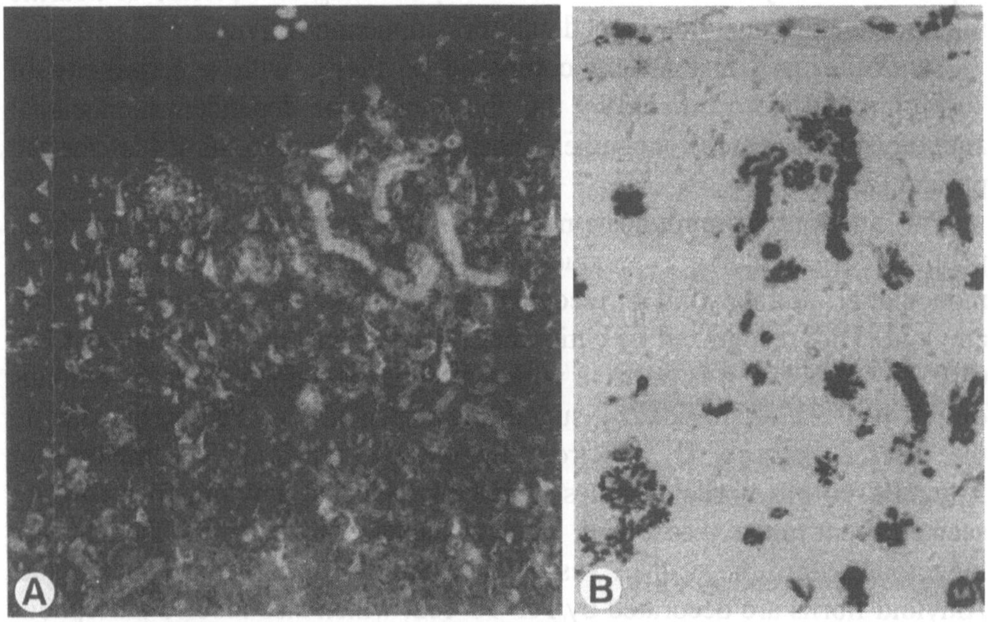

Figure 1. Neocortex from a patient affect by PrP-CAA. A) Amyloid angiopathy and neurofibrillary tangles are shown in a thioflavin S preparation. B) Numerous capillaries are immunolabeled with antibody 3F4. Magnifications: A. x177, B. x188

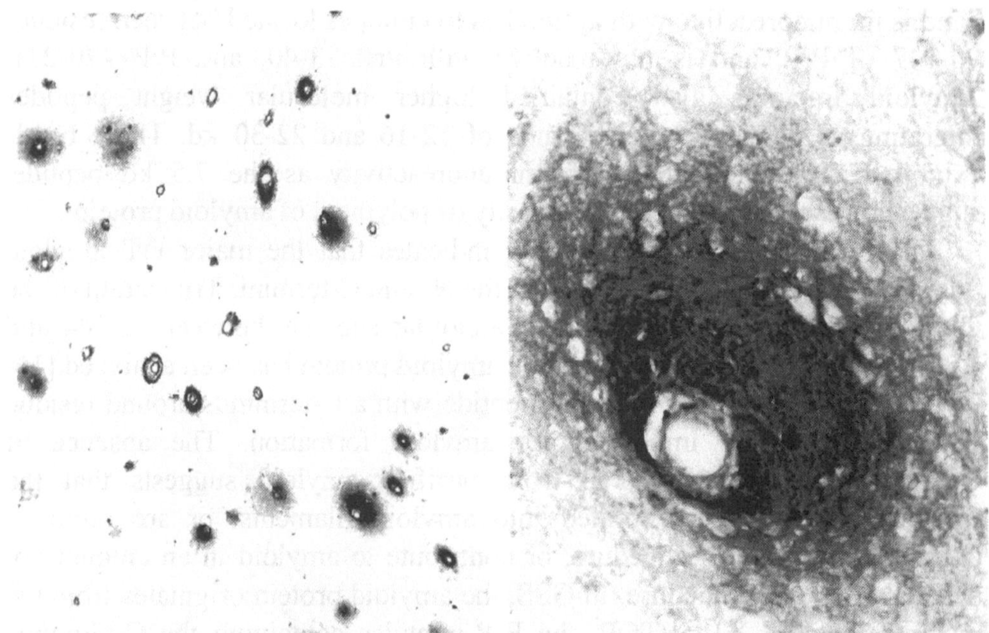

Figure 2. Neocortex from a patient affected by FBD. A) Perivascular plaques specifically labeled with antibody 338 (anti-C-terminal ABri antibody; 1:2,000). B) High power view of a perivascular deposit. Magnifications: A. x135, B. x1345

3. FAMILIAL BRITISH DEMENTIA

Familial British dementia (FBD) is an autosomal dominant form of CAA clinically characterized by progressive dementia, spastic tetraparesis and cerebellar ataxia, with an age of onset in the fourth to fifth decade. An extensive pedigree with the disease occurring in multiple generations has been reported [29]. Neuropathologically, there is severe and widespread amyloid angiopathy of the brain and spinal cord with perivascular amyloid plaque formation, periventricular white matter changes resembling Binswanger's leucoencephalopathy, neuritic and non-neuritic amyloid plaques affecting cerebellum, hippocampus, amygdala and occasionally cerebral cortex, and neurofibrillary degeneration of hippocampal neurons. Magnetic resonance brain scans are useful for early detection of the white matter changes, which are usually accompanied by ventricular dilatation. In spite of the extensive amyloid deposition in the CNS vasculature, large intracerebral hemorrhage is not a common feature of the disease.

FBD was first reported in 1933 by Worster-Drought in two siblings [30] and the detailed neuropathological description was published 7 years later [31]. The pedigree has been followed and expanded since then [29]. At present, the Worster-Drought kindred comprises 343 individuals over nine generations dating back to ~1780. Although this pedigree is large, it is not complete since the descendants of 22 individuals from early generations who were at risk of the disease could not be traced. FBD has been interpreted as an atypical form of familial Alzheimer's disease (AD) [32], as an example of a spongiform encephalopathy [33-34] and also regarded as a specific form of primary congophilic angiopathy [35]. Due to the extensive cerebrovascular involvement, the disorder has been previously designated as familial cerebral amyloid angiopathy - British type [29] and cerebrovascular amyloidosis - British type [36].

The biochemical nature of the amyloid fibrils extracted from leptomeningeal deposits in FBD was recently uncovered [37]. The amyloid subunit, named ABri, is composed of 34 amino acids (pEASNCFAIRHFENKFAVETLICSRTVKKNIIEEN) and has no sequence identity when compared to known amyloid proteins. The isolated amyloid has a molecular mass of 3,937 Da and some degree of N- and C-terminal heterogeneity. It is devoid of glycine, methionine, proline, aspartic acid, tryptophane, tyrosine and glutamine, while it features pyroglutamate at the N-terminus. Two cysteine residues at positions 5 and 22 may be of importance for polymerization and fibrillization. The novel peptide has a predicted isoelectric point of 7.0, suggestive of low solubility properties at physiologic pH; in fact, synthetic peptides homologous to the full length ABri spontaneously polymerize and aggregate in solution.

Antibodies against the last 10 residues of the ABri sequence (Ab 338 [37]) specifically recognize the non-neuritic and perivascular plaques as well as vascular amyloid deposits (Figure 2). Immunoreactivity was completely abolished by absorption of the antibody with ABri synthetic peptide. Antibody 338 did not immunolabel parenchymal and cerebrovascular amyloid deposits in brain sections of sporadic CAA, sporadic Alzheimer disease, Down syndrome, hereditary cerebral hemorrhage with amyloidosis-Dutch type, hereditary cerebral hemorrhage with amyloidosis - Icelandic type, Hungarian transthyretin cerebral amyloidosis, and systemic cases of light chain amyloidosis (kidney), light chain deposition disease (kidney) and amyloid A (heart).

This new amyloid protein is a degradation product of a 277 amino acids precursor molecule with a primary structure that resembles a type II single-spanning transmembrane protein. The precursor protein is encoded by a single gene *BRI* located on the long arm of chromosome 13. In patients with FBD, a single nucleotide substitution (TGA to AGA at codon 267) results in the presence of an arginine residue in place of the stop codon normally occurring in the wild type precursor molecule and a longer open-reading frame of 277 amino acids instead of 266. The ABri amyloid peptide is formed by the last 34 amino acids of the mutated precursor protein, presumably as a result of a furin-like processing [38]. A consensus sequence for a number of calcium-dependent serine proprotein convertases (PCs) is present at positions 238 to 243 of the precursor protein. Many of these enzymes (i.e. furin) localize to the trans-Golgi network by virtue of their transmembrane domains feature that provides access to many precursor molecules that move to the cell surface via constitutive vesicles [39]. Further degradation by local amino- and carboxyl-peptidases may take place *in situ* producing the N- and C- terminal heterogeneity observed in the isolated amyloid. The post-translational modification observed at position 1 (pyroglutamate) has been previously found in other brain amyloids, i.e. peptides derived from Alzheimer's Aβ [40-42]. The N-terminal pyroglutamate would be expected to offer some protection against *in vivo* proteolysis and to increase the β-pleated sheet content of the ABri peptide and its tendency to aggregate and polymerize [43].

ABri is structurally unrelated to all known amyloids, including those deposited in the brain. It is the first example of an amyloid molecule created by the abolishment of a stop codon in its precursor protein. However, immunohistochemical and electron microscopical studies demonstrated that the neurofibrillary tangles in these patients are identical to those seen in the patient with PrP-CAA; in fact, they immunoreact with phosphorylation-dependent and phosphorylation-independent anti-tau antibodies [20, 36]. Thus, the cytoskeletal pathology in these patients is indistinguishable from that seen in other neurodegenerative conditions, including AD and GSS [15, 36, 44-45]. Amyloid peptides may be of primary importance in the initiation of neurodegeneration; however, the exact mechanism by which different amyloid peptides could trigger similar neuropathological changes leading to neuronal loss and dementia remains to be determined.

REFERENCES

1. Gudmundsson G., Hallgrimsson, J., Jonasson, T., and Bugiani, O. (1972) Hereditary cerebral hemorrhage with amyloidosis, *Brain* **95**, 387-404.
2. Luyendijk W. and Schoen, J, (1964) Intracerebral haematomas. A clinical study of 40 surgical cases, *Psychiat. Neurol. Neurochir.* **67**, 445-468.
3. Vidal R., Garzuly, F., Budka, H., Lalowski, M., Linke, R., Brittig, F., Frangione, B., and Wisniewski, T. (1996) Meningocerebrovascular amyloidosis associated with a novel Transthyretin mis-sense mutation at codon 18 (TTRD18G), *Am. J. Pathol.* **148**, 361-366.
4. Petersen R., Goren, H., Cohen, M., Richardson, S., Tresser, N., Lynn, A., Gali, M., Estes, M., and Gambetti, P., (1997) Transthyretin amyloidosis: A new mutation associated with dementia, *Ann. Neurol.* **41**, 307-313.
5. Kiuru S, Salonen O, and Haltia M. (1999) Gelsolin-related spinal and cerebral amyloid angiopathy. *Ann. Neurol.* **45,** 305-311.
6. Ghiso J., Jensson, O., and Frangione, B. (1986) Amyloid fibrils in hereditary cerebral hemorrhage with amyloidosis of Icelandic type is a variant of γ-trace basic protein (cystatin-C). *Proc. Natl. Acad. Sci.USA* **83**, 2974-2978.
7. Levy E., Lopez-Otin, C., Ghiso, J., Geltner, D., and Frangione, B., (1989) Stroke in Icelandic patients with hereditary amyloid angiopathy is related to a mutation in the cystatin-C gene, an inhibitor of cysteine proteases. *J. Exp. Med.* **169**, 1771-1778.
8. Van Duinen S., Castaño, E., Prelli, F., Bots, G., Luyenkijk, W., and Frangione, B. (1987) Hereditary cerebral hemorrhage with amyloidosis in patients of Dutch origin is related to Alzheimer's disease. *Proc. Natl. Acad. Sci. USA* **84**, 5991-5994.
9. Prelli F., Levy, E., Van Duinen, S., Bots, G., Luyendijk, W., and Frangione, B. (1990) Expression of a normal and variant Alzheimer's β-protein gene in amyloid of hereditary cerebral hemorrhage, Dutch type: DNA and protein diagnostic assays, *Biochem. Biophys. Res. Commun.* **170**, 301-307.
10. Levy E., Carman, M., Fernandez-Madrid, I., Power, M., Lieberburg, I., Van Duinen, S., Bots, G., Luyendijk, W., and Frangione, B. (1990) Mutation of the Alzheimer's disease amyloid gene in hereditary cerebral hemorrhage, Dutch type, *Science* **248**, 1124-1126.
11. Ghiso J, Haltia M, Prelli F, Novello J, and Frangione B. (1990) Gelsolin variant (Asn-187) in familial amyloidosis Finnish type. *Biochem. J* **272,** 827-830.
12. Maury, C.P. and Baumann M. (1990) Isolation and characterization of cardiac amyloid in familial amyloid polyneuropathy type IV (Finnish): relationship of the amyloid protein to variant gelsolin. *Biochim. Biophys. Acta* **1096,** 84-86.
13. Levy E, Haltia M, Fernandez-Madrid I, Kouvinen O, Ghiso J, Prelli F, and Frangione B. (1990) Mutation in gelsolin gene in Finnish hereditary amyloidosis. *J. Exp. Med.* **172,** 1865-1867
14. de la Chapelle A, Kere J, Sack GH Jr, Tolvanen R, Maury CP. (1992) Familial amyloidosis, Finnish type: G654T -a mutation of the gelsolin gene in Finnish families and an unrelated American family *Genomics* **13,** 898-901.

15. Ghetti, B., Dlouhy, S. R., Giaccone, G., Bugiani, O., Frangione, B., Farlow, M. R., Tagliavini, F. (1995) Gerstmann-Sträussler-Scheinker disease and the Indiana kindred, *Brain Pathol.* **5**, 61-75.
16. Tagliavini F., Prelli F., Ghiso J., Bugiani O., Serban D., Prusiner S.B., Farlow M.R., Ghetti B., Frangione B. (1991) Amyloid protein of Gerstmann-Sträussler-Scheinker disease (Indiana kindred) is an 11 kd fragment of prion protein with an N-terminal glycine at codon 58. *EMBO J.* **10**, 513-519.
17. Tagliavini, F., Prelli, F., Porro, M., Rossi, G., Giaccone, G., Farlow, M. R., Dlouhy, S. R., Ghetti, B., Bugiani, O., Frangione, B. (1994) Amyloid fibrils in Gerstmann-Sträussler-Scheinker disease (Indiana and Swedish Kindreds) express only PrP peptides encoded by the mutant allele, *Cell* **79**, 695-703.
18. Tagliavini, F., Prelli, F., Porro, M., Rossi, G., Giaccone, G., Bird, T. D., Dlouhy, S. R., Young, K., Piccardo, P., Ghetti, B., Bugiani, O., Frangione, B. (1995) Only mutant PrP participates in amyloid formation in Gerstmann-Sträussler-Scheinker disease with Ala|Val substitution at codon 117. *J. Neuropath. Exp. Neurol.* **54**, 416.
19. Kitamoto, T., Iizuka, R., Tateishi, J. (1993) An amber mutation of prion protein in Gerstmann-Sträussler syndrome with mutant PrP plaques, *Biochem. Biophys. Res. Commun.* **192**, 525-531.
20. Ghetti B., Piccardo P., Spillantini M.G., Ichimiya Y., Porro M., Perini F., Kitamoto T., Tateishi J., Seiler C., Frangione B., Bugiani O., Giaccone G., Prelli F., Goedert M., Dlouhy S.R., Tagliavini F. (1996) Vascular variant of prion protein cerebral amyloidosis with tau-positive neurofibrillary tangles: The phenotype of the stop codon 145 mutation in *PRNP. Proc. Natl. Acad. Sci. USA* **93**, 744-748.
21. Wolozin B.L., Pruchnicki A., Dickson D.W., Davies P. (1986) A neuronal antigen in the brains of Alzheimer patients. *Science* **232**, 648-650.
22. Goedert M., Spillantini M.G., Jakes R. (1991) Localization of the Alz-50 epitope in recombinant human microtubule-associated protein tau. *Neurosci. Lett.* **126**, 149-154.
23. Carmel G., Mager E.M., Binder L.I., Kuret J. (1996) The structural basis of monoclonal antibody Alz50's selectivity for Alzheimer's disease pathology. *J. Biol. Chem.* **277**, 32789-32795.
24. Goedert M., Jakes R., Vanmechelen E. (1995) Monoclonal antibody AT8 recognizes tau protein phosphorylated at both serine 202 and threonine 205. *Neurosci. Lett.* **189**, 167-170.
25. Otvos L., Feiner L., Lang E., Szendrei G.I., Goedert M., Lee V.M.Y. (1994) Monoclonal antibody PHF-1 recognizes tau protein phosphorylated at serine residues 396 and 404. *J. Neurosci. Res.* **39**, 669-673.
26. Piccardo, P., Ghetti, B., Dickson, D. W., Vinters, H. V., Giaccone, G., Bugiani, O., Tagliavini, F., Young, K., Dlouhy, S. R., Seiler, C., Jones, C. K., Lazzarini, A., Golbe, L. I., Zimmerman, T. R., Perlman, S. L., McLachlan, D. C., St George-Hyslop, P. H., Lennox, A. (1995) Gerstmann-Straussler-Scheinker disease (PRNP P102L): amyloid deposits are best recognized by antibodies directed to epitopes in PrP region 90-165, *J. Neuropath. Exp. Neurol.* **54**, 790-801
27. Kascsak, R. J., Rubenstein, R., Merz, P. A., Tonna-DeMasi, M.; Fersko, R.; Carp, R. I.; Wisniewski, H. M.; Diringer, H. (1987) Mouse polyclonal and monoclonal antibody to scrapie-associated fibril proteins, *J. Virol.* **61**, 3688-3693
28. Rogers, M., Serban, D., Gyuris, T., Scott, M., Torchia, T., Prusiner, S. B. (1991) Epitope mapping of the Syrian hamster prion protein utilizing chimeric and mutant genes in a vaccinia virus expression system, *J. Immunol.* **147**, 3568-3574

29. Plant, G.T., Révész, T., Barnard, R.O., Harding, A.E. and Gautier-Smith, P.C. (1990). Familial cerebral amyloid angiopathy with nonneuritic plaque formation, *Brain* **113**, 721-747
30. Worster-Drought, C., Hill, T. R.and McMenemey, W. H. (1933) Familial presenile dementia with spastic paralysis. *J. Neurol. Psychopathol.* **14**, 27-34.
31. Worster-Drought, C., Greenfield, J.G and McMenemey, W.H. (1940) A form of familial presenile dementia with spastic paralysis (including the pathological examination of a case). *Brain* **63**, 237-254.
32. Corsellis, J. and Brierley, J.B. (1954) An unusual type of presenile dementia: (atypical Alzheimer's disease with amyloid vascular change), *Brain* **77**, 571-587.
33. Keohane, C., Peatfield, R. and Duchen, L. W. (1985) Subacute spongiform encephalopathy (Creutzfeldt-Jakob disease) with amyloid angiopathy. *J. Neurol. Neurosurg. Psych.* **48**, 1175-1178.
34. Pearlman, R. L., Towfighi, J., Pezeshkpour, G. H., Tenser, R. B. and Turel, A.P. (1988) Clinical significance of types of cerebellar amyloid plaques in human spongiform encephalopathies, *Neurology* **38**, 1249-1254.
35. Vinters, H. (1987) Cerebral amyloid angiopathy: a critical review, *Stroke* 18, 311-324.
36. Révész, T., Holton, J. L., Doshi, B., Anderton, B. H., Scaravilli, F. & Plant, G.T.(1999) Cytoskeletal pathology in familial amyloid angiopathy (British type) with non-neuritic plaque formation. *Acta Neuropath.* **97**, 170-176.
37. Vidal, R., Frangione, B., Rostagno, A., Mead, S., Révész, T., Plant, G., and Ghiso, J. (1999) A stop-codon mutation in the BRI gene associated with familial British dementia, *Nature* **399**, 776-781.
38. Kim S.H., Wang, R., Gordon, D.J., Bass, J., Steiner, D.F., Tinakaran, G., Meredith,S.C. and Sisodia, S.S. (1999) Furin mediates enhanced production of fibrillogenic ABri peptide in familial British dementia. *Nature Neurosci.* **2**, 984-988.
39. Zhou, A., Webb, G., Zhu, X., and Steiner, D.F. (1999) Proteolytic processing in the secretory pathway, *J. Biol. Chem.* **274**, 20745-20748.
40. Mori, H., Takio, K., Ogawara, M., and Selkoe, D. (1992) Mass spectrometry of purified amyloid beta protein in Alzheimer's disease, *J. Biol. Chem.* **267**, 17082-17086.
41. Saido, T., Yamao-harigaya, W., Iwatsubo, T., and Kawashima, S. (1996) Amino- and carboxyl-terminal heterogeneity of beta-amyloid peptides deposited in human brain, *Neurosci. Lett.* **13**, 173-176.
42. Tekirian, T., Saido, T., Markesbery, W., Russell, M., Wekstein, D., Patel, E. and Geddes J. (1998) N-terminal heterogeneity of parenchymal and cerebrovascular Aβ deposits, *J. Neuropathol. Exp. Neurol.* **57**, 76-94.
43. He, W., and Barrow, C.J. (1999) The Aβ 3-pyroglutamyl and 11-pyrogluramyl peptides found in senile plaques have greater β-sheet formation and aggregation propensities *in vitro* than full-length Aβ, *Biochemistry* **38**, 10871-10877.
44. Dickson, D. (1997) Neurodegenerative diseases with cytoskeletal pathology: a biochemical classification, *Ann.Neurol.* **42**, 541-544.
45. Giaccone, G., Tagliavini, F., Verga, L., Frangione, B., Farlow, M., Bugiani, O. and Ghetti, B. (1990) Neurofibrillary tangles in the Indiana kindred of Gerstmann-Sträussler-Scheinker disease share antigenic determinants with those of Alzheimer's disease, *Brain Res.* **530**, 325-329.

SECTION IV

IN VITRO AND ANIMAL MODELS OF CAA

Chapter 15

AMYLOID β PROTEIN INTERNALIZATION AND PRODUCTION BY CANINE SMOOTH MUSCLE CELLS

REINHARD PRIOR and BRITTA URMONEIT
Department of Neurology, University of Duesseldorf, Duesseldorf, Germany

Aging dogs show progressive β-amyloid (Aβ) deposition within the walls of small cortical and leptomeningeal arteries and develop cerebral amyloid angiopathy (CAA) that is morphologically identical to human CAA associated with Alzheimer's disease and with aging. The canine and the human Aβ amino acid sequences are identical, which is an essential requirement for an animal model of β-amyloidosis, because subtle changes within the Aβ peptide sequence may impredictibly change its aggregation properties. Since cerebrovascular Aβ deposits are always closely associated with vascular smooth muscle cells (SMCs) or the SMC-related parenchymal pericytes, primary canine cerebrovascular SMC cultures have been used to investigate the molecular mechanisms underlying the development of cerebrovascular Aβ deposition. This review summarizes the results obtained with canine SMC cultures and discusses their potential implications for the pathogenesis of CAA.

1. SMC CULTURE

Primary cultures of canine cerebrovascular SMCs are obtained by enzymatic disaggregation of small arterioles dissected from canine leptomeninges [1,2]. They grow under standard cell culture conditions and can be propagated for a consistent number of passages after which proliferation decreases and dedifferentiation occurs. Morphologically, the cells are spindle-shaped or spread (Figure 1A) resembling SMCs derived from the media of large vessels such as the aorta. Cerebrovascular SMCs show two types of growth patterns: until confluency, monolayers are formed, whereas thereafter cells tend to cluster in threedimensional multilayered nodules, according to a "hill and valley" growth pattern also described for

M.M. Verbeek et al. (eds.),
Cerebral Amyloid Angiopathy in Alzheimer's Disease and Related Disorders, 251–264.

aortic SMCs [3] (Figure 1C,D). The identity of cultured SMCs can be confirmed by immunostaining for smooth muscle actin: in typical cultures, between 20 and 90% of the cells are positive for smooth muscle α-actin, depending on growth state and culture density (Figure 1A).

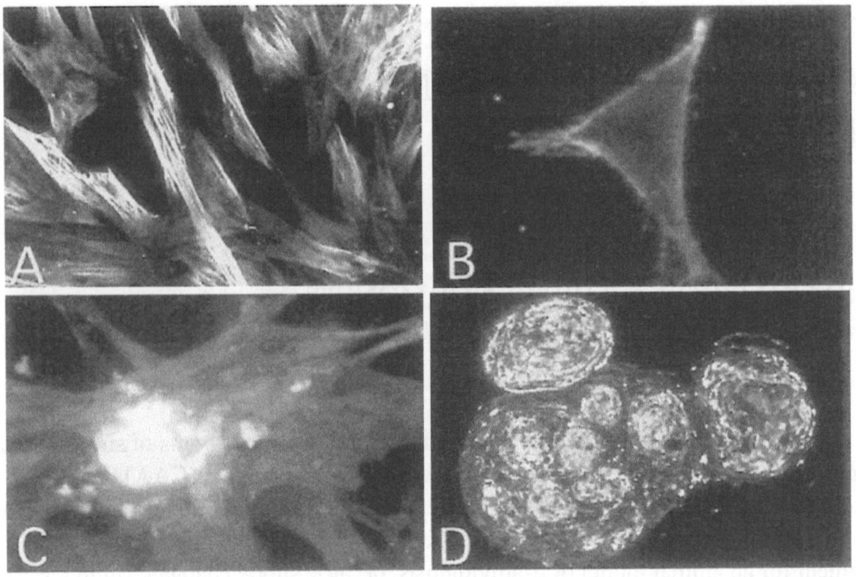

Figure 1. Morphology of primary canine cerebrovascular SMC cultures and binding of fluorescein-Aβ to SMCs ($A\beta_{1\text{-}40}$, 3 μM, at 4 °C in B,C or added to the culture medium in D). Cells are spindle-shaped or spread and form both monolayers and multilayered nodules. A: Immunofluorescence for smooth muscle actin confirms SMC differentiation. B: Surface binding of Aβ. C: Strong Aβ-binding to multilayered nodules. D: Cryosection of an SMC-nodule grown for 1 week under non-adherent conditions in the presence of fluorescein-conjugated $A\beta_{1\text{-}40}$. Aβ is integrated in a distinctive pattern that probably reflects the distribution of extracellular matrix components

When monolayer cultures of canine SMCs are incubated at 4 °C with fluorescein-conjugated Aβ peptides ($A\beta_{1\text{-}40}$ and $A\beta_{1\text{-}42}$), Aβ binds to the cell surface [2] (Figure 1B). The most intense Aβ binding however is found on multilayered SMC nodules suggesting that the binding of Aβ depends on the SMC growth state and that Aβ has a strong affinity for specific components of the extracellular matrix produced by SMCs once they grow in multilayered nodules (Figure 1C). When SMCs are cultured under non-adherent conditions to produce cell spheroids with addition of fluorescein-conjugated Aβ-peptides, Aβ peptides are continuously integrated in the extracellular matrix of growing nodules and appear deposited within distinctive rounded structures on spheroid sections (Figure 1D).

2. INTERNALIZATION OF Aβ BY CANINE SMCS VIA MULTIPLE LIPOPROTEIN PATHWAYS

After incubation of SMCs at 37 °C for various time periods with fluorescein-conjugated Aβ-peptides ($A\beta_{1-40}$ and $A\beta_{1-42}$), Aβ is localized to endosomes and later to lysosomes as was evident by colocalization with the lysosome-associated membrane protein LAMP-1 [2] (figure 2).

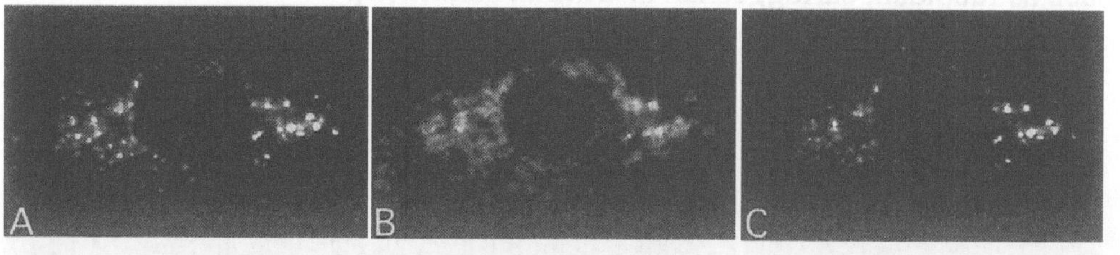

Figure 2. Colocalization between fluorescein-conjugated $A\beta_{1-40}$ (1 μM) after 12 h of incubation of SMCs at 37 °C (A) and immunofluorescence for LAMP-1 (B) reveals the lysosomal localization of internalized Aβ. Images were obtained by confocal laser scanning microscopy, the colocalization is shown in C.

On the cell surface and within early endosomes, fluorescein-conjugated Aβ was colocalized to apolipoprotein E (ApoE, Figure 3A-C) and its receptor, the low density lipoprotein receptor-related protein (LRP, Figure 3D-F) suggesting that Aβ is at least partially internalized by receptor-mediated endocytosis following a lipoprotein pathway and that ApoE containing lipoproteins may mediate the internalization of Aβ into SMCs. Aβ internalization via receptor-mediated endocytosis is also supported by additional experimental evidence derived from treatment of SMCs with trypsin, cycloheximide or Brefeldin A in order to degrade cell surface receptors, to inhibit receptor expression or inhibit intracellular transport, respectively. All of these treatments prevented the internalization of Aβ by SMCs. The internalization of Aβ via a lipoprotein pathway is further supported by coincubation of SMCs with receptor associated protein (RAP), which is a specific antagonist for LRP and which inhibited Aβ internalization by SMCs. Internalization of Aβ was also abolished when serum-free or lipid-free culture media were used suggesting that the binding of Aβ to lipoproteins contained within the serum is essential to its subsequent internalization [2].

The internalization of Aβ by cultured SMCs appears to involve multiple lipoprotein receptors or multiple pathways, because RAP only partially blocked Aβ-internalization [2]. In addition to LRP, endosomal colocalization was also demonstrated for Aβ with scavenger receptors. These are involved in the removal of modified lipoproteins and lipid accumulation within macrophages and SMCs of arteriosclerotic lesions and have been extensively investigated in the context of atherosclerosis (reviewed in [4,5]). Both class A and class B scavenger receptors are colocalized to Aβ within the endosomes of canine SMCs cultured at 37 °C in the presence of fluorescein-conjugated Aβ peptides. Moreover, the internalization of Aβ is strongly enhanced in rabbit SMCs transfected with class A scavenger receptors and was partially blocked by different types of scavenger receptor antagonists such as fucoidan, carraggeenan or dextran sulfate [6].

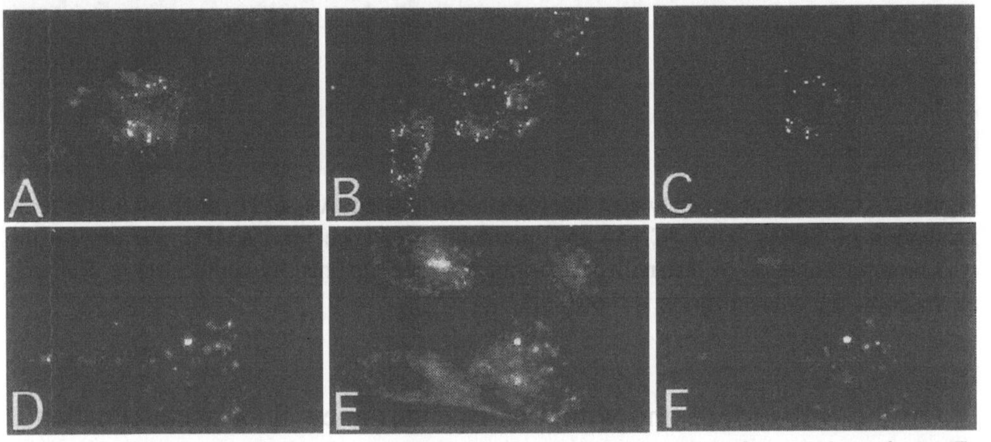

Figure 3. A-C: Colocalization between fluorescein-conjugated $A\beta_{1-40}$ (A) and ApoE-immunofluorescence (B). Images were obtained by confocal laser scanning microscopy, the colocalization is shown in C. D-F: Colocalization between fluorescein-conjugated $A\beta_{1-40}$ (D) and LRP-immunofluorescence (E); colocalization is shown in F. Images were obtained after binding of 1 μM $A\beta_{1-40}$ at 4 °C to the cell surface and subsequent incubation for 20 min at 37 °C.

The internalization of Aβ by SMCs may also be mediated by heparan sulfate proteoglycans (HSPGs), since HSPGs are produced by SMCs and have been shown to be involved in SMC lipoprotein uptake [7,8]. In addition, there may also be a HSPG-mediated receptor-independent pathway as it has been demonstrated in studies of remnant lipoprotein metabolism (reviewed in [9]). Interestingly, the receptor-independent HSPG-mediated pathway relies on the presence of ApoE-containing lipoproteins and is the only pathway responsible for isoform-specific differences observed for cellular lipid uptake when the ApoE3 and ApoE4 isoforms are compared [9].

Internalization of Aβ by canine SMCs is regulated and is affected by the presence of growth factors in the culture medium. Both platelet-derived growth factor (PDGF) and transforming growth factor β1 (TGF-β1) increase the intracellular accumulation of Aβ peptides contained in the culture medium in a dose-dependent manner (Figure 4A,B). The latter observation is intriguing, because transgenic mice overexpressing TGF-β1 develop CAA [10]. The expression of TGF-β1 is also increased in Alzheimer brains [10] and TGF-β1 may affect the Aβ-processing by cerebrovascular SMCs in a way that leads to CAA development. In addition, PDGF and TGF-β1 are both involved in the pathogenesis of atherosclerosis and recent epidemiological studies suggest an association between the presence of atherosclerosis and the incidence of CAA [11]. A pathogenetic relationship between CAA and atherosclerosis is also supported by the fact that both conditions are favored by the presence of the ApoE4 genotype. Since ApoE functions as a cholesterol and lipid transporter, altered cellular uptake of ApoE- and Aβ-containing lipoproteins may have a central role during CAA development.

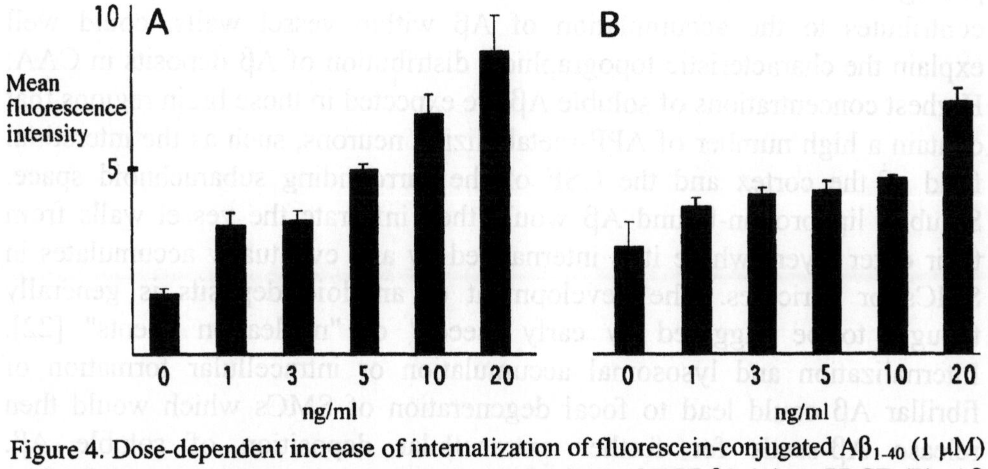

Figure 4. Dose-dependent increase of internalization of fluorescein-conjugated $A\beta_{1\text{-}40}$ (1 μM) by canine SMCs treated with increasing concentrations of TGF-β1 (A) or PDGF (B). Aβ internalization is expressed in arbitrary units representing the mean fluorescence signal obtained by flow cytometry after 2h incubation at 37 °C. Each bar represents mean and standard deviation of three parallel experiments measuring a total of 3x1000 cells.

3. THE PATHOPHYSIOLOGICAL SIGNIFICANCE OF INTERNALIZATION OF Aβ BY SMCS

Internalization of Aβ into cerebrovascular SMCs or into pericytes associated with cortical capillaries may occur *in vivo*, where small cortical and leptomeningeal arteries which represent the predilection site for CAA deposits are surrounded by cerebrospinal fluid (CSF) that contains Aβ peptides bound to lipoproteins. CSF contains high concentrations of ApoE and CSF-derived ApoE binds Aβ [12,13]; it appears therefore likely that CSF-Aβ is actively internalized by SMCs of cerebral vessels via ApoE-mediated lipoprotein pathways. CSF contains a heterogeneous set of Aβ peptides with the more soluble $A\beta_{1-40}$ isoform being a major Aβ species within the CSF [14]. Accordingly, the accumulation of $A\beta_{1-40}$ was selectively found in brains with prominent CAA by various studies [15-18]. Interestingly, another study found a sharp peak of vessel-associated soluble Aβ between ages 50 and 70 even in the absence of CAA-pathology [17]. These data strongly suggest that cerebral vessels internalize and accumulate CSF-dervived Aβ *in vivo*.

Internalization of Aβ may lead to intracellular accumulation of Aβ within lysosomes of SMCs which may favour the development of CAA. It has been shown in fibroblasts [19,20] and in microglia [21] that Aβ internalized into lysosomes is resistant to degradation for prolonged time periods. A pathogenetic model, in which internalization of CSF-derived Aβ by SMCs contributes to the accumulation of Aβ within vessel walls, could well explain the characteristic topographical distribution of Aβ deposits in CAA. Highest concentrations of soluble Aβ are expected in those brain regions that contain a high number of APP-metabolizing neurons, such as the interstitial fluid of the cortex and the CSF of the surrounding subarachnoid space. Soluble, lipoprotein-bound Aβ would then infiltrate the vessel walls from their outer layers where it is internalized by and eventually accumulates in SMCs or pericytes. The development of amyloid deposits is generally thought to be triggered by early "seeds" or "nucleation events" [22]. Internalization and lysosomal accumulation or intracellular formation of fibrillar Aβ could lead to focal degeneration of SMCs which would then release Aβ-seeds for further extracellular deposition of soluble Aβ. Alternatively, the natural turnover of SMCs as it has been described in coronary arteries [23], may lead to cell death of single SMCs which would then leave intracellular Aβ-precipitates in the extracellular space. Using organ cultures of canine leptomeninges [24,25], it was shown that physiological CSF concentrations of Aβ are sufficient to sustain the growth of extracellular CAA deposits, once these deposits have been formed. It has been shown that early non fibrillar Aβ-deposits are detected within SMCs or as granules between myocytes [26]. However, intracellular Aβ deposits are only rarely detected with conventional (immuno)staining methods and are not detected on histological routine specimens of CAA. Special fixation methods or pretreatment of tissue sections may be necessary to reveal the presence of intracellular fibrillar Aβ precipitates *in situ*.

4. APOE4 AS A RISK FACTOR FOR CAA: REDUCED Aβ CLEARANCE AS A POTENTIAL PATHOGENETIC MECHANISM

As an alternative to the hypothesis proposed above, the internalization of soluble CSF-derived and lipoprotein-bound Aβ by brain SMCs may be physiological and may represent a protective mechanism of local Aβ clearance. Such a physiological mechanism may, however, become saturated leading to intra- or extracellular Aβ concentrations that exceed the solubility limit of Aβ resulting in Aβ-fibril formation. When cerebrovascular Aβ deposition is viewed as a consequence of reduced or insufficient Aβ clearance from the extracellular space, recently published findings may explain why CAA deposits form in the vascular extracellular space and why the ApoE4 genotype is a risk factor for CAA: there are several recent findings that indicate a less efficient lipid transport and cellular uptake by the ApoE4 isoform. Binding of Aβ to ApoE was shown to be less efficient for ApoE4 compared to ApoE3 when ApoE was produced by mammalian cells in two independent studies suggesting a less efficient ApoE-mediated Aβ-clearance in carriers of the ε4 genotype [27,28]. Also, intracellular accumulation or clearance of ApoE in hepatocytes [29], neuronal cells [29-31] and CHO cells [32] was significantly lower for ApoE4 when compared to ApoE3. Most importantly, one study [29] showed that the isoform-specific effect was only observed in the presence of cell surface HSPGs, which is abundantly expressed also by SMCs and has been convincingly implicated as an important cofactor in the development of cerebral and vascular Aβ-amyloid deposits [33-35]. The strong binding of fluorescein-Aβ to multilayered SMC nodules (Figure 1C,D) is highly suggestive of an interaction between Aβ and HSPGs or other specific extracellular matrix components, although this remains to be investigated.

Less efficient clearance of CSF-derived Aβ by SMCs could be a pathogenetic factor explaining the increased risk to develop CAA in carriers of the ApoE4 genotype. ApoE mediated Aβ-clearance by cerebrovascular SMCs may be a physiological and protective mechanism within a complex balance between Aβ-production, Aβ-clearance and Aβ-solubility, the latter being also affected by Aβ-binding to other CSF proteins or by changes in Aβ-peptide structure as in hereditary cerebral hemorrhage with amyloidosis-Dutch type. Absolute or relative deficiency of Aβ-clearance may be associated to atherosclerosis, aging, or the ApoE4 genotype and may lead to increased Aβ-concentrations within the vascular extracellular matrix, where specific components such as HSPGs ultimately promote deposition and growth of CAA deposits on the surface of SMCs. Cell surface precipitation of fibrillar Aβ has already been shown in a human cerebrovascular SMC model [36] when $A\beta_{1-42}$ or the mutated Dutch-variant Aβ-peptide were used at micromolar concentrations, which are consistently higher than the physiological nanomolar CSF concentrations of Aβ. Increased extracellular concentrations of Aβ therefore appear necessary to initiate extracellular fibril formation and reduced vascular Aβ-clearance could be one of several factors contributing to the development of CAA.

5. THE ORIGIN OF CEREBROVASCULAR Aβ

Based on experimental findings in squirrel monkeys which were intravenously injected with radiolabeled Aβ [37], it has been suggested that the Aβ deposited within vessel walls affected by CAA may originate from the blood and therefore derive from the vessel lumen. A blood origin of the amyloid protein appears likely for other rare forms of cerebrovascular amyloidoses such as the amyloidoses caused by the deposition of mutated transthyretin [38], gelsolin [39], or cystatin C [40] in which - in contrast to CAA caused by Aβ-deposition - there is also involvement of various systemic organs or of the peripheral nervous system. CAA with Aβ-deposition is always solely confined to the central nervous system, which strongly argues against a systemic origin of the amyloid protein. Moreover, extravascular and cerebral origin of cerebrovascular Aβ is supported by the abluminal localization of early Aβ-deposits in Aβ-CAA and the almost exclusive distribution of CAA to cortical and leptomeningeal vessels together with its absence in white matter where neurons are myelinated and therefore do not secrete Aβ to the extracellular space. Wisniewski and coworkers have proposed that cerebrovascular SMCs produce the Aβ peptides that are deposited within CAA-affected vessel walls *in situ*. This hypothesis is based on ultrastructural findings showing amyloid fibrils close to cytoplasmic vesicles and within the extracellular matrix produced by SMCs [26] and, more importantly, on the detection of Aβ-immunoreactivity within cytoplasmic granules of canine leptomeningeal SMCs in culture [41,42]. Analyzing various conditions that influenced the fraction of Aβ-immunoreactive cells, the same authors have also reported that cytoplasmic accumulation of Aβ is induced by the addition of serum [43] or ApoE [44] to the culture medium. No sequencing data are available, however, to exclude that the reported Aβ-immunoreactivity represents antibody cross-reactivity with the amyloid precursor protein (APP) which is also detected within cytoplasmic vesicles of canine SMCs. Another study has shown that the presence of $A\beta_{1-42}$ enhances the generation of carboxyl terminal APP-fragments in cultured human SMCs while causing SMC degeneration [45]. It is possible, that SMCs accumulate and secrete Aβ-peptides, the amount of

Aβ produced by canine SMCs, however, appears very low or undetectable when compared to the amount of Aβ secreted into the supernatant by Aβ-transfected neuronal cells (Figure 5). Moreover, *in situ* production of Aβ by vascular SMCs hardly explains the total absence of vascular Aβ deposits in systemic vessels and their rare appearance in cerebral white matter vessels. Finally, an extravascular and probably neuronal and CSF-origin of the cerebrovascular Aβ deposits is supported by recent findings in transgenic mice where prominent CAA develops in the absence of vascular APP expression [46].

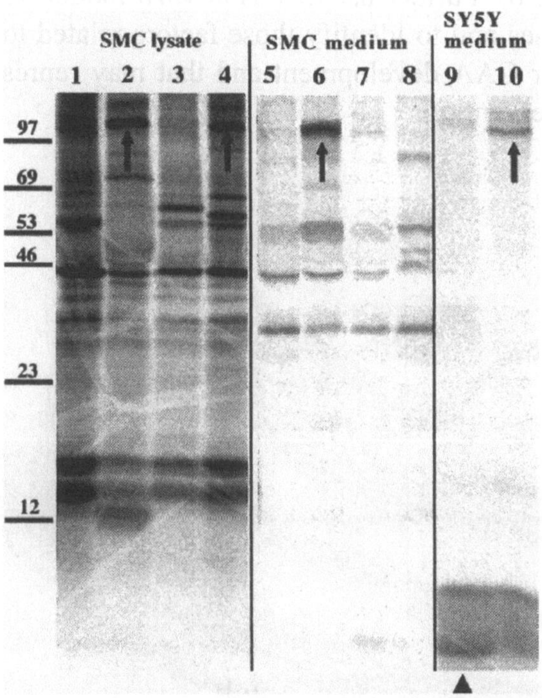

Figure 5. SDS-PAGE (8-12%) analysis of expression and processing of the amyloid precursor protein (APP) in primary canine SMC lysate (lanes 1-4), SMC medium (lanes 5-8), and in the medium of APP-transfected neuronal SY5Y cells (lanes 9,10). SMCs were kept 3 weeks in culture, labeled with 35S-methionine and the cell lysate or the medium was immunoprecipitated with normal rabbit serum (1,5), a polyclonal APP-antibody (2,6), a polyclonal Aβ-antiserum (3,7,9) and the widely used monoclonal antibody 4G8 (anti-Aβ; 4,8,10). SMCs and SY5Y cells express APP, which is recognized by the polyclonal APP-antiserum, but also by the monoclonal 4G8 Aβ-antibody (arrows). Aβ is not detected within SMC lysate or SMC medium, but only in the medium of control cells (9,10; full-length and p3-fragment of Aβ; arrowhead).

In summary, since both early and late Aβ-deposits are always closely associated with SMCs or with pericytes, the development of CAA appears closely related to SMC differentiation and biology. The experimental setup of cultured SMCs that internalize exogenous Aβ from the culture medium reflects the *in vivo* situation, where vascular SMCs are surrounded by CSF containing brain derived Aβ peptides of extravascular, mostly neuronal origin. The observation of Aβ-internalization by SMCs indicates that SMCs may process CSF-derived Aβ peptides also *in vivo* and that the CSF may be the origin of the cerebrovascular Aβ deposits. The internalization and processing of extravascular Aβ by cerebrovascular SMCs may represent a physiological and protective mechanism which may become impaired during the pathogenesis of CAA with the result of intra- or extracellular Aβ-precipitation (Figure 6). Further use of this *in vitro* model will allow to test the current hypotheses and to identify those factors related to SMC biology that critically favour CAA development and that may represent targets for pharmacological intervention.

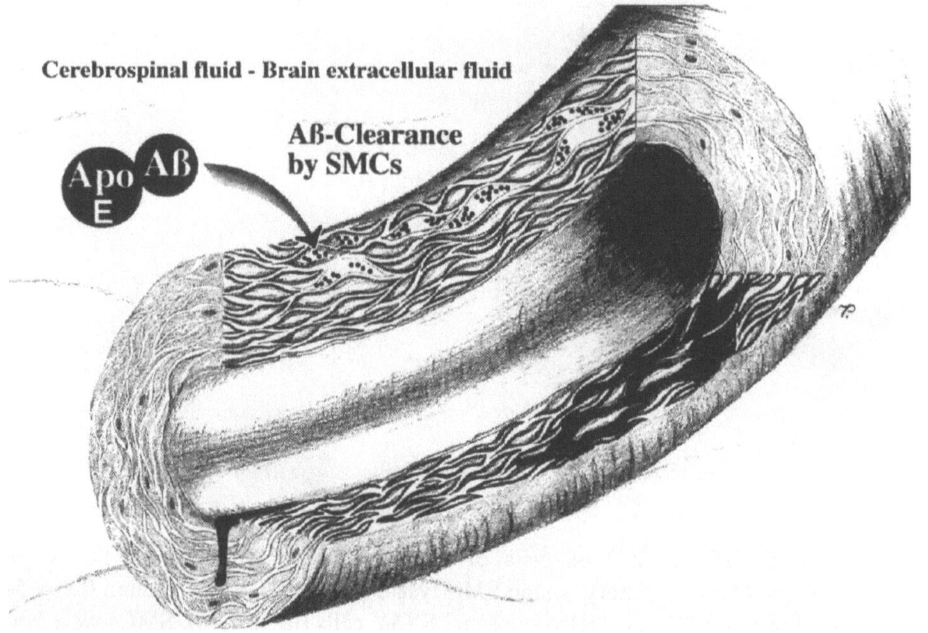

Figure 6: Possible pathophysiological mechanisms in CAA. Potential Aβ-precipitation sites are intracellular lysosomes or, alternatively, the extracellular matrix between SMCs (indicated by black dots and black areas respectively). Leptomeningeal or cortical arterioles are surrounded by CSF or extracellular brain fluid. SMCs of the vessel wall internalize ApoE- and Aβ-containing lipoproteins from the CSF. Normally, Aβ is degraded within SMC-lysosomes and the internalization of Aβ by SMCs represents a physiological mechanism for local Aβ-clearance. When lysosomal breakdown or the cellular internalization mechanisms become overloaded or less efficient, Aβ-precipitates may form within lysosomes or within the extracellular space between SMCs where specific extracellular matrix components such as HSPGs promote Aβ-fibril formation.

REFERENCES

1.Frackowiak, J., Mazur-Kolecka, B., Wegiel, J., Kim, K. S., and Wiesniewski, H. M. (1995). Culture of canine vascular myocytes as a model to study production and accumulation of β-protein by cells involved in amyloidogenesis, in K. Iqbal, J.A. Mortimer, B. Winblad, and H.M. Wisniewski (eds.), *Research Advances in Alzheimer's Disease and Related Disorders*, John Wiley & Sons, Chichester, pp 747-754

2. Urmoneit, B., Prikulis, I., Wihl, G., D'Urso, D., Frank, R., Heeren, J., Beisiegel, U., and Prior, R. (1997) Cerebrovascular smooth muscle cells internalize Alzheimer amyloid beta protein via a lipoprotein pathway: implications for cerebral amyloid angiopathy, *Lab. Invest.* **77**, 157-166.

3. Bjorkerud, S. (1985) Cultivated human arterial smooth muscle displays heterogeneous pattern of growth and phenotypic variation, *Lab. Invest.* **53**, 303-310.

4. Greaves, D. R., Gough, P. J., and Gordon, S. (1998) Recent progress in defining the role of scavenger receptors in lipid transport, atherosclerosis and host defence, *Curr. Opin. Lipidol.* **9**, 425-432.

5. Yamada, Y., Doi, T., Hamakubo, T., and Kodama, T. (1998) Scavenger receptor family proteins: roles for atherosclerosis, host defence and disorders of the central nervous system, *Cell. Mol. Life Sci.* **54**, 628-640.

6. Prior, R., Wihl, G., and Urmoneit, B. Apolipoprotein E, smooth muscle cells and the pathogenesis of cerebral amyloid angiopathy: the potential role of impaired cerebrovascular Aβ-clearance, *Ann. NY Acad. Sci.* (in press).

7. Berrou, E., Quarck, R., Fontenay-Roupie, M., Levy-Toledano, S., Tobelem, G., and Bryckaert, M. (1995) Transforming growth factor-beta 1 increases internalization of basic fibroblast growth factor by smooth muscle cells: implication of cell- surface heparan sulphate proteoglycan endocytosis, *Biochem. J.* **311**, 393-399.

8. Weaver, A. M., Lysiak, J. J., and Gonias, S. L. (1997) LDL receptor family-dependent and -independent pathways for the internalization and digestion of lipoprotein lipase-associated beta- VLDL by rat vascular smooth muscle cells, *J. Lipid Res.* **38**, 1841-1850.

9. Mahley, R. W., and Ji, Z. S. (1999) Remnant lipoprotein metabolism: key pathways involving cell-surface heparan sulfate proteoglycans and apolipoprotein E, *J. Lipid Res.* **40**, 1-16.

10. Wyss-Coray, T., Masliah, E., Mallory, M., McConlogue, L., Johnson-Wood, K., Lin, C., and Mucke, L. (1997) Amyloidogenic role of cytokine TGF-beta1 in transgenic mice and in Alzheimer's disease, *Nature* **389**, 603-606.

11. Ellis, R. J., Olichney, J. M., Thal, L. J., Mirra, S. S., Morris, J. C., Beekly, D., and Heyman, A. (1996) Cerebral amyloid angiopathy in the brains of patients with Alzheimer's disease: the CERAD experience, Part XV, *Neurology* **46**, 1592-1596.

12. Strittmatter, W. J., Saunders, A. M., Schmechel, D., Pericak-Vance, M., Enghild, J., Salvesen, G. S., and Roses, A. D. (1993) Apolipoprotein E: high-avidity binding to beta-amyloid and increased frequency of type 4 allele in late-onset familial Alzheimer disease, *Proc. Natl. Acad. Sci. USA* **90**, 1977-1981.

13. Wisniewski, T., Golabek, A., Matsubara, E., Ghiso, J., and Frangione, B. (1993) Apolipoprotein E: binding to soluble Alzheimer's beta-amyloid, *Biochem. Biophys. Res. Commun.* **192**, 359-365.

14. Vigo-Pelfrey, C., Lee, D., Keim, P., Lieberburg, I., and Schenk, D. B. (1993) Characterization of beta-amyloid peptide from human cerebrospinal fluid, *J. Neurochem.* 61, 1965-1968.

15. Alonzo, N. C., Hyman, B. T., Rebeck, G. W., and Greenberg, S. M. (1998) Progression of cerebral amyloid angiopathy: accumulation of amyloid- beta40 in affected vessels, *J. Neuropathol. Exp. Neurol.* **57**, 353-359.
16. Gravina, S. A., Ho, L., Eckman, C. B., Long, K. E., Otvos, L., Jr., Younkin, L. H., Suzuki, N., and Younkin, S. G. (1995) Amyloid beta protein (A beta) in Alzheimer's disease brain. Biochemical and immunocytochemical analysis with antibodies specific for forms ending at A beta 40 or A beta 42(43), *J. Biol. Chem.* **270**, 7013-7016.
17. Shinkai, Y., Yoshimura, M., Ito, Y., Odaka, A., Suzuki, N., Yanagisawa, K., and Ihara, Y. (1995) Amyloid beta-proteins 1-40 and 1-42(43) in the soluble fraction of extra- and intracranial blood vessels, *Ann. Neurol.* **38**, 421-428.
18. Suzuki, N., Iwatsubo, T., Odaka, A., Ishibashi, Y., Kitada, C., and Ihara, Y. (1994) High tissue content of soluble beta 1-40 is linked to cerebral amyloid angiopathy, *Am. J. Pathol.* **145**, 452-460.
19. Burdick, D., Kosmoski, J., Knauer, M. F., and Glabe, C. G. (1997) Preferential adsorption, internalization and resistance to degradation of the major isoform of the Alzheimer's amyloid peptide, A beta 1-42, in differentiated PC12 cells, *Brain Res.* **746**, 275-284.
20. Knauer, M. F., Soreghan, B., Burdick, D., Kosmoski, J., and Glabe, C. G. (1992) Intracellular accumulation and resistance to degradation of the Alzheimer amyloid A4/beta protein, *Proc. Natl. Acad. Sci. USA* **89**, 7437-7441.
21. Paresce, D. M., Chung, H., and Maxfield, F. R. (1997) Slow degradation of aggregates of the Alzheimer's disease amyloid beta- protein by microglial cells, *J. Biol. Chem.* **272**, 29390-29397.
22. Jarrett, J. T., Berger, E. P., and Lansbury, P. T., Jr. (1993) The carboxy terminus of the beta amyloid protein is critical for the seeding of amyloid formation: implications for the pathogenesis of Alzheimer's disease, *Biochemistry* **32**, 4693-4697.
23. Newby, A. C., and George, S. J. (1996) Proliferation, migration, matrix turnover, and death of smooth muscle cells in native coronary and vein graft atherosclerosis, *Curr. Opin. Cardiol.* **11**, 574-582.
24. Prior, R., D'Urso, D., Frank, R., Prikulis, I., and Pavlakovic, G. (1995) Experimental deposition of Alzheimer amyloid beta-protein in canine leptomeningeal vessels, *Neuroreport* **6**, 1747-1751.
25. Prior, R., D'Urso, D., Frank, R., Prikulis, I., Wihl, G., and Pavlakovic, G. (1996) Canine leptomeningeal organ culture: a new experimental model for cerebrovascular beta-amyloidosis, *J. Neurosci. Methods* **68**, 143-148.
26. Frackowiak, J., Zoltowska, A., and Wisniewski, H. M. (1994) Non-fibrillar beta-amyloid protein is associated with smooth muscle cells of vessel walls in Alzheimer disease, *J. Neuropathol. Exp. Neurol.* **53**, 637-645.
27. Aleshkov, S., Abraham, C. R., and Zannis, V. I. (1997) Interaction of nascent ApoE2, ApoE3, and ApoE4 isoforms expressed in mammalian cells with amyloid peptide beta (1-40). Relevance to Alzheimer's disease, *Biochemistry* **36**, 10571-10580.
28. LaDu, M. J., Falduto, M. T., Manelli, A. M., Reardon, C. A., Getz, G. S., and Frail, D. E. (1994) Isoform-specific binding of apolipoprotein E to beta-amyloid, *J. Biol. Chem.* **269**, 23403-23406.
29. Ji, Z. S., Pitas, R. E., and Mahley, R. W. (1998) Differential cellular accumulation/retention of apolipoprotein E mediated by cell surface heparan sulfate proteoglycans. Apolipoproteins E3 and E2 greater than E4, *J. Biol. Chem.* **273**, 13452-13460.

30. Beffert, U., Aumont, N., Dea, D., Lussier-Cacan, S., Davignon, J., and Poirier, J. (1999) Apolipoprotein E isoform-specific reduction of extracellular amyloid in neuronal cultures, *Brain Res. Mol. Brain Res.* **68**, 181-185.
31. Jordan, J., Galindo, M. F., Miller, R. J., Reardon, C. A., Getz, G. S., and LaDu, M. J. (1998) Isoform-specific effect of apolipoprotein E on cell survival and beta- amyloid-induced toxicity in rat hippocampal pyramidal neuronal cultures, *J. Neurosci.* **18**, 195-204.
32. Yang, D. S., Small, D. H., Seydel, U., Smith, J. D., Hallmayer, J., Gandy, S. E., and Martins, R. N. (1999) Apolipoprotein E promotes the binding and uptake of beta-amyloid into Chinese hamster ovary cells in an isoform-specific manner, *Neuroscience* **90**, 1217-1226.
33. Castillo, G. M., Ngo, C., Cummings, J., Wight, T. N., and Snow, A. D. (1997) Perlecan binds to the beta-amyloid proteins (A beta) of Alzheimer's disease, accelerates A beta fibril formation, and maintains A beta fibril stability, *J. Neurochem.* **69**, 2452-2465.
34. Snow, A. D., Kinsella, M. G., Parks, E., Sekiguchi, R. T., Miller, J. D., Kimata, K., and Wight, T. N. (1995) Differential binding of vascular cell-derived proteoglycans (perlecan, biglycan, decorin, and versican) to the beta-amyloid protein of Alzheimer's disease, *Arch. Biochem. Biophys.* **320**, 84-95.
35. Snow, A. D., Sekiguchi, R., Nochlin, D., Fraser, P., Kimata, K., Mizutani, A., Arai, M., Schreier, W. A., and Morgan, D. G. (1994) An important role of heparan sulfate proteoglycan (Perlecan) in a model system for the deposition and persistence of fibrillar A beta-amyloid in rat brain, *Neuron* **12**, 219-234.
36. Van Nostrand, W. E., Melchor, J. P., and Ruffini, L. (1998) Pathologic amyloid beta-protein cell surface fibril assembly on cultured human cerebrovascular smooth muscle cells, *J. Neurochem.* **70**, 216-223.
37. Mackic, J. B., Weiss, M. H., Miao, W., Kirkman, E., Ghiso, J., Calero, M., Bading, J., Frangione, B., and Zlokovic, B. V. (1998) Cerebrovascular accumulation and increased blood-brain barrier permeability to circulating Alzheimer's amyloid beta peptide in aged squirrel monkey with cerebral amyloid angiopathy, *J. Neurochem.* **70**, 210-215.
38. Ushiyama, M., Ikeda, S., and Yanagisawa, N. (1991) Transthyretin-type cerebral amyloid angiopathy in type I familial amyloid polyneuropathy, *Acta Neuropathol.* **81**, 524-528.
39. Kiuru, S. (1998) Gelsolin-related familial amyloidosis, Finnish type (FAF), and its variants found worldwide, *Amyloid* **5**, 55-66.
40. Levy, E., Lopez-Otin, C., Ghiso, J., Geltner, D., and Frangione, B. (1989) Stroke in Icelandic patients with hereditary amyloid angiopathy is related to a mutation in the cystatin C gene, an inhibitor of cysteine proteases, *J. Exp. Med.* **169**, 1771-1778.
41. Frackowiak, J., Mazur-Kolecka, B., Wisniewski, H. M., Potempska, A., Carroll, R. T., Emmerling, M. R., and Kim, K. S. (1995) Secretion and accumulation of Alzheimer's beta-protein by cultured vascular smooth muscle cells from old and young dogs, *Brain. Res.* **676**, 225-230.
42. Wisniewski, H. M., Frackowiak, J., and Mazur-Kolecka, B. (1995) *In vitro* production of beta-amyloid in smooth muscle cells isolated from amyloid angiopathy-affected vessels, *Neurosci. Lett.* **183**, 120-123.
43. Mazur-Kolecka, B., Frackowiak, J., Carroll, R. T., and Wisniewski, H. M. (1997) Accumulation of Alzheimer amyloid-beta peptide in cultured myocytes is enhanced by serum and reduced by cerebrospinal fluid, *J. Neuropathol. Exp. Neurol.* **56**, 263-272.
44. Mazur-Kolecka, B., Frackowiak, J., Krzeslowska, J., Ramakrishna, N., Haske, T., Emmerling, M. R., Zhang, W., Kim, K. S., and Wisniewski, H. M. (1999) Apolipoprotein E alters metabolism of AbetaPP in cells engaged in beta- amyloidosis, *J. Neuropathol. Exp. Neurol.* **58**, 288-295.

45. Van Nostrand, W. E., Davis-Salinas, J., and Saporito-Irwin, S. M. (1996) Amyloid beta-protein induces the cerebrovascular cellular pathology of Alzheimer's disease and related disorders, *Ann. N Y Acad. Sci.* **777**, 297-302.
46. Calhoun, M. E., Burgermeister, P., Phinney, A. L., Stalder, M., Tolnay, M., Wiederhold, K. H., Abramowski, D., Sturchler-Pierrat, C., Sommer, B., Staufenbiel, M., and Jucker, M. Neuronal overexpression of mutant APP results in prominent deposition of cerebrovascular amyloid (submitted).

Chapter 16

DEGENERATION OF HUMAN CEREBROVASCULAR SMOOTH MUSCLE CELLS AND PERICYTES CAUSED BY AMYLOID β PROTEIN

MARCEL M. VERBEEK*[†], WILLIAM E. VAN NOSTRAND[‡] and ROBERT M.W. DE WAAL[†]
*Departments of *Neurology and [†]Pathology, University Hospital Nijmegen, The Netherlands and [‡]Departments of Medicine and Pathology, Health Sciences Center, State University of New York, Stony Brook, NY, USA.*

The deposition of amyloid β protein in the cerebral vasculature (also known as cerebral amyloid angiopathy, CAA, or congophilic angiopathy) is one of the neuropathological hallmarks of Alzheimer's disease and several related disorders, such as hereditary cerebral hemorrhage with amyloidosis of the Dutch type. CAA can also occur independently of Alzheimer's disease and is common in the brain of elederly individuals. When this condition develops in an advanced stage, it may be associated with cerebral hemorrhages and stroke. CAA-related brain hemorrhages often occur in association with brain microvascular alterations that include fibrinoid necrosis, microaneurysm formation and smooth muscle cell and pericyte death. In this review we summarize the relationship between amyloid β peptides and degeneration of smooth muscle cells and pericytes that have been studied *in vitro*. In particular, the effects of specific Aβ isoforms and the role of peptide fibril assembly in the induction of cellular degeneration and in increased cell-associated amyloid precursor protein has been extensively investigated. Evidence will be presented to demonstrate a differential response to different Aβ assembly states that mediate the toxic effect of the peptide on cerebrovascular cells on the one hand, and neurons on the other. These data suggest that there may be different recognition and/or pathogenic mechanisms for cells in the cerebrovascular compartment compared to neuronal cells.

M.M. Verbeek et al. (eds.),
Cerebral Amyloid Angiopathy in Alzheimer's Disease and Related Disorders, 265–279.

1. INTRODUCTION

Cerebral amyloid angiopathy (CAA) is one of the neuropathological hallmarks of patients with Alzheimer's disease (AD), the others being senile plaques, neurofibrillary tangles (NFTs), neuropil threads and neuronal cell and synapse loss. The incidence of CAA in brains from individuals over 60 years of age is 36% [1], but in association with AD this incidence is much higher, approximately 85% [2,3]. The ε4 allele of apolipoprotein E is a risk factor for the development of AD, whereas the ε2 allele may have a protective effect. On the other hand, both the ε2 and ε4 alleles have been identified as risk factors for the development of CAA (see McCarron and Nicoll, this volume). The amyloid β protein (Aβ) is the major component of both senile plaques and CAA. Both lesions are found in brains from patients with AD, Down syndrome or hereditary cerebral hemorrhage with amyloidosis-Dutch type (HCHWA-D). In HCHWA-D patients, the cerebrovascular Aβ deposition is more severe, leading to small infarcts and often to fatal intracerebral hemorrhages in the brains of these patients. A mutation at position 22 of the Aβ sequence, substituting a glutamine for a glutamic acid [4], is directly related to the deposition of Aβ in diffuse senile plaques and CAA (see Maat-Schieman *et al.*, this volume). The Aβ peptide may occur in different isoforms varying both at its C- and N-terminus, resulting in peptides ending either at residue 40 or 42 and starting at amino acid 1, 3, 11 or 17. Immunohistochemical data indicate that CAA in AD and HCHWA-D contains both Aβ1-40 and Aβ1-42 [5-8]. In advanced stages of CAA, Aβ1-40 may become the predominant species, but Aβ1-42 is the isoform that is initially deposited (see Yamaguchi and Maat-Schieman, this volume).

In this review we summarize recent studies in which the role of cerebrovascular cells, in particular smooth muscle cells and pericytes, was investigated in relation to the formation of CAA. The focus of this review will be on the relationship between Aβ and degeneration of cerebrovascular smooth muscle cells and pericytes.

2. CELLULAR DEGENERATION IN CAA.

In comparison with brains from elderly non-demented patients, the capillary system in AD brains shows several pathological changes. This microangiopathy comprises a reduced microvascular density, an increased number of fragmented vessels with fewer intact branches and with atrophic string vessels [9]. Furthermore, the outer capillary surface appears irregular and marked changes in the capillary diameter (both augmented and attenuated) are observed [10]. However, these microvascular changes are not restricted to AD and are also found in other neurodegenerative disorders [9].

Cerebrovascular pathological changes in AD are more obvious in association with CAA. Using immunohistochemical analysis, it was found that capillaries with collapsed or degenerated endothelium frequently occurred in AD and Down's syndrome brains with Aβ deposits [11]. Using electron microscopy, it was observed that the endothelium in amyloid-laden vessels is degenerated [12]. In the same study, it was observed that pericytes also show an advanced stage of degeneration in amyloid-laden vessels. Scanning electron microscopy of AD capillaries confirmed that both terminal arterioles and capillaries had an irregular shape, with focal constrictions and dilatations [13]. Pericytes also demonstrated signs of degeneration in capillaries. Leptomeningeal vessels with Aβ deposition appeared to have a thinner wall and the smooth muscle cells of these vessels exhibited signs of degeneration [14]. Moreover, smooth muscle atrophy, measured as the ratio of muscle layer/total vessel thickness, was consistently reduced in AD brains independent of Aβ deposition in the vessel wall [15]. Finally, a loss of vessel wall viability was observed in isolated canine leptomeningeal vessel segments with CAA [16].

Besides these degenerative cellular changes observed in vessels affected by CAA, several other vascular abnormalities have been observed in association with CAA including hyalinization or fibrosis and microaneurysm formation (see Vinters and Vonsattel, this volume). Secondary microvascular degenerative and inflammatory changes are particularly prominent in CAA of patients with HCHWA-D and may contribute to CAA-associated stroke [17]. In HCHWA-D the number of cerebrovascular lesions correlates with the severity of these CAA-associated microvasculopathies [18].

3. DEGENERATION OF CULTURED SMOOTH MUSCLE CELLS AND PERICYTES CAUSED BY Aβ

3.1 Characteristics of cerebrovascular smooth muscle cells and pericytes in culture

Smooth muscle cells (SMCs) can be rapidly isolated by collagenase/dispase treatment of leptomeningeal vessels dissected from human brains obtained at the time of autopsy [19,20]. SMCs display a typical spindle-shaped or polygonal appearance and robustly express smooth muscle cell α-actin. Pericytes have been isolated after collagenase treatment from isolated human brain capillaries [21]. Similar to cultured SMCs, pericytes display a polygonal shape with overlapping processes indicating the absence of contact inhibition. Only a small fraction of cultured pericytes express smooth muscle cell α-actin, which is consistent with the absence of smooth muscle cell α-actin expression in cerebral capillaries *in vivo* [22]. Furthermore, pericytes express a specific panel of cell surface markers, including the high molecular weight-melanoma associated antigen, vascular cell adhesion molecule-1 and intercellular adhesion molecule-1 [23], but lack endothelial cell markers [21]. Both cell types produce and secrete Aβ and amyloid precursor protein (APP) [24,25].

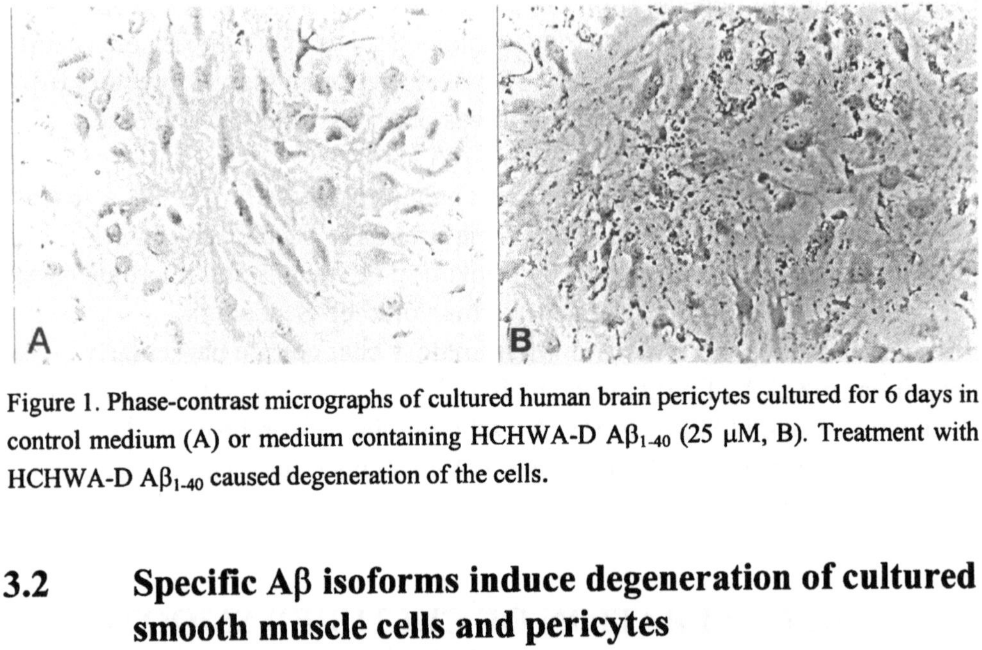

Figure 1. Phase-contrast micrographs of cultured human brain pericytes cultured for 6 days in control medium (A) or medium containing HCHWA-D $A\beta_{1\text{-}40}$ (25 μM, B). Treatment with HCHWA-D $A\beta_{1\text{-}40}$ caused degeneration of the cells.

3.2 Specific Aβ isoforms induce degeneration of cultured smooth muscle cells and pericytes

In a number of studies several Aβ isoforms were investigated for their effects on the morphology and viability of cultured SMCs and pericytes. These studies revealed that SMCs or pericytes exhibit pronounced signs of cellular degeneration when incubated for a prolonged period of time (6-12 days) with freshly resuspended wild-type $A\beta_{1\text{-}42}$ at 25 μM under serum-free conditions. Extensive degeneration was visible by loss of the characteristic morphology of the cells, disappearance of cell contours and evidence of cellular atrophy [24,25] (Figure 1). Despite this extensive degeneration, cells remained attached to the culture dish for prolonged incubation times. In contrast, it was observed that shorter isoforms of wild-type Aβ (e.g. $A\beta_{1-39}$ or $A\beta_{1-40}$) did not induce morphological degeneration (Figure 2). However, a neurotoxic fragment of Aβ, $A\beta_{25-35}$, also induced degeneration of SMCs, albeit at a higher concentration (100 μM). The degenerative effects of wild-type $A\beta_{1-42}$ on cultured SMCs / pericytes were both time- and dose-dependent.

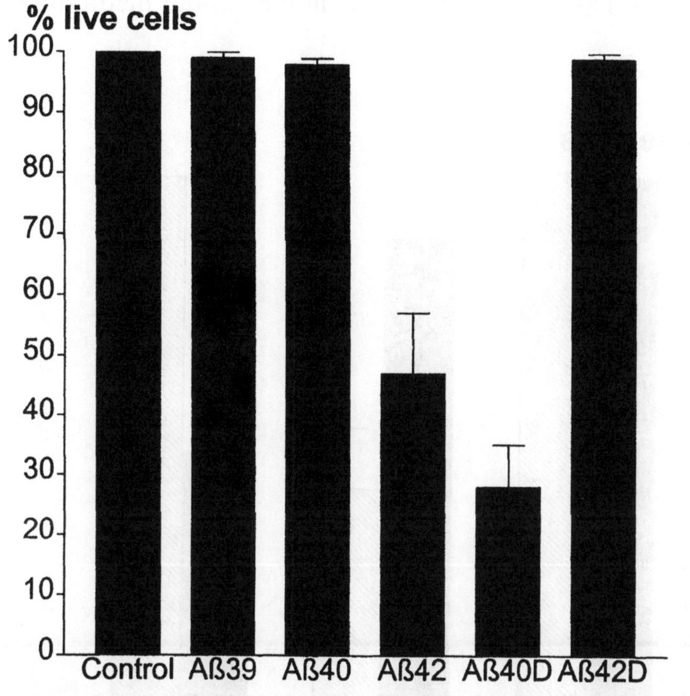

Figure 2: Viability of cultured smooth muscle cells after incubation with various Aβ peptides. Cultured smooth muscle cells were incubated with 25 μM Aβ for 12 days after which the number of live cells was determined. Both $A\beta_{42}$ and $A\beta_{40}D$ strongly reduced the number of live cells. AβD peptides carry the HCHWA-D mutation.

In contrast to these findings, when wild-type $A\beta_{1-42}$ was assembled into fibrils in solution prior to addition to cultures of SMCs, the degenerative effect of the peptide was completely abolished [26]. SMC cultures treated with freshly resuspended wild-type $A\beta_{1-42}$ exhibited a robust decrease in viability whereas in cultures treated with assembled, fibrillar wild-type $A\beta_{1-42}$ no loss in cell viability was observed (Figure 3).

An intriguing set of observations was obtained by applying Aβ peptides carrying the E22Q mutation found in HCHWA-D patients. When HCHWA-D $A\beta_{1-40}$ was applied to cultures of SMCs or pericytes a robust decline in viability of the cultures was observed after 6 days of incubation [25,27]. This effect was even stronger than that observed with wild-type $A\beta_{1-42}$ (Figure 2). By comparison with SMC cultures, pericytes appeared to be particularly sensitive to HCHWA-D $A\beta_{1\text{-}40}$ treatment, since after only two days of incubation, a significant decrease in cell viability was observed, whereas in SMC cultures this effect was only observed after 6 days of incubation. Furthermore, after removal of the peptide cellular viability continued to decrease. SMCs were better able to recover from short-term treatment with HCHWA-D $A\beta_{1\text{-}40}$ relative to pericyte cultures [24]. HCHWA-D $A\beta_{1\text{-}42}$ was found to be inactive towards SMCs and pericytes, however; this was attributed to its rapid assembly into fibrils in the culture medium [24,27] (Figure 2).

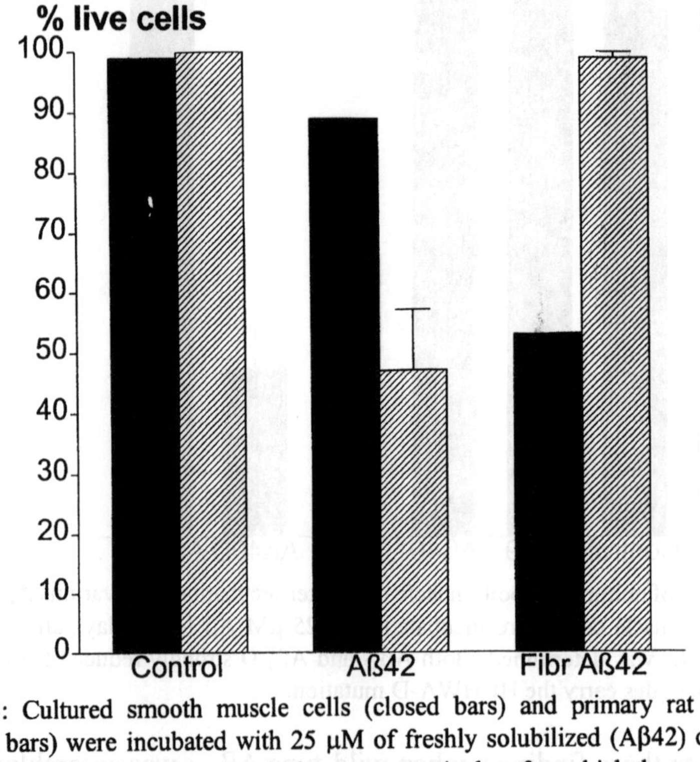

Figure 3: Cultured smooth muscle cells (closed bars) and primary rat cortical neurons (hatched bars) were incubated with 25 μM of freshly solubilized (Aβ42) or pre-aggregated $A\beta_{1\text{-}42}$ (Fibr Aβ42) for 12 days or 24 hours, respectively, after which the number of live cells was determined.

Although solution-assembled Aβ fibrils were not toxic for cultured SMCs, evidence for the involvement of Aβ fibril assembly came from immmunohistochemical and electron microscopical identification of fibril assembly at the cell surface of cultured SMCs and pericytes [28] (and unpublished observations). Further evidence was obtained in experiments in which Congo red was applied to the cells together with Aβ. Congo red inhibits neurotoxicity of Aβ by either binding to preformed fibrils or, alternatively, by inhibiting fibril formation [29]. It was found that Congo red protected cultured SMCs and pericytes from cellular degeneration by

HCHWA-D $A\beta_{1-40}$ (Figure 4). It was suggested that both the interaction of Aβ with the cell surface of SMCs or pericytes and/or its assembly into fibrils at the cell surface is essential for the degenerative effects of the peptide. The recognition and assembly rate of each specific Aβ isoform at the cell surface may then determine its pathogenic effect on SMCs and pericytes. Wild-type $A\beta_{1-42}$ fibrillizes faster than wild-type $A\beta_{1-40}$ [30], and in accordance, the former is pathogenic for cultured pericytes and SMCs, whereas the latter is not. Furthermore, Aβ peptides containing the HCHWA-D mutation have been reported to assemble into fibrils more rapidly than wild-type Aβ [31,32]. In line with these observations, HCHWA-D $A\beta_{1-40}$ causes degeneration of cultured pericytes and SMCs, whereas wild-type $A\beta_{1-40}$ is non-toxic. In these studies HCHWA-D $A\beta_{1-42}$ formed fibrils so rapidly that it likely negated its ability to properly interact with the cell surface of SMCs or pericytes.

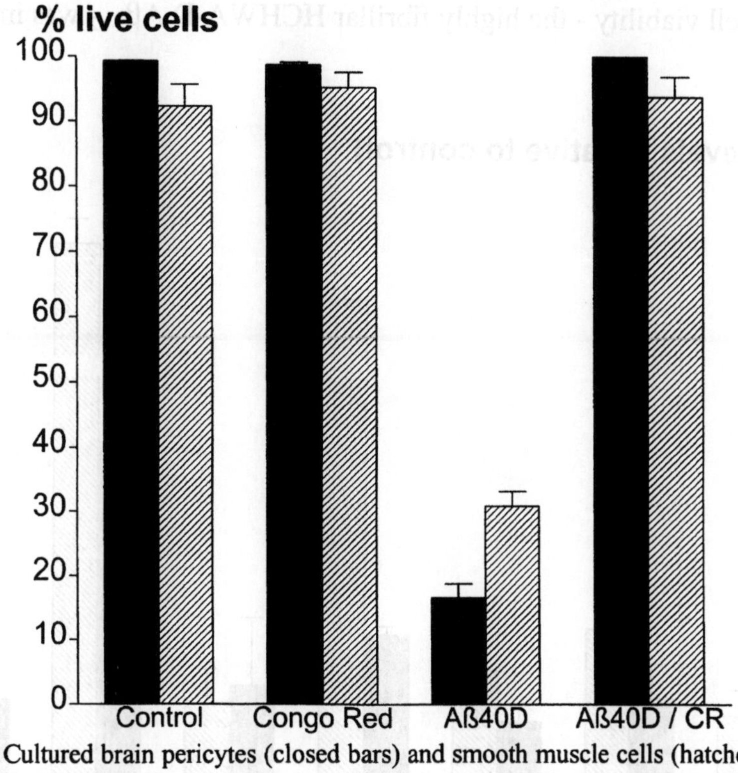

Figure 4: Cultured brain pericytes (closed bars) and smooth muscle cells (hatched bars) were incubated 25 μM HCHWA-D $A\beta_{40}$ and / or 25 μM Congo Red for 12 days after which the number of live cells were determined. Congo red completely inhibits the degenerative effects of incubation with HCHWA-D $A\beta_{40}$. Abbreviations: D, HCHWA-D; CR, Congo Red.

3.3 Aβ enhances production and secretion of APP and Aβ in cultured smooth muscle cells and pericytes

In parallel with the effects of pathogenic forms of Aβ peptides on the viability of cultured SMCs and pericytes, these peptides also affect the levels of cell-associated and secreted APP. Pathogenic wild-type $A\beta_{1\text{-}42}$ caused a striking 10- to 15-fold increase in the amount of cell-associated APP [24,25,27], whereas wild-type $A\beta_{1\text{-}40}$ did not affect cell-associated APP levels (Figure 5). Moreover, assembled fibrillar wild-type $A\beta_{1\text{-}42}$ was also incapable of changing cell-associated APP levels [26]. A much smaller increase in secreted APP, however, was observed after treatment with wild-type $A\beta_{1\text{-}42}$. In contrast, HCHWA-D $A\beta_{1\text{-}40}$ had a more pronounced effect on both secreted and cell-associated APP compared to wild-type $A\beta_{1\text{-}42}$ (Figure 5). This peptide caused a 25-fold increase in cell-associated APP and 4-fold increase in secreted APP [24,27], whereas – in line with the absence of any effect on cell viability - the highly fibrillar HCHWA-D $A\beta_{1\text{-}42}$ was inactive.

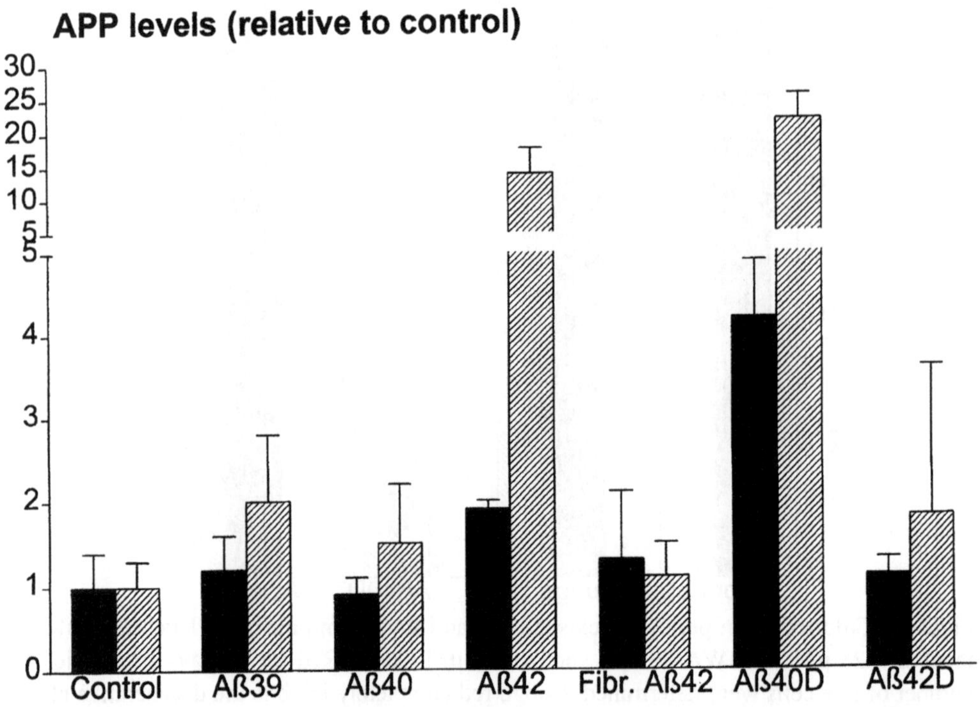

Figure 5: Smooth muscle cells were incubated with 25 μM of Aβ peptides for 6 days after which cell-associated (closed bars) and secreted APP (hatched bars) levels were determined. Both $A\beta_{42}$ and HCHWA-D $A\beta_{40}$ strongly increased APP expression. Abbreviations: Fibr., fibrillar; D, HCHWA-D.

The effects on APP of both pathogenic Aβ peptides were both dose- and time-dependent. For both wild-type $A\beta_{1-42}$ and HCHWA-D $A\beta_{1-40}$ cell-associated levels of APP were already elevated at 10 μM with a maximal effect at 25-50 μM peptide. Maximal induction of cell-associated APP was observed after 6 – 9 days incubation and then decreased due to an increase in the number of dead cells [27]. It was demonstrated that the presence of pathogenic Aβ peptides was directly related to a change in cell-associated APP levels [24]. APP levels decreased in response to removal of HCHWA-D $A\beta_{1-40}$, whereas cell viability continued to decrease when the peptide was removed. Furthermore, it was shown that induction of cell death was preceded by the increased production of both cellular and secreted APP. In addition to this increase in the production and secretion of its own precursor molecule, pathogenic Aβ also enhanced the secretion of soluble Aβ itself in SMC cultures [25]. Finally, in line with its effect on cell viability, Congo red completely abolished the increase in both cell-associated and secreted APP when co-incubated with HCHWA-D $A\beta_{1-40}$ [24], suggesting the involvement of Aβ fibril assembly in initiating a pathological cascade that also includes alterations in the production of APP.

4. DIFFERENTIAL PATHOGENIC EFFECTS OF Aβ ON NEURONAL CELLS AND ON VASCULAR CELLS

The presence of neuronal degeneration in AD brains is reflected in the formation of intraneuronal NFTs, dystrophic neurites and neuropil threads. Several studies have indicated that neuronal degeneration may be related to the presence of Aβ, but this molecule may also exert protective effects on neuronal cells: soluble Aβ, added in low doses to cultures of freshly isolated neurons, may enhance their rate of survival [33]. However, reports describing a toxic effect of Aβ on neuronal cells are more numerous. Injection of isolated human amyloid cores containing fibrillar Aβ into rat brains induced neurodegeneration [34], suggesting that the peptide has a direct toxic effect. Insights into the pathogenic activity of Aβ have come from *in vitro* studies showing that Aβ can spontaneously assemble into β-pleated sheet containing fibrils [35,36], and that when this fibrillar Aβ is administered to cultured neurons it induces neurotoxicity, loss of presynaptic terminals and the development of dystrophic neurites [29,37]. The neurotoxic effect of pre-assembled, fibrillar or soluble Aβ can be blocked by the amyloid-binding dye Congo red [29], which is probably based on its inhibition of fibril formation or binding to preformed fibrils. These *in vitro* data on Aβ-mediated neuronal toxicity correlate well with pathologic changes observed in AD brain. Brain tissue containing diffuse senile plaques that are composed of non-fibrillar Aβ usually lack evidence of associated neuronal degeneration, whereas in classic senile plaques Aβ fibrils and neurodegenerative changes (i.e. "neurites") coincide. Together, these *in vivo* and *in vitro* observations have led to the general hypothesis that the pathogenic activity of Aβ is dependent on its transformation into a β-pleated sheet, fibrillar conformation similar to that found in senile plaques.

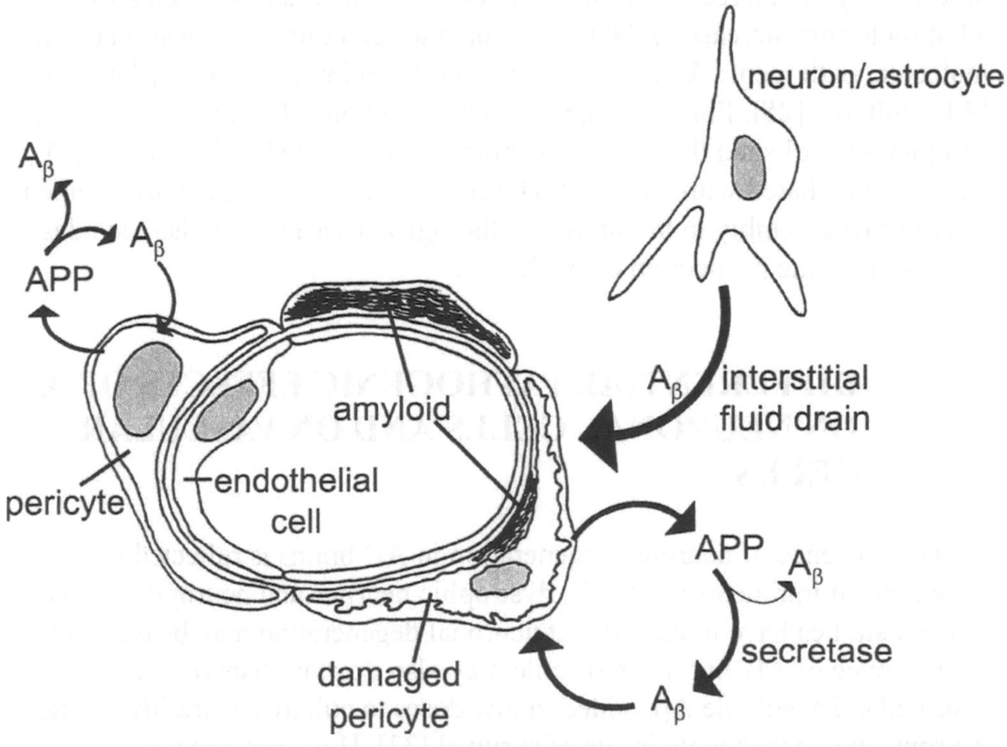

Figure 6. Schematic diagram illustrating pathways that may be involved in the evolution of CAA. Pericytes and smooth muscle cells (SMCs) produce small amounts of Aβ that are likely not toxic to the producing cells. Aβ produced by parenchymal cells is drained via the interstitial fluid pathway along the vasculature. Aβ may interact with the cell surface of pericytes and SMCs via mechanisms that are not yet well understood. Aβ accumulating at the cell surface aggregates into fibrils and in response, cells degenerate and produce increased amounts of APP and Aβ, which may accelerate the subsequent deposition of Aβ and associated cellular degeneration.

In contrast to fibrillar Aβ-induced neuronal toxicity, cultured SMCs and pericytes are completely unresponsive to assembled, fibrillar forms of Aβ [26]. Pathogenic forms of Aβ peptide such as wild-type $A\beta_{1\text{-}42}$ or HCHWA-D $A\beta_{1\text{-}40}$ must be introduced in a soluble, non β-pleated sheet conformation to these cerebrovascular cells in order to induce cellular degeneration and increased levels of cell-associated APP [27,28] (Figure 3). However, it has been shown that assembly of soluble pathogenic forms of Aβ fibrils on the surface of these cultured cerebrovascular cells coincided with the induction of pathologic responses [28]. Congo red blocked both the assembly of cell surface fibrils and pathologic responses in SMCs and pericytes further implicating their importance in Aβ-induced cerebrovascular cell toxicity. These findings suggest that neuronal cells and cerebrovascular cells may possess different recognition mechanisms for pathogenic forms of Aβ. Furthermore, this observation may provide insight into the different or overlapping etiologies of senile plaques and CAA [38].

Several lines of evidence suggest that the neurotoxic action of Aβ is mediated by reactive oxygen species, such as hydrogen peroxide, causing lipid peroxidation, and can be inhibited by free radical scavengers or anti-oxidants [39-41]. In contrast to these data, however, a recent study suggests that neuronal cell death induced by Aβ is not mediated by lipid peroxidation and could not be prevented by anti-oxidants [42]. Furthermore, *in vivo* and *in vitro* studies have suggested that neuronal cells may die via apoptotic mechanisms when incubated with Aβ peptides [43]. Similarly, recent studies have indicated that SMCs may undergo an apoptotic mechanism of cell death in response to pathogenic Aβ [44]. However, the possible involvement of free radicals and reactive oxygen species in this process have not yet been identified in cultures of SMCs or pericytes.

5. DISCUSSION: RELEVANCE OF THESE DATA FOR CAA / ALZHEIMER'S DISEASE

Together, these studies provide several lines of evidence that cultured primary cerebrovascular SMCs and pericytes are useful *in vitro* paradigms to further investigate how Aβ interacts with cerebrovascular cells to cause the pathological consequences of CAA observed *in vivo* (Figure 6). First, pathogenic forms of Aβ assemble into fibrils on the surface of SMCs and pericytes leading to their degeneration and eventual death. This parallels the *in vivo* finding that cerebrovascular fibrillar Aβ deposition is associated with cellular degeneration and death of these cells within the vessel wall. Second, pathogenic Aβ cell surface fibril assembly on cultured SMCs and pericytes is accompanied by a striking increase in the levels of cell-associated APP. Increased levels of APP have been observed in cerebral vessels containing Aβ deposits [45]. Third, Aβ containing the HCHWA-D mutation was found to be much more toxic than wild-type Aβ to cultured SMCs and pericytes. Patients afflicted with HCHWA-D develop CAA earlier and to a much more severe degree than most AD patients, further supporting the notion that this mutation in Aβ targets the pathogenicity of the peptide for the cerebral vasculature. Finally, *in vitro* studies have shown that the type and fibrillar structure of Aβ that is pathogenic for cultured SMCs and pericytes can be quite different than what is toxic for cultured neuronal cells. This suggests that there may be different recognition and/or pathogenic mechanisms for cells in the cerebrovascular compartment compared to neuronal cells. This would appear to warrant a detailed analysis of the specific mechanisms that mediate Aβ-induced degeneration of smooth muscle cells and pericytes.

ACKNOWLEDGEMENTS

This work has been financially supported by grants from the Internationale Stichting Alzheimer Onderzoek (ISAO) to M.M. Verbeek and by NIH grants NS35781, HL49566, HL03229, and AG16223 to W.E. Van Nostrand. The authors thank Irene Otte-Höller for immunohistochemistry and preparation of the illustrations.

REFERENCES

1. Vinters, H.V. and Gilbert, J.J. (1983) Cerebral amyloid angiopathy: incidence and complications in the aging brain. II. The distribution of amyloid vascular changes, *Stroke* **14**, 924-928.
2. Glenner, G.G., Henry, J.H., and Fujihara, S. (1981) Congophilic angiopathy in the pathogenesis of Alzheimer's degeneration, *Ann. Pathol.* **1**, 120-129.
3. Ellis, R.J., Olichney, J.M., Thal, L.J., Mirra, S.S., Morris, J.C., Beekly, D., and Heyman, A. (1996) Cerebral amyloid angiopathy in the brains of patients with Alzheimer's disease: the CERAD experience, Part XV, *Neurology* **46**, 1592-1596.
4. Levy, E., Carman, M.D., Fernandez-Madrid, I.J., Power, M.D., Lieberburg, I., van Duinen, S.G., Bots, G.T.A.M., Luyendijk, W., and Frangione, B. (1990) Mutation of the Alzheimer's disease amyloid gene in hereditary cerebral hemorrhage, Dutch type, *Science* **248**, 1124-1126.
5. Iwatsubo, T., Odaka, A., Suzuki, N., Mizusawa, H., Nukina, N., and Ihara, Y. (1994) Visualization of Aβ42(43) and Aβ40 in senile plaques with end-specific monoclonals: Evidence that an initially deposited species is Aβ42(43), *Neuron* **13**, 45-53.
6. Savage, M.J., Kawooya, J.K., Pinsker, L.R., Emmons, T.L., Mistretta, S., Siman, R., and Greenberg, B.D. (1995) Elevated Aβ levels in Alzheimer's disease brain are associated with selective accumulation of $A\beta_{42}$ in parenchymal amyloid plaques and both $A\beta_{40}$ and $A\beta_{42}$ in cerebrovascular deposits, *Amyloid: Int. J. Exp. Clin. Invest.* **2**, 234-240.
7. Mann, D.M.A., Iwatsubo, T., Ihara, Y., Cairns, N.J., Lantos, P.L., Bogdanovic, N., Lannfelt, L., Winblad, B., Maat-Schieman, M.L.C., and Rossor, M.N. (1996) Predominant deposition of Amyloid-$\beta_{42(43)}$ in plaques in cases of Alzheimer's disease and hereditary

cerebral hemorrhage associated with mutations in the amyloid precursor protein gene, *Am. J. Pathol.* **148**, 1257-1266.
8. Gravina, S.A., Ho, L., Eckman, C.B., Long, K.E., Otvos Jr., L., Younkin, L.H., Suzuki, N., and Younkin, S.G. (1995) Amyloid β protein (Aβ) in Alzheimer's disease brain, *J. Biol. Chem.* **270**, 7013-7016.
9. Buée, L., Hof, P.R., Bouras, C., Delacourte, A., Perl, D.P., Morrison, J.H., and Fillit, H.M. (1994) Pathological alterations of the cerebral microvasculature in Alzheimer's disease and related dementing disorders, *Acta. Neuropathol.* **87**, 469-480.
10. Perlmutter, L.S., Chui, H.C., Saperia, D., and Athanikar, J. (1990) Microangiopathy and the colocalization of heparan sulfate proteoglycan with amyloid in senile plaques of Alzheimer's disease, *Brain Res.* **508**, 13-19.
11. Kalaria, R.N. and Hedera, P. (1995) Differential degeneration of the cerebral microvasculature in Alzheimer's disease, *NeuroReport* **6**, 477-480.
12. Wisniewski, H.M., Wegiel, J., Wang, K.C., and Lach, B. (1992) Ultrastructural studies of the cells forming amyloid in the cortical vessel wall in Alzheimer's disease, *Acta. Neuropathol.* **84**, 117-127.
13. Kimura, T., Hashimura, T., and Miyakawa, T. (1991) Observations of microvessels in the brain with Alzheimer's disease by the scanning electron microscope, *Jpn. J. Psychiatr. Neurol.* **45**, 671-676.
14. Kawai, M., Kalaria, R.N., Cras, P., Siedlak, S.L., Velasco, M.E., Shelton, E.R., Chan, H.W., Greenberg, B., and Perry, G. (1993) Degeneration of vascular muscle cells in cerebral amyloid angiopathy of Alzheimer disease, *Brain Res.* **623**, 142-146.
15. Perry, G., Smith, M.A., McCann, C.E., Siedlak, S.L., Jones, P.K., and Friedland, R.P. (1998) Cerebrovascular muscle atrophy is a feature of Alzheimer's disease, *Brain Res.* **791**, 63-66.
16. Prior, R., D'Urso, D., Frank, R., Prikulis, I., and Pavlakovic, G. (1996) Loss of vessel wall viability in cerebral amyloid angiopathy, *Neuroreport* **7**, 562-564.
17. Vinters, H.V., Natte, R., Maat-Schieman, M.L.C., van Duinen, S.G., Hegeman, K., I, Welling-Graafland C., Haan, J., and Roos, R.A. (1998) Secondary microvascular degeneration in amyloid angiopathy of patients with hereditary cerebral hemorrhage with amyloidosis, Dutch type (HCHWA-D), *Acta Neuropathol. Berl.* **95**, 235-244.
18. Natte, R., Vinters, H.V., Maat-Schieman, M.L.C., Bornebroek, M., Haan, J., Roos, R.A., and van Duinen, S.G. (1998) Microvasculopathy is associated with the number of cerebrovascular lesions in hereditary cerebral hemorrhage with amyloidosis, Dutch type, *Stroke* **29**, 1588-1594.
19. Van Nostrand, W.E., Rozemuller, J.M., Chung, R., Cotman, C.W., and Saporito-Irwin, S.M. (1994) Amyloid β-protein precursor in cultured leptomeningeal smooth muscle cells, *Amyloid: Int. J. Exp. Clin. Invest.* **1**, 1-7.
20. Urmoneit, B., Prikulis, I., Wihl, G., D'Urso, D., Frank, R., Heeren, J., Beisiegel, U., and Prior, R. (1997) Cerebrovascular smooth muscle cells internalize Alzheimer amyloid beta protein via a lipoprotein pathway: implications for cerebral amyloid angiopathy, *Lab. Invest.* **77**, 157-166.
21. Verbeek, M.M., Otte-Höller, I., Wesseling, P., Ruiter, D.J., and de Waal, R.M.W. (1994) Induction of α-smooth muscle actin expression in cultured human brain pericytes by TGFβ1, *Am. J. Pathol.* **144**, 372-382.
22. Nehls, V. and Drenckhahn, D. (1991) Heterogeneity of microvascular pericytes for smooth muscle type alpha-actin, *J. Cell. Biol.* **113**, 147-154.

23. Verbeek, M.M., Westphal, J.R., Ruiter, D.J., and de Waal, R.M.W. (1995) T lymphocyte adhesion to human brain pericytes is mediated via VLA-4/VCAM-1 interactions, *J. Immunol.* **154**, 5876-5884.
24. Verbeek, M.M., de Waal, R.M.W., Schipper, J.J., and Van Nostrand, W.E. (1997) Rapid degeneration of cultured human brain pericytes by amyloid β protein, *J. Neurochem.* **68**, 1135-1141.
25. Davis-Salinas, J., Saporito-Irwin, S.M., Cotman, C.W., and Van Nostrand, W.E. (1995) Amyloid β-protein induces its own production in cultured degenerating cerbrovascular smooth muscle cells, *J. Neurochem.* **65**, 931-934.
26. Davis-Salinas, J. and Van Nostrand, W.E. (1995) Amyloid beta-protein aggregation nullifies its pathologic properties in cultured cerebrovascular smooth muscle cells, *J. Biol. Chem.* **270**, 20887-20890.
27. Davis, J. and Van Nostrand, W.E. (1996) Enhanced pathologic properties of Dutch-type mutant amyloid β-protein, *Proc. Natl. Acad. Sci. USA* **93** , 2996-3000.
28. Van Nostrand, W.E., Melchor, J.P., and Ruffini, L. (1998) Pathologic amyloid β-protein cell surface assembly on cultured human cerebrovascular smooth muscle cells, *J. Neurochem.* **70**, 216-223.
29. Lorenzo, A. and Yankner, B.A. (1994) β-Amyloid neurotoxicity requires fibril formation and is inhibited by Congo red, *Proc. Natl. Acad. Sci. USA* **91**, 12243-12247.
30. Jarrett, J.T., Berger, E.P., and Lansbury, P.T. (1993) The carboxy terminus of the β amyloid protein is critical for the seeding of amyloid formation: Implications for the pathogenesis of Alzheimer's disease, *Biochemistry* **32**, 4693-4697.
31. Clements, A., Walsh, D.M., Williams, C.H., and Allsop, D. (1993) Effects of the mutations Glu22 to Gln and Ala21 to Gly on the aggregation of a synthetic fragment of the Alzheimer's amyloid β/A4-peptide, *Neurosci. Lett.* **161**, 17-20.
32. Wisniewski, T., Ghiso, J., and Frangione, B. (1991) Peptides homologous to the amyloid protein of Alzheimer's disease containing a glutamine for glutamic acid substitution have accelerated amyloid fibril formation, *Biochem. Biophys. Res. Commun.* **179**, 1247-1254.
33. Yankner, B.A., Duffy, L.K., and Kirschner, D.A. (1990) Neurotrophic and neurotoxic effects of amyloid β protein: Reversal by tachykinin neuropeptides, *Science* **250**, 279-282.
34. Kowall, N.W., Beal, M.F., Busciglio, J., Duffy, L.K., and Yankner, B.A. (1991) An *in vivo* model for the neurodegenerative effects of β amyloid and protection by substance P, *Proc. Natl. Acad. Sci. USA* **88**, 7247-7251.
35. Burdick, D., Soreghan, B., Kwon, M., Kosmoski, J., Knauer, M., Henschen, A., Yates, J., Cotman, C., and Glabe, C. (1992) Assembly and agrregation properties of synthetic Alzheimer's A4/β amyloid peptide analogs, *J. Biol. Chem.* **267**, 546-554.
36. Kirschner, D.A., Inouye, H., Duffy, L.K., Sinclair, A., Lind, M., and Selkoe, D.J. (1987) Synthetic peptide homologous to β protein from Alzheimer disease forms amyloid-like fibrils *in vitro*, *Proc. Natl. Acad. Sci. USA* **84**, 6953-6957.
37. Pike, C.J., Burdick, D., Walencewicz, A.J., Glabe, C.G., and Cotman, C.W. (1993) Neurodegeneration induced by β-amyloid peptides *in vitro*: The role of peptide assembly state, *J. Neurosci.* **13**, 1676-1687.
38. Verbeek, M.M., Eikelenboom, P., and de Waal, R.M.W. (1997) Differences between the pathogenesis of senile plaques and congophilic angiopathy in Alzheimer's disease, *J. Neuropathol. Exp. Neurol.* **56**, 751-761.
39. Harris, M.E., Hensley, K., Butterfield, D.A., Leedle, R.A., and Carney, J.M. (1995) Direct evidence of oxidative injury produced by the Alzheimer's β-amyloid peptide (1-40) in cultured hippocampal neurons, *Exp. Neurol.* **131**, 193-202.

40. Bruce, A.J., Malfroy, B., and Baudry, M. (1996) beta-Amyloid toxicity in organotypic hippocampal cultures: protection by EUK-8, a synthetic catalytic free radical scavenger, *Proc. Natl. Acad. Sci. U. S. A.* **93**, 2312-2316.
41. Behl, C., Davis, J.B., Lesley, R., and Schubert, D. (1994) Hydrogen peroxide mediates amyloid beta protein toxicity, *Cell* **77**, 817-827.
42. Pike, C.J., Ramezan, A.N., and Cotman, C.W. (1997) Beta-amyloid neurotoxicity *in vitro*: evidence of oxidative stress but not protection by antioxidants, *J. Neurochem.* **69**, 1601-1611.
43. Yang, F., Sun, X., Beech, W., Teter, B., Wu, S., Sigel, J., Vinters, H.V., Frautschy, S.A., and Cole, G.M. (1998) Antibody to caspase-cleaved actin detects apoptosis in differentiated neuroblastoma and plaque-associated neurons and microglia in Alzheimer's disease, *Am. J. Pathol.* **152**, 379-389.
44. Davis, J., Cribbs, D.H., Cotman, C.W., and Van Nostrand, W.E. (1999) Pathogenic amyloid β-protein induces apoptosis in cultured human cerebrovascular smooth muscle cells, *Amyloid: Int. J. Exp. Clin. Invest.* in press
45. Rozemuller, A.J.M., Roos, R.A.C., Bots, G.T.A.M., Kamphorst, W., Eikelenboom, P., and Van Nostrand, W.E. (1993) Distribution of β/A4 protein and amyloid precursor protein in hereditary cerebral hemorrhage with amyloidosis-Dutch type and Alzheimer's disease, *Am. J. Pathol.* **142**, 1449-1457.

Chapter 17

VASOACTIVITY OF AMYLOID β PEPTIDES

DANIEL PARIS, TERRENCE TOWN and MICHAEL MULLAN
Roskamp Institute, University of South Florida, Tampa, FL, USA

It is becoming increasingly recognized that risk factors for vascular disease increase the risk for developing Alzheimer's disease (AD), suggesting that the vasculature may be a contributing factor to the pathophysiology of AD. Furthermore, in the majority of AD cases, vascular deposits of β-amyloid (Aβ) peptides are observed in the condition known as cerebral amyloid angiopathy, suggesting that Aβ peptides may be biologically active at the vascular level. We present data demonstrating that low doses (in the nM range) of freshly solubilized Aβ peptides (1-40 and 1-42) greatly enhance the vasoconstriction induced by endothelin-1 (ET-1) in isolated, intact vessels. We also show that freshly solubilized Aβ peptides transduce vasoactivity via stimulation of a pro-inflammatory pathway involving activation of cytosolic phospholipase A_2, cyclooxygenase-2, and 5-lipoxygenase. Our data demonstrate that soluble forms of Aβ peptides (at doses similar to those found in AD patient plasma) are able to trigger a pro-inflammatory response in vessels, suggesting that, in AD, which has a significant inflammatory component, Aβ peptides may exert pro-inflammatory effects prior to deposition as insoluble Aβ aggregates.

1. Aβ PEPTIDES, ALZHEIMER'S DISEASE, AND CEREBRAL AMYLOID ANGIOPATHY

Vascular pathology is the norm in advanced cases of Alzheimer's Disease (AD), with cerebral amyloid angiopathy (CAA) being one of the commonest abnormalities detected at autopsy in carefully standardized examination (83% of AD cases as assessed by CERAD [1]). It is becoming increasingly accepted that risk factors for vascular pathology, such as the apolipoprotein E ε4 allele, elevated cholesterol levels, and hypertension, are also risk factors for AD [2]. However, we are concerned here not with the effects of anatomic pathological changes on these outcomes but with the investigation of the possibility that β-amyloid (Aβ) peptide-induced

M.M. Verbeek et al. (eds.),
Cerebral Amyloid Angiopathy in Alzheimer's Disease and Related Disorders, 281–294.

functional abnormalities occur in the microvasculature in AD, and that these functional changes may contribute to the clinical picture, pathology and progression of the disease. This hypothesis arises from the growing evidence, reviewed here, that soluble forms of Aβ peptides have vasoactive effects in isolated mammalian vessels and in transgenic animal models where Aβ peptides are overexpressed. At the clinical level, the possibility arises that these "physiologic" effects of soluble Aβ contribute to hypoperfusion and perhaps ischemia in AD brains, thereby amplifying the AD pathological process. This hypothesis includes the tenet that Aβ peptides exert these effects in their soluble form and that deposition of insoluble amyloid is not required to produce these effects [3].

With regard to hypoperfusion, both SPECT and PET studies confirm reduction in cerebral blood flow (CBF) in AD [4,5]. However, glucose hypometabolism is also observed in AD cases [6], suggesting two possibilities: first, that reduced CBF leads to lowered brain glucose levels, and second, that glucose metabolism is downregulated as a result of reduced neuronal metabolic demand. Although any potential vasoactive effect of Aβ has not been demonstrated in humans, data from transgenic animals overexpressing Aβ peptides supports the hypothesis that soluble Aβ peptides can induce vasoconstriction and oppose normal vasorelaxation during the development of AD and CAA. For instance, transgenic mice which overexpress Aβ peptides exhibit decreased CBF in response to vasodilators [7] prior to the formation of amyloid plaques, suggesting that the isolated vessel bath system may be an accurate model of the pre-clinical effects of soluble Aβ levels on the microcerebrovasculature in AD. Other transgenic models of AD also overexpressing Aβ do poorly after cerebral ischaemia compared to their non-transgenic littermates. In particular, after middle cerebral artery occlusion there is enlarged infarct size, reduced blood flow in the penumbra of the infarct and reduced response to vasodilators in the same area [8]. These data support the possibility of a vasoactive role for soluble Aβ peptides, as their levels reach supra-physiologic amounts around the microcerebrovasculature in the brain parenchyma.

However, in persons affected with AD there has been no direct testing of the existence or clinical importance of any Aβ-induced vasoactive effects similar to those observed *in vitro* or in transgenic animals. Indeed, a primary role of hypoperfusion is not established and, as mentioned, thought to be largely consequent to reduced metabolic demand of dysfunctional neurons. For instance, in a small study Jagust and colleagues [9] suggested that there was no evidence to propose that the prominent perfusion changes seen in AD were primary and that AD cases had similar increases in perfusion compared to normals when challenged by hypocapnia. By contrast, Nagata and colleagues [10] observed increased oxygen extraction from cerebral vessels into the brain and decreased CBF in AD cases compared to controls, suggesting a primary role for the vasculature in limiting perfusion to the brain. Interestingly, the same study suggested no change in vascular reactivity, at least in relation to changes in $PaCO_2$. When reviewing these data, there arise important differences in studies of blood flow in AD cases and transgenic models of AD. Some of these differences may be due to the different experimental paradigms between human and animal studies and, in particular, the restrictions inherent in human studies that preclude challenge of the appropriate vasoactive mechanisms triggered by Aβ. For instance, although vascular reactivity is maintained in AD patients under hypercapnic conditions, it may not be maintained when these patients are challenged with compounds that directly impinge upon the signal transduction pathway leading to Aβ-induced vasoactivity. The experimental design of studies that would provide an answer to this question is not obvious, but much needed. As a first step to elucidating the contribution of Aβ's vasoactive effects in transgenic models of AD and in AD itself, we must determine exactly how Aβ mediates its vasoactive effects, and this is the subject of this chapter.

2. CHARACTERISTICS OF Aβ VASOACTIVITY *IN VITRO*

In order to investigate the vasoactive properties of Aβ peptides, we employed freshly dissected rat aortic rings in a tissue bath system. The tissue bath system consists of a series of glass chambers containing 7 mL of Krebs buffer, which is continuously oxygenated (95% O_2, 5% CO_2) and maintained at physiologic pH (7.5) and temperature (37°C). Rat aortae are suspended in Krebs buffer via small hooks, which are connected to an isometric transducer linked to a MacLab system. Aortic rings are equilibrated in the tissue bath system for 2 h, with the Krebs buffer changed every 30 min prior to beginning the vasoactivity assays. Data obtained from these experiments were analyzed using analysis of variance (ANOVA), with all independent variables for each experiment included in a single model (i.e., vasotension = $\beta_1 factor_1 + \beta_1 factor_2 + \ldots \ldots$ + all interactive terms + constant, where $factor_{1-n}$ = ET-1 dose, Aβ, MAFP, melittin, etc.). Alternatively, data were analyzed using one-way ANOVA followed by post-hoc comparisons by means of Bonferonni's method.

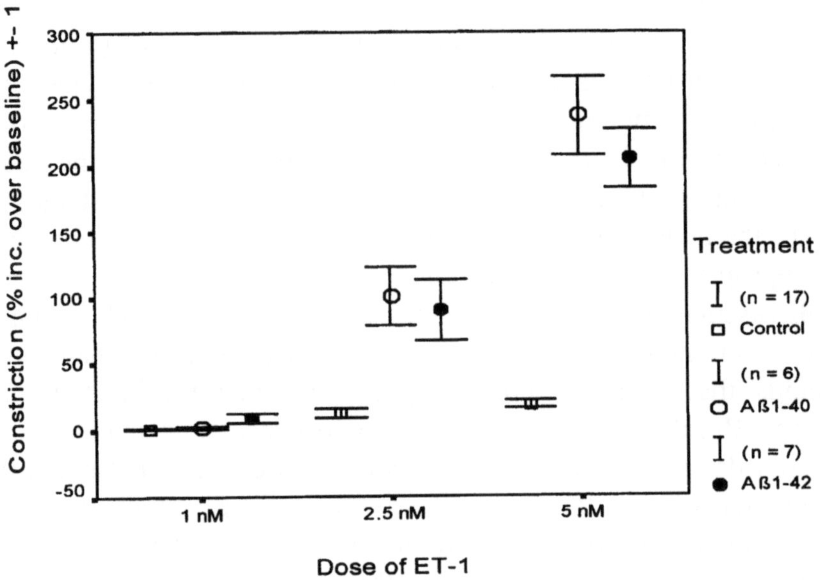

Figure 1: Freshly solublilized $A\beta_{1-40}$ or $A\beta_{1-42}$ enhance ET-1-induced vasoconstriction. Aortic rings were treated with 1 μM of $A\beta_{1-40}$ or $A\beta_{1-42}$ 5 minutes prior to the addition of a dose range of ET-1. ANOVA showed significant treatment effects of ET-1 dose ($p < .001$), $A\beta_{1-40}$ ($p < .001$), and $A\beta_{1-42}$ ($p < .001$). ANOVA also revealed significant interactive terms between ET-1 dose and either $A\beta_{1-40}$ ($p < .001$) or $A\beta_{1-42}$ ($p = .001$), indicating ET-1 dose-dependent enhancement of vasoconstriction by Aβ.

$A\beta_{1-40}$ has previously been shown to exhibit vasoactive properties, and it has been suggested that this effect is mediated by free radical production. Thomas and colleagues [11] showed that treatment of intact aortic rings with Aβ prior to the addition of the vasoconstrictor phenylephrine (PE) enhances the vasoconstriction normally induced by PE and diminishes relaxation in response to acetylcholine. Interestingly, Aβ vasoactivity appears to be related to the conformation adopted by the peptide in solution [3], and can be observed with soluble but not with aggregated forms of Aβ. To determine the impact of Aβ on endogenous vasoconstrictors, we examined the effect of endothelin-1 (ET-1) on vessels pretreated with Aβ. ET-1 is one of the most potent endogenous vasoconstrictors known in the brain, and, together with NO, controls cerebral vasoregulation [12]. Our data show that freshly solubilized Aβ peptides (1 μM, 1-40 or 1-42) synergistically enhance ET-1-induced vasoconstriction to a similar extent (Figure 1). Furthermore, the response of Aβ to ET-1-induced vasoconstriction is much more potent than with PE, and can be observed within minutes, whereas Aβ enhancement of PE-induced vasoconstriction is minor, as PE alone induces a much greater vasoconstrictive response than ET-1 alone. Specifically, ET-1 induces a small constrictive event in rat aortae (an approximate 25 to 50% increase in vasotension over baseline at the 5 nM dose of ET-1). However, while we do not observe intrinsic vasoconstriction with Aβ, the effect of Aβ on ET-1 induced vasoconstriction is dramatic, and is characterized by a synergistic (greater than 2-fold) increase in ET-1-induced vasoconstriction. This observation led us to focus on the ET-1 system as an assay for Aβ vasoactivity.

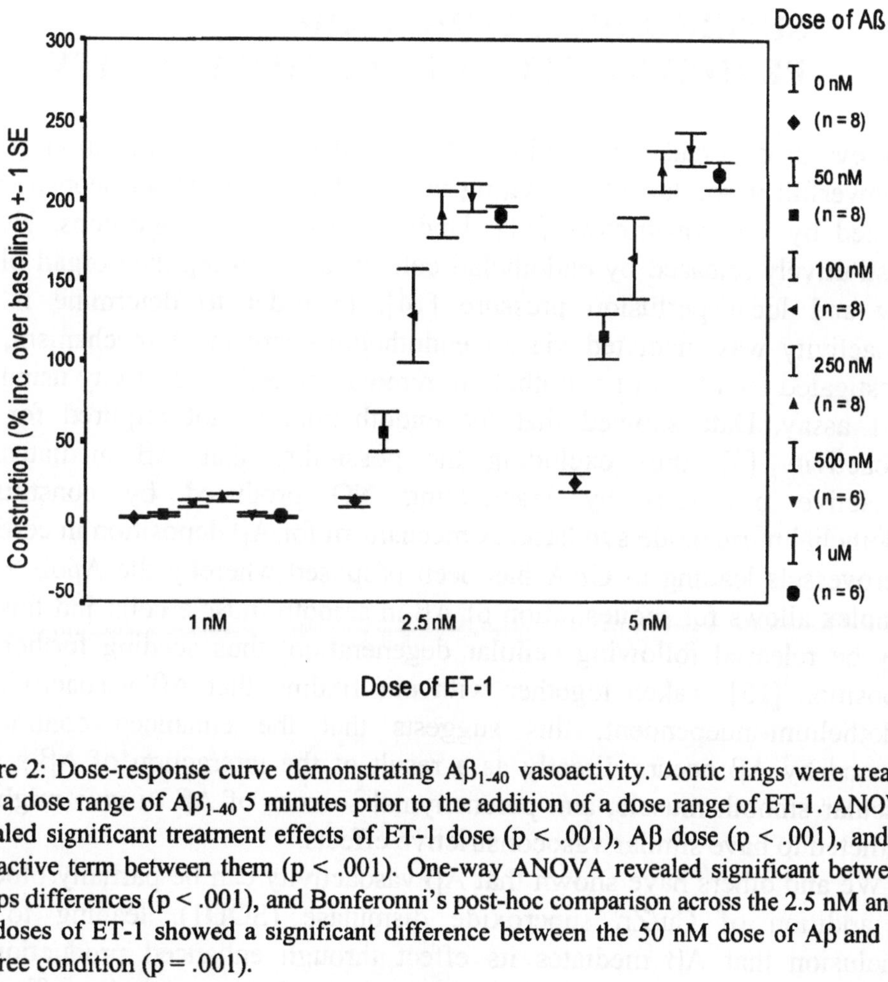

Figure 2: Dose-response curve demonstrating $A\beta_{1\text{-}40}$ vasoactivity. Aortic rings were treated with a dose range of $A\beta_{1\text{-}40}$ 5 minutes prior to the addition of a dose range of ET-1. ANOVA revealed significant treatment effects of ET-1 dose ($p < .001$), Aβ dose ($p < .001$), and an interactive term between them ($p < .001$). One-way ANOVA revealed significant between-groups differences ($p < .001$), and Bonferonni's post-hoc comparison across the 2.5 nM and 5 nM doses of ET-1 showed a significant difference between the 50 nM dose of Aβ and the Aβ-free condition ($p = .001$).

The initial experiments showing the vasoactive response of Aβ employed relatively high doses of the peptides. However, the physiologic relevance of such supraphysiologic doses is questionable. We now show Aβ vasoactivity within minutes, with nanomolar doses of freshly solubilized Aβ (with a maximum effect at 250 nM, Figure 2), doses which are much closer to blood-plasma levels of Aβ in AD patients (on average, 50 nM in AD and 9 nM in age-matched control plasma [13]). This dose-response curve (Figure

2) shows that the vasculature is extremely sensitive to the effects of soluble Aβ, and that, with 250 nM of Aβ, the Aβ pathway leading to vasoconstriction is maximally stimulated. These data raise the possibility that, in AD and CAA as perivascular soluble Aβ levels rise, the cerebrovasculature may play an important role in the pathogenesis of the disease prior to the formation of Aβ aggregates.

3. SUPEROXIDE, NITRIC OXIDE, PEROXYNITRITE AND Aβ VASOACTIVITY

Previous data had suggested that Aβ may decrease the production of NO (a powerfull vasorelaxant), resulting in potentiation of the vasoconstriction induced by vasoconstrictors [11]. Under physiological conditions, NO is constitutively released by endothelial cells in order to regulate organ blood flow and local perfusion pressure [14]. In order to determine if Aβ vasoactivity was mediated via an endothelium-dependent mechanism, we investigated the effect of endothelium removal on Aβ vasoactivity using the ET-1 assay. Data showed that the endothelium is not required for Aβ vasoactivity [3], thus excluding the possibility that Aβ mediates its vasoactive properties by inactivating NO produced by constitutive endothelial nitric oxide synthase. A mechanism for Aβ deposition in cerebral microvessels leading to CAA has been proposed whereby the ApoE + Aβ complex allows for sequestration of Aβ in smooth muscle cells and this Aβ may be released following cellular degeneration, thus seeding further Aβ deposition [15]. Taken together with our finding that Aβ vasoactivity is endothelium-independent, this suggests that the enhanced contraction induced by Aβ occurs directly as a result of the interaction of Aβ's with vascular smooth muscle, and parenchymal sources of Aβ *in vivo* might be predicted to have similar vasoconstrictive effects.

We and others have shown that Aβ vasoactivity can be partially blocked by addition of Cu/Zn superoxide dismutase (SOD1), leading to the conclusion that Aβ mediates its effect through enhanced production of superoxide [13,16,17]. The hypothesized mechanism of the effect of superoxide was via a direct interaction with NO, resulting in the formation of peroxynitrite, which would make NO less available to oppose vasoconstriction. This inactivation of NO by superoxide may have explained the diminished response to vasorelaxants and the enhanced effect of vasoconstrictors in response to Aβ stimulation. Yet, the SOD mimetic Mn(III)tetrakis(4-benzoic acid)porphyrin chloride (MnTBAP), which can clear superoxide both intracellularly and extracellularly and scavenge peroxynitrite is inefficient at abolishing the vasoconstriction induced by

Aβ [18]. Moreover, we have shown that peroxynitrite displays vasorelaxant properties when added alone to intact aortae and addition of peroxynitrite to Aβ-treated vessels does not result in a statistically interactive effect [18], showing that peroxynitrite is not a critical mediator of Aβ vasoactivity.

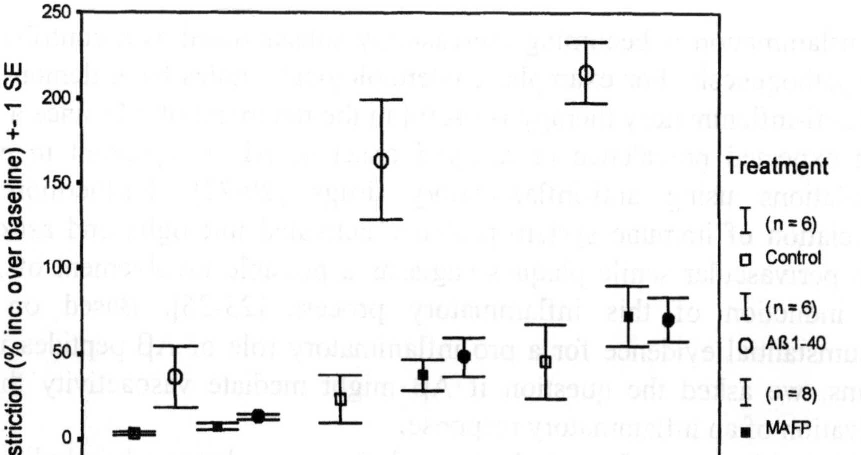

Figure 3: Effect of $cPLA_2$ inhibition on Aβ vasoactivity. Aortic rings were treated with 1 μM of freshly solubilized $A\beta_{1\text{-}40}$, 1 μM of MAFP, MAFP + Aβ, or untreated (control) 5 minutes prior to the addition of a dose range of ET-1. ANOVA showed significant treatment effects of ET-1 dose ($p < .001$), Aβ ($p < .001$), and MAFP ($p < .001$). There were also significant interactive terms between ET-1 dose and Aβ ($p < .01$), and among ET-1, Aβ and MAFP ($p < .01$). One-way ANOVA revealed significant between-groups differences ($p < .001$), and Bonferonni's post hoc test across the 2.5 nM and 5 nM doses of ET-1 did not reveal a significant difference between control and Aβ + MAFP treated vessels ($p = 1.00$), indicating complete blockade of Aβ

Having shown that both superoxide and peroxynitrite are unlikely to mediate Aβ vasoactivity, we went on to test the possibility that NO production could be decreased in response to Aβ stimulation. Thus, we investigated the effect of nitric oxide synthase (NOS) inhibition in our vessel bath system. Data show that NOS inhibition does not mimic Aβ vasoactivity, but rather it results in only a slight increase in ET-1-induced vasoconstriction. Moreover, NOS inhibition together with Aβ does not result in a statistically interactive effect [19]. These data show that Aβ vasoactivity is not due to a reduction in NO production via a decreased activity of NOS, and lead us to postulate an alternate mechanism for the vasoactivity induced by Aβ.

4. MECHANISM OF Aβ VASOACTIVITY

Inflammation is becoming increasingly substantiated as a contributor to AD pathogenesis. For example, epidemiological studies have demonstrated that anti-inflammatory therapy is useful in the treatment of AD since a lower than expected prevalence or delayed onset of AD is apparent in patient populations using anti-inflammatory drugs [20-22]. Furthermore, the association of immune system proteins, activated microglia and astrocytes with perivascular senile plaques suggests a possible involvement of Aβ in the induction of this inflammatory process [23-26]. Based on such circumstantial evidence for a pro-inflammatory role of Aβ peptides in AD brains, we asked the question if Aβ might mediate vasoactivity through activation of an inflammatory response.

Since cleavage of arachidonic acid from membrane phospholipids by phospholipase A_2 (PLA_2) initiates a classical inflammatory cascade, we first investigated the effect of blocking PLA_2 on Aβ vasoactivity. Data show that methyl arachidonyl fluorophosphonate (MAFP), a selective irreversible inhibitor of both calcium-dependent and calcium-independent cytosolic PLA_2s ($cPLA_2$s [27]), is able to completely abolish Aβ vasoactivity (Figure 3). Furthermore, stimulation of $cPLA_2$ using the small (26 amino acid residue) peptide melittin [28] mimics Aβ vasoactivity and results in statistical interaction when added in combination with Aβ (Figure 4).

Activation of $cPLA_2$ results in an increase in arachidonic acid (AA) production since $cPLA_2$ displays strict substrate specificity for AA-containing phospholipids. Thus, we went on to investigate the contribution of AA metabolism to Aβ vasoactivity. AA can be metabolized via two distinct pathways involving the cyclooxygenases (COX) and lipoxygenases (LOX). The COX pathway gives rise to various prostaglandins and thromboxanes, while the LOX pathway results in the production of leukotrienes and lipoxins. All of these eicosanoids are ubiquitous lipid mediators, with a broad range of physiologic activities, including modulation of inflammation and arterial blood pressure [29]. We find that NS-398, a specific COX-2 inhibitor [30], almost completely inhibits Aβ vasoactivity in a statistically interactive manner (Figure 5), showing that COX-2 activity is required to mediate Aβ-induced vasoconstriction. We then tested the effect of MK-886, a compound that impairs the translocation of 5-LOX and its subsequent activation by 5-LOX-activating protein [31], and observed that

MK-886 partially inhibits Aβ vasoactivity in a statistically interactive manner (Figure 6). These data show that Aβ vasoactivity is also mediated via the 5-LOX pathway. Simultaneous inhibition of COX-2 and 5-LOX results in complete inhibition of Aβ vasoactivity (Bonferonni's post-hoc comparison of control to Aβ + MK-886 + NS-398 treated vessels across the 2.5 nM and 5 nM doses of ET-1 does not reveal a significant difference, $p = 1.00$, data not shown), showing that both COX-2 and 5-LOX activities are required to transduce Aβ vasoactivity.

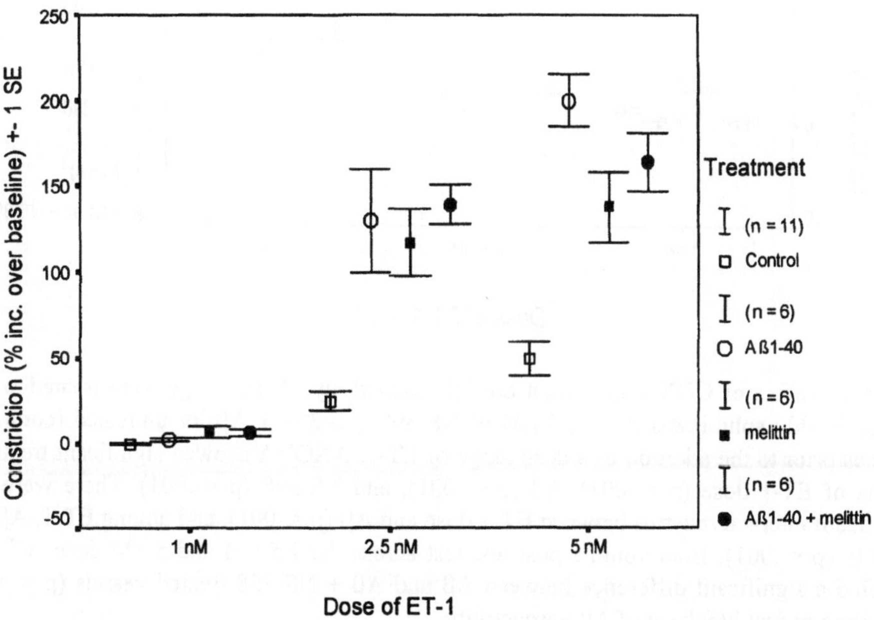

Figure 4: Effect of $cPLA_2$ stimulation on Aβ vasoactivity. Aortic rings were treated with 1 μM of freshly solubilized $A\beta_{1\text{-}40}$, 1 μM of melittin, melittin + Aβ, or untreated (control) 5 minutes prior to the addition of a dose range of ET-1. ANOVA showed significant treatment effects of ET-1 dose ($p < .001$), Aβ ($p < .001$) and melittin ($p < .001$), as well as significant interactive terms between ET-1 dose and either Aβ ($p < .001$) or melittin ($p < .05$). Furthermore, there was a significant interactive term among ET-1 dose, Aβ and melittin ($p < .01$). One-way ANOVA revealed significant between-groups differences ($p < .001$), but Bonferonni's post-hoc test across the 2.5 nM and 5 nM doses of ET-1 did not reveal significant differences between Aβ and melittin ($p = .266$), or between Aβ and Aβ + melittin ($p = 1.00$), indicating that melittin mimics Aβ vasoactivity.

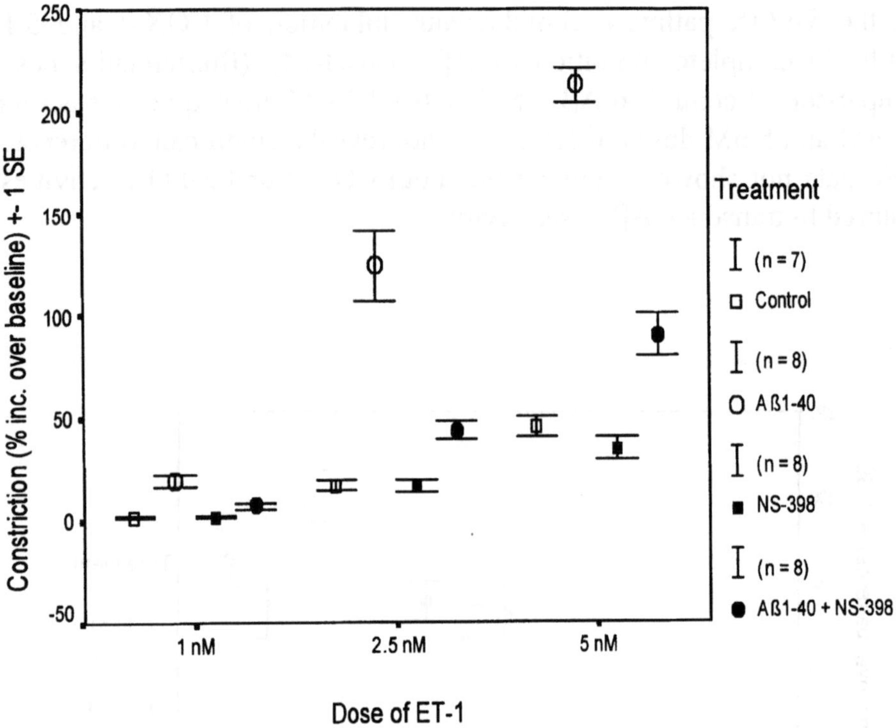

Figure 5: Effect of COX-2 inhibition on Aβ vasoactivity. Aortic rings were treated with 1 μM of freshly solubilized $A\beta_{1\text{-}40}$, 5 μM of NS-398, NS-398 + Aβ, or untreated (control) 5 minutes prior to the addition of a dose range of ET-1. ANOVA showed significant treatment effects of ET-1 dose ($p < .001$), Aβ ($p < .001$), and NS-398 ($p < .001$). There were also significant interactive terms between ET-1 dose and Aβ ($p < .001$), and among ET-1, Aβ and NS-398 ($p < .001$). Bonferonni's post hoc test across the 2.5 nM and 5 nM doses of ET-1 revealed a significant difference between Aβ and Aβ + NS-398 treated vessels ($p < .001$), indicating partial blockade of Aβ vasoactivity.

5. DISCUSSION

We have shown that the use of the tissue bath system constitutes a rapid and efficient means of delineating specific signal transduction pathways triggered by Aβ. Ultimately, these effects can be modulated by manipulating particular targets in such signal transduction pathways, providing the basis for novel therapeutic intervention. By using this system, we have found that stimulation of a primitive inflammatory response is necessary for transduction of Aβ's vasoactivity. Specifically, we show that Aβ's vasoconstrictive effect is mediated by a stimulation of $cPLA_2$ resulting in an increased production of arachidonic acid which is further metabolized into various eicosanoid endproducts via COX and LOX pathways. Our data suggest that activation of $cPLA_2$, COX-2 and 5-LOX is sufficient to bring about Aβ vasoactivity. Moreover, we demonstrate that soluble Aβ peptides trigger this inflammatory response, suggesting that, in AD where inflammation is a key factor in the pathogenesis of the disease, soluble forms of the peptide might initiate this process prior to the formation of Aβ aggregates. Furthermore, our data showing that soluble forms of Aβ activate a pro-inflammatory pathway suggest that inflammation may occur during the pre-clinical phases of AD and CAA, thereby paving the way for future exacerbation of this inflammatory process as Aβ peptides continue to be deposited in brain parenchyma and between vascular smooth muscle cells.

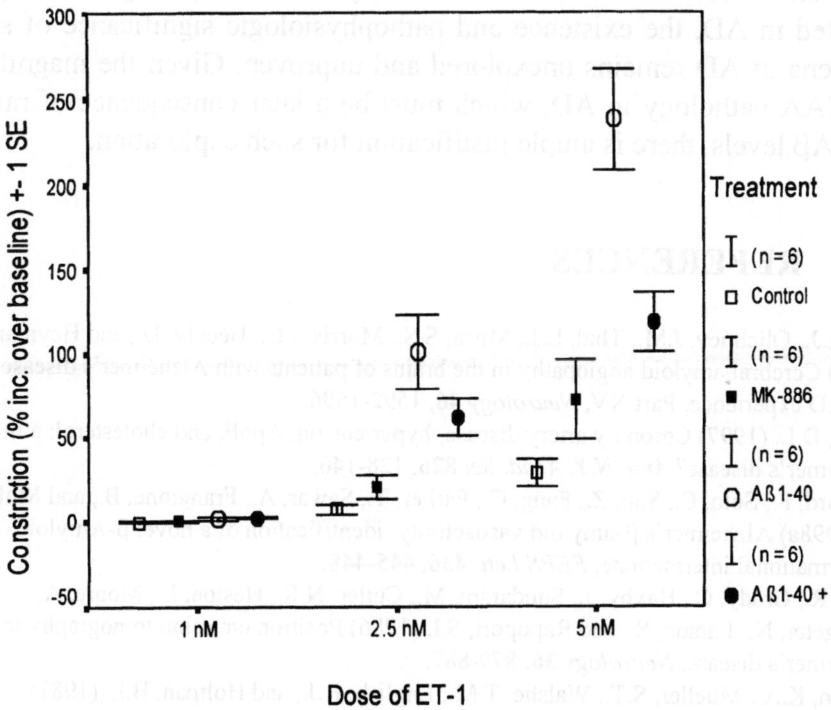

Figure 6: Effect of 5-LOX inhibition in Aβ vasoactivity. Aortic rings were treated with 1 μM of freshly solubilized $A\beta_{1-40}$, 1 μM of MK-886, MK-886 + Aβ, or untreated (control) 5 minutes prior to the addition of a dose range of ET-1. ANOVA showed significant treatment effects of ET-1 dose ($p < .001$), Aβ ($p < .001$), and MK-886 ($p < .05$). There were also significant interactive terms between ET-1 dose and Aβ ($p < .01$), and among ET-1, Aβ and MK-886 ($p = .001$). Bonferonni's post hoc test across the 2.5 nM and 5 nM doses of ET-1 revealed a significant difference between Aβ and Aβ + MK-886 treated vessels ($p = .01$), indicating partial blockade of Aβ vasoactivity.

There are several implications of these findings for AD and CAA. It has been suggested that Aβ peptides are the primary pathogenic molecules in AD, and CAA is an integral part of the AD process, occurring in the majority (83%) of AD patients, particularly in cerebral arterioles, pre-capillaries, and capillaries [1]. Microscopically, Aβ deposition is also associated with degeneration of smooth muscle cells in CAA [32,33]. Increased amounts of soluble Aβ have been demonstrated in isolated vessels from AD subjects compared to age-matched controls [34], and Aβ vasoactivity has been observed both *in vitro* in peripheral and *in vivo* in cerebral microvessels [13,17,35], suggesting that the juxtaposition of Aβ peptides with large and small cerebral vessels results in aberrant vasoactivity. The evidence, which we have presented here, suggests that this effect is mediated by a pro-inflammatory mechanism, although free radical concentrations can modify these effects, probably indirectly. Although modulation of cerebrovascular resistance by soluble Aβ might be highly anticipated in AD, the existence and pathophysiologic significance of such phenomena in AD remains unexplored and unproven. Given the magnitude of the CAA pathology in AD, which must be a later consequence of raised soluble Aβ levels, there is ample justification for such exploration.

REFERENCES

1. Ellis, R.J., Olichney, J.M., Thal, L.J., Mirra, S.S., Morris, J.C., Beekly, D., and Heyman, A. (1996) Cerebral amyloid angiopathy in the brains of patients with Alzheimer's disease: the CERAD experience, Part XV, *Neurology* **46**, 1592-1596.
2. Sparks, D.L. (1997) Coronary artery disease, hypertension, ApoE, and cholesterol: a link to Alzheimer's disease? *Ann. N.Y. Acad. Sci* **826**, 128-146.
3. Crawford, F., Soto, C., Suo, Z., Fang, C., Parker, T., Sawar, A., Frangione, B., and Mullan, M. (1998a) Alzheimer's β-amyloid vasoactivity: identification of a novel β-Amyloid conformational intermediate, *FEBS Lett.* **436**, 445-448.
4. Duara, R., Grady, C., Haxby, J., Sundaram, M., Cutler, N.R., Heston, L., Moore, A., Schlageter, N., Larson, S., and Rapoport, S.I. (1986) Positron emission tomography in Alzheimer's disease, *Neurology* **36**, 879-887.
5. Johnson, K.A., Mueller, S.T., Walshe, T.M., English, R.J., and Holman, B.L. (1987) Cerebral perfusion imaging in Alzheimer's disease. Use of a single photon emission computed tomography and iofetamine hydrochloride I 123, *Arch. Neurol.* **44**, 165-168.
6. Friedland, R.P., Budinger, T.F., Ganz, E., Yano, Y., Mathis, C.A., Koss, B., Ober, B.A., Huesman, R.H., and Derenzo, S.E. (1983) Regional cerebral metabolic alterations in dementia of the Alzheimer type: positron emission tomography with [18F]fluorodeoxyglucose, *J. Comput. Assist. Tomogr.* **7**, 590-598.
7. Iadecola, C., Zhang, F., Niwa, K., Eckman, C., Turner, S., Fischer, E., Younkin, S., Borchelt, D., Hsaio, K., and Carlson, G. (1999) SOD1 rescues cerebral endothelial dysfunction in mice overexpressing amyloid precursor protein, *Nat. Neurosci.* **2**, 157-161.

8. Zhang, F., Eckman, C., Younkin, S., Hsiao, K.K., and Iadecola, C. (1997) Increased susceptibility to ischemic brain damage in mice overexpressing the amyloid precursor protein, *J. Neurosci.* **17**, 7655-7661.
9. Jagust, W.J., Eberling, J.L., Reed, B.R., Mathis, C.A., and Budinger, T.F. (1997) Clinical studies of cerebral blood flow in Alzheimer's disease, *Ann. N. Y. Acad. Sci.* **826**, 254-262.
10. Nagata, K., Buchan, R.J., Yokoyama, E., Kondoh, Y., Sato, M., Terashi, H., Sato, Y., Watahiki, Y., Senova, M., Hirata, Y., and Hatazawa, J. (1997) Misery perfusion with preserved vascular reactivity in Alzheimer's disease, *Ann. N. Y. Acad. Sci.* **826**, 272-281.
11. Thomas, T., Thomas, G., McLendon, C., Sutton, T., and Mullan, M. (1996) β-Amyloid mediated vasoactivity and vascular endothelial damage, *Nature* **380**, 168-171.
12. Douglas, S.A. and Ohlstein, E.H. (1997) Signal transduction mechanisms mediating the vascular actions of endothelin, *J. Vas. Res.* **34**, 152-164.
13. Kuo, Y.M., Emmerling, M.R., Lampert, H.C., Hempelman, S.R., Kokjohn, T.A., Woods, A.S., Cotter, R.J., and Roher, A.E. (1999) High level of circulating Aβ42 sequestered by plasma proteins in Alzheimer's disease. *Biochem. Biophys. Resch. Com.* **257**, 787-791.
14. Moncada, S., Palmer, R., and Higgs, E. (1991) Nitric oxide: physiology, pathophysiology, and pharmacology, *Pharmacol. Rev.* **43**, 109-141.
15. Urmoneit, B., Prikulis, I., Wihl, G., D'Urso, D., Frank, R., Heeren, J., Beisiegel, U., and Prior, R. (1997) Cardiovascular smooth muscle cells internalize Alzheimer amyloid beta protein via a lipoprotein pathway: implications for cerebral amyloid angiopathy, *Lab. Invest.* **77**, 157-166.
16. Bonafi, L., Thomas, S.R., Hill, R.G., and Longmore, J. (1998) β-Amyloid inhibits endothelial-dependent relaxations in rabbit isolated aorta: an interaction between superoxide radicals and nitric oxide?, *Alzh. Reports* **5**, 297-302.
17. Crawford, F., Suo, Z., Fang, C., and Mullan, M. (1998b) Characteristics of the in vitro vasoactivity of β-amyloid peptides, *Exp. Neurol.* **150**, 159-168.
18. Paris, D., Parker, T.A., Town, T., Suo, Z., Fang, C., Humphrey, J., Crawford, F., and Mullan, M. (1998) Role of Peroxynitrite in the vasoactive and cytotoxic effects of Alzheimer's β-amyloid peptide, *Exp. Neurol.* **152**, 116-122.
19. Paris, D., Town, T., Parker, T.A., Humphrey, J., Tan, J., Crawford, F., and Mullan, M. (1999) Inhibition of Alzheimer's β-amyloid induced vasoactivity and pro-inflammatory response in microglia by a cGMP-dependent mechanism, *Exp. Neurol.* **157**, 211-221.
20. McGeer PL, Schulzer M, and McGeer EG. (1996) Arthritis and anti-inflammatory agents as possible protective factors for Alzheimer's disease: a review of 17 epidemiologic studies, *Neurology* **47**: 425-432.
21. Stewart, W.F., Kawas, C., Corrada, M., and Metter, E.J. (1997) Risk of Alzheimer's disease and duration of NSAID use, *Neurology* **48**, 626-632.
22. Rogers, J., Kirby, L.C., and Hempielman, S.R. (1993) Clinical trial of indomethacin in Alzheimer's disease, *Neurology* **43**, 1609-1611.
23. Coria, F., Moreno, A., Rubio, I., Garcia, M.A., Morato, E., and Mayor, F. (1993) The cellular pathology associated with Alzheimer β-amyloid deposits in non-demented aged individuals, *Neuropathol. Appl. Neurobiol.* **19**, 261-268.
24. Griffin WST, Sheng JG, Roberts GW, and Mrak RE. (1995) Interleukin-1 expression in different plaque types in Alzheimer's disease: significance in plaque evolution, *J. Neuropathol. Exp Neurol.* **54**, 276-281.
25. Itagaki S, McGeer PL, Akiyama H, Zhu S, and Selkoe D. (1989) Relationship of microglia and astrocytes to amyloid deposits of Alzheimer disease, *J. Neuroimmunol.* **24**, 173-182.

26. Lue, L.F., Brachova, L., Civin, W.H., and Rogers, J. (1996) Inflammation, Abeta deposition, and neurofibrillary tangle formation as correlates of Alzheimer's disease neurodegeneration, *J. Neuropathol. Exp. Neurol.* **55**, 1083-1088.
27. Lio, Y.C., Reynolds, L.J., Balsinde, J., and Dennis, E.A. (1996) Irreversible inhibition of Ca(2+)-independent phospholipase A2 by methyl arachidonyl fluorophosphonate, *Biochem. Biophys. Acta.* **1302**, 55-60.
28. Wu, Y.L., Jiang, X.R., Newland, A.C., and Kelsey, S.M. (1998) Failure to activate cytosolic phospholipase A2 causes TNF resistance in human leukemic cells, *J. Immunol.* **160**, 5929-5935.
29. Gurwitz, J.G., Avorn, J., Bohn, R.L., Glynn, R.J., Monane, M., and Mogun, H. (1994) Initiation of antihypertensive treatment during nonsteroidal anti-inflammatory drug therapy, *JAMA* **272**, 781-786.
30. Futaki, N., Takahashi, S., Yokoyama, M., Arai, I., Higuchi, S., and Otomo, S. (1994) NS-398, a new anti-inflammatory agent, selectively inhibits prostaglandin G/H synthase/cyclooxygenase (COX-2) activity in vitro, *Prostaglandins* **47**, 55-59.
31. Lepley, R.A., Muskardin, D.T., and Fitzpatrick, F.A. (1996) Tyrosine kinase activity modulates catalysis and translocation of cellular 5-lipoxygenase, *J. Biol. Chem.* **271**, 6179-6184.
32. Kalaria., R.N. (1997) Cerebrovascular degeneration is related to amyloid-beta protein deposition in Alzheimer's disease. *Ann. N.Y. Acad. Sci.* **826**, 263-271.
33. Kawai., M., Kalaria., R.N., Cras., P., Siedlak., S.L., Velasco, M.E., Shelton, E.R., Chan, H.W., Greenberg, B.D., Perry, G. (1993) Degeneration of vascular muscle cells in cerebral amyloid angiopathy of Alzheimer disease. *Brain. Res.* **623**, 142-146.
34. Kalaria, R.D., Bhatti, S., Lust, W., and Perry, G. (1993) The blood brain barrier and cerebral microcirculation in Alzheimer's disease, *Cerebrovas. Brain Metab. Rev.* **4**, 226-260.
35. Suo, Z., Humphrey, J., Kundtz, A., Sethi, F., Placzek, A., Crawford, F., and Mullan, M. (1998) Soluble Alzheimer's β-amyloid constricts the cerebral vasculature in vivo, *Neurosci. Lett.* **257**, 77-80.

Chapter 18

CAA IN TRANSGENIC MOUSE MODELS OF ALZHEIMER'S DISEASE

What Can We Learn from APP Transgenic Mouse Models?

GREG M. COLE and FUSHENG YANG

Department of Medicine and Neurology, UCLA and Sepulveda VAMC, GRECC11E, North Hills, CA, USA

The development of mouse lines with high level expression of mutant β-amyloid precursor protein (APP) trangenes has afforded researchers convenient animal models with CNS β-amyloidosis. Two well-characterized lines bearing neuronally expressed human APP with "Swedish" double mutations with dramatically increased production of Aβ$_{1-40}$ and Aβ$_{1-42}$, develop significant neuritic plaque and vascular amyloidosis. Other transgenics with the " London" APP mutant transgene, which results in an increased percentage of APP processed to Aβ$_{1-42}$ but not Aβ$_{1-40}$, show very high levels of primarily plaque amyloid suggesting that the site of deposition depends on the relative levels of the different Aβ species. The roles of transforming growth factor β1 and apolipoprotein E in both CAA and plaque amyloid deposition are also being addressed in transgenic mouse models. From a review of the literature on amyloidosis in transgenic mice, we develop several hypotheses about how Aβ accumulates in CAA and discuss possible applications of APP transgenics to research on the prevention and treatment of CAA.

1. INTRODUCTION.

After the isolation and sequencing of amyloid β-protein (Aβ) from vascular amyloid [1], it remained unclear whether senile plaque amyloid involved the same peptide until purified plaque amyloid was sequenced [2]. Unlike the case with vascular amyloid, there was initial difficulty in sequencing the plaque amyloid form because of blocked and ragged N-termini. Plaque amyloid was found to contain a mixture of peptides with numerous posttranslational modifications including a high percentage of pyroglutamate at the N-terminus, isoaspartates and other alterations. The plaque amyloid peptide also was found to involve a higher percentage of the

M.M. Verbeek et al. (eds.),
***Cerebral Amyloid Angiopathy in Alzheimer's Disease and Related Disorders*, 295–311.**

42 than 40 amino acid form which predominated in vascular amyloid [3,4]. The percentage of Aβ_{40} in brain has been found to correlate with the cerebral amyloid angiopathy (CAA) [5,6]; cases which have essentially no vascular amyloid have been reported to contain nearly 95% A$\beta_{x\text{-}42}$ in the brain [3]. Cultured cells of many different types have been found to produce a greater percentage of Aβ_{40} (around 80%), but Aβ_{42} aggregates much more readily than shorter peptides, which may explain the higher percentage of A$\beta_{x\text{-}42}$ in plaques. It also accounts for the inheritance of AD in individuals carrying APP717 or presenilin mutations which result in a higher percentage of A$\beta_{1\text{-}42}$ formation, but no increase in total Aβ production [7,8]. Vascular β-amyloid with its abundant A$\beta_{1\text{-}40}$ presumably involves other pathogenetic mechanisms. In Dutch variant amyloidosis (HCHWA-D), where the primary cause is a change in internal Aβ sequence believed to cause increased peptide aggregation, the vascular amyloid is still predominantly Aβ_{40} [9]. Non-genetic factors promoting CAA could include either an increase in co-factors involved in amyloidosis or an increase in Aβ_{40} production sufficient to induce aggregation of amyloid fibrils. Since vascular β-amyloid is quite variable in extent in AD cases, there must be independent factors regulating vascular amyloid formation. Consistent with the involvement of independent mechanisms, spontaneous age-related, primary vascular β-amyloidosis occurs in some higher primates, notably the squirrel monkey [10]. However, it is convenient to have small animal models for vascular β-amyloidosis in order to work out the cause and consequences of this pathology. The objective of this review is to briefly describe the APP transgenics and vascular amyloid occuring in transgenic models related to AD, and discuss the implications of these observations and potential of the models.

2. A BRIEF SUMMARY OF TRANSGENIC LITERATURE RELEVANT TO VASCULAR β-AMYLOIDOSIS.

2.1 APP TRANSGENIC MICE.

Human β-amyloid protein precursor (huAPP) transgenic mouse lines have been established by a number of groups in an effort to develop convenient animal models for Alzheimer's disease (AD) [11-14]. In general, transgenics with huAPP expression levels below 2-3 fold the endogenous mouse APP levels have failed to develop amyloid deposits while those with 5-10 fold or greater elevated expression of huAPP developed classical amyloid deposits by 12 months of age. The first transgenic mouse model with significant β-amyloid deposition, used a platelet derived growth factor promoter to drive very high level expression of a human APP717F mutant transgene (PDAPP line 109) [12]. In addition to the very high level of APP expression, the APP717 mutations result in an increased percentage of $A\beta_{1-42}$ production. These animals have some, relatively limited vascular amyloid. Biochemical data showed that as the animals aged and amyloid accumulated, the percentage of $A\beta_{42}$ in brain rose from 27% at 4 months to 99% at 10 months and remained about 90% $A\beta_{1-42}$. Further, immunocytochemistry showed limited amyloid angiopathy by 18 months and the $A\beta_{40}$ end-specific immunoreactivity was primarily seen in compact plaques rather than increasing amyloid angiopathy [15,16]. In contrast, high level expression of human APP670/671NL "Swedish" mutant APP (huAPPsw, line Tg2576) resulted in nearly 60% $A\beta_{x-40}$ at 10-12 months of age [13]. As shown below, these mice have abundant vascular amyloid and $A\beta_{1-40}$ immunoreactivity in both compact plaques and vascular amyloid. A similar, but even higher expressing APPsw line (APP23) developed at Novartis by Staufenbiel's group [14] also exhibits significant vascular amyloid as described by M. Jucker and discussed below.

2.2 VASCULAR AMYLOID IN TG2576 APPsw TRANSGENIC MICE.

The PrP promoter drives predominantly neuronal expression of huAPPsw in Tg2576 mice [17]. At 12 months (n=6), most of the Aβ staining was plaque-associated [13], but variable amounts of limited, patchy leptomeningeal staining were evident (Figure 1A). This early vascular Aβ staining was blocked by preabsorption with peptide antigen (Figure 1B). The early deposits were labeled with affinity purified end-specific antibodies to $A\beta_{42}$ (Figure 1C). Vascular Aβ was a common finding affecting all of the animals, but to a variable extent, which appeared to increase along with the plaque deposition as the animals aged to 24 months. By 16 months (n=20), leptomeningeal deposits were no longer patchy, but circumvented most affected vessels (Figs. 1D,E). By this age, all transgene-positive mouse brains exhibited leptomeningeal Aβ deposits which were more evident in animals with abundant plaque amyloid. However, as shown in Figure 2, in some mice, regional variation within the brain was pronounced. Piriform and entorhinal/ perirhinal regions with significant plaque labeling also had cortical large and small vessel Aβ deposits (Figure 2A). In contrast, more frontal regions of the same section showed predominantly cortical large vessel and leptomeningeal Aβ (Figure 2B). The reasons underlying significant regional differences in the pattern of cortical amyloid deposition in the same animals are unknown. In addition to "within animal" variation, marked differences existed between different outbred mice. Occasionally (~10%), the major lesion was vascular amyloid (Figure 2C). In these animals $A\beta_{40}$ end-specific antibodies (Figure 2D) labeled almost every plaque as well as all of the vascular amyloid. However, in most animals as in humans, $A\beta_{40}$ antibodies more selectively stained the vascular amyloid and a variable subset of plaques. The most consistent and prominent $A\beta_{40}$ staining was always in the leptomeningeal vessels, but deeper cortical, hippocampal sulcus and thalamic vessel labeling was also present (Figure 2D).

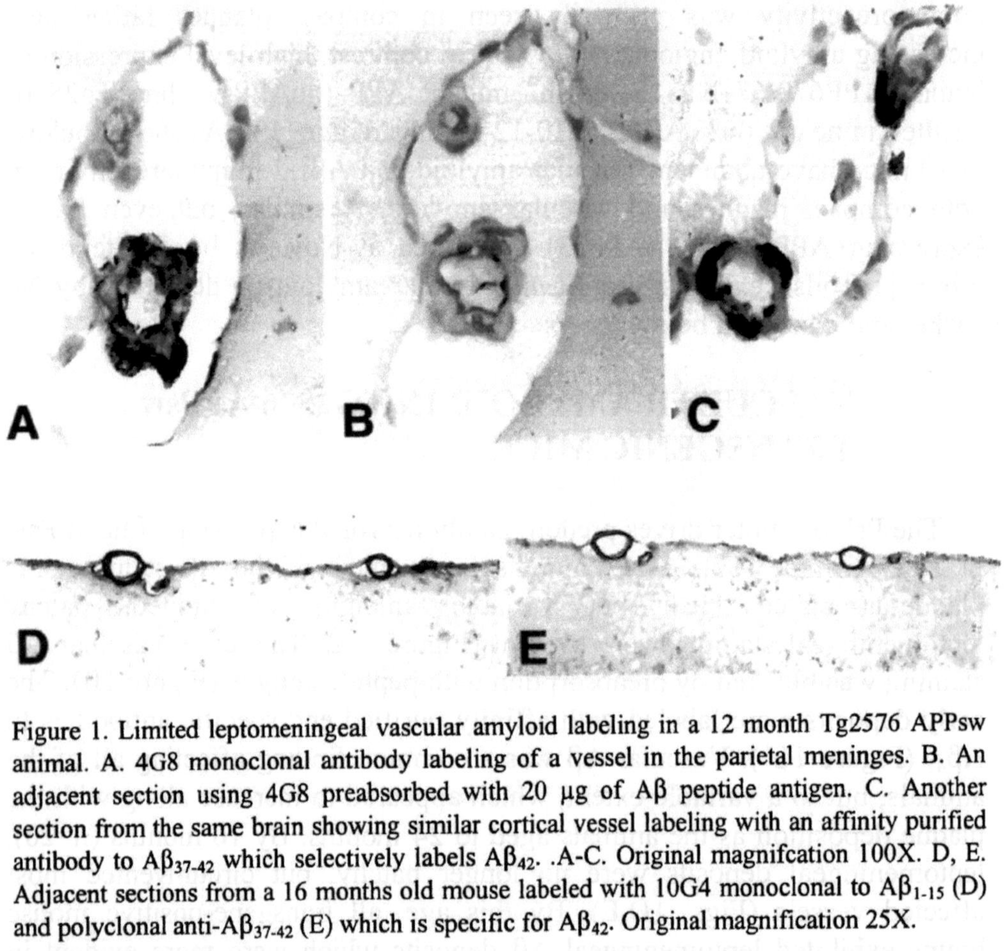

Figure 1. Limited leptomeningeal vascular amyloid labeling in a 12 month Tg2576 APPsw animal. A. 4G8 monoclonal antibody labeling of a vessel in the parietal meninges. B. An adjacent section using 4G8 preabsorbed with 20 μg of Aβ peptide antigen. C. Another section from the same brain showing similar cortical vessel labeling with an affinity purified antibody to $A\beta_{37-42}$ which selectively labels $A\beta_{42}$. .A-C. Original magnifcation 100X. D, E. Adjacent sections from a 16 months old mouse labeled with 10G4 monoclonal to $A\beta_{1-15}$ (D) and polyclonal anti-$A\beta_{37-42}$ (E) which is specific for $A\beta_{42}$. Original magnification 25X.

To confirm the presence of amyloid we employed classical thioflavin S fluorescent labeling. Thioflavin S staining (Figure 3) of 16 month Tg2576 APPsw mice revealed numerous large vessels, notably in the leptomeninges and cortex. The hippocampal sulcus was also a frequent site of vascular deposits. In the thalamus there was typically prominent CAA, but very few plaques. In general, all of the vascular Aβ staining seen with rabbit polyclonals corresponded to thioflavin S labeled vessels, but occasional brains had areas with suboptimal perfusion and artefactual, thioflavin negative vessel labeling when monoclonals to Aβ were detected with biotinylated anti-mouse secondary antibodies.

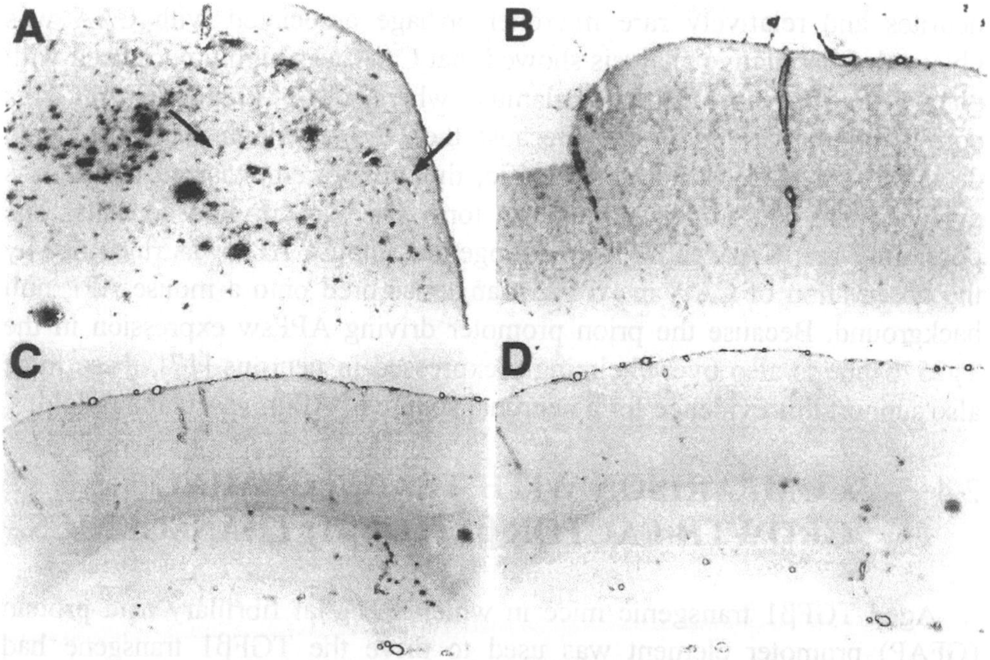

Figure 2. Variation in Aβ and $A\beta_{40}$ immunostaining in 16 month old Tg2576 APPsw mice. A. Anti-$A\beta_{1\text{-}13}$ antisera (DAE) labeled many plaques and vessels in the entorhinal perirhinal region of 16 month mice. Arrows indicate small vessel cortical amyloid. B. Vascular amyloid demonstrated with the same antibody in frontal cortex from the same section (DAE). C. DAE labeling for Aβ in a section from another 16 month old mouse with primarily vascular amyloid. D. An adjacent section labeled for $A\beta_{40}$ with anti-$A\beta_{34\text{-}40}$ antisera. Original magnifications 10X.

We have yet to see evidence of hemorrhage in relation to vascular amyloid in the Tg2576 mice, but have only examined a few mice greater than two years of age. Whether or not there is significant damage to the vascular smooth muscle or endothelial cells or neighboring neurons and their processes is yet to be determined.

2.3 COMPARISON WITH CAA IN APP23 TRANSGENIC MICE.

Very similar Aβ immunoreactive, thioflavin S and Congo red positive CAA has been previously reported in leptomeninges, cortex, thalamus and hippocampus in the APP23 line by Jucker *et al.* [18]. Ultrastructural examination showed that vessel amyloid fibrils were associated with the microvascular basal lamina. Qualitative evidence of neuron loss, dystrophic neurites and relatively rare microhemorrhage associated with CAA was obtained. Quantitative analysis showed that CAA was highly correlated with plaque load except in the thalamus, where CAA predominated. This correlation was driven by age. Because the Thy1 promoter element used to drive the transgene is neuron specific, the authors emphasized that this is evidence that neuron-derived Aβ can form vascular amyloid deposits. The possibility that CAA came from endogenous mouse APP was ruled out by the observation of CAA in APP23 trangenics bred onto a mouse APP null background. Because the prion promoter driving APPsw expression in the Tg2576 line is also overwhelmingly expressed in neurons [17], these mice also support this evidence for a neuronal source of Aβ in vascular amyloid.

2.4 COMPARISON WITH TRANSFORMING GROWTH FACTOR β1 (TGFβ1) TRANSGENICS.

Aged TGFβ1 transgenic mice in which the glial fibrillary acid protein (GFAP) promoter element was used to drive the TGFβ1 transgene had prominent perivascular astrocyte TGFβ1 expression and vascular β-amyloid at 16-18 months [19]. The amyloid deposition was TGFβ1 dose-dependent. Furthermore, a cross between PDAPP mice and the TGFβ1 transgenics resulted in accelerated vascular amyloid deposition in the bigenic mice which was already evident at 10 weeks of age, much earlier than amyloid deposition in PDAPP or any of the other APP lines. As with the prion promoter in Tg2576 and the Thy1 promoter in the APP23 line, the PDGF promoter results in predominantly neuron-specific APP CNS transgene expression [12], again demonstrating the potential for neuronal Aβ to deposit in vessels. These same authors also found that TGFβ1 expression was elevated in AD cases with significant CAA suggesting that TGFβ1 might also be involved in the vascular deposition of neuron-derived Aβ in Alzheimer's disease.

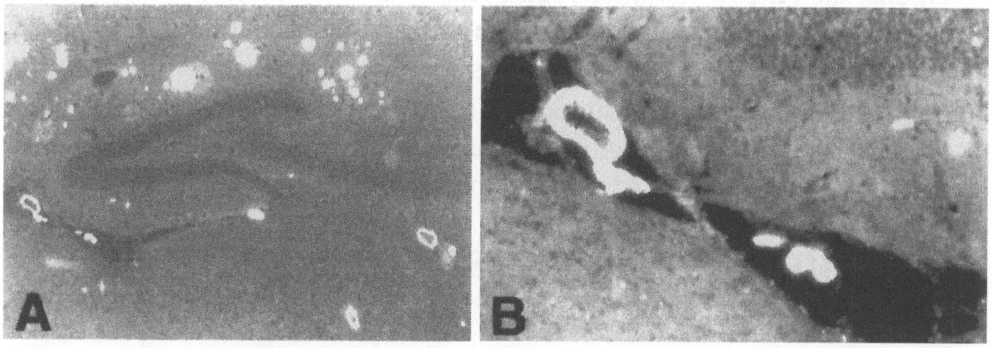

Figure 3. Thioflavin S labeling of plaque and vascular amyloid in a 16 month old Tg2576 APPsw mouse. A. Hippocampus, ventricle and thalamus. Original magnification 10X. B. Higher magnification of same field showing vascular amyloid in ventricle and hippocampus. Original magnification 100X.

3. DISCUSSION.

Several interesting points are raised by these observations:

3.1 VASCULAR AMYLOID CAN HAVE A NEURONAL SOURCE.

As noted by Jucker *et al.*, the transgenic data argue that authentic vascular β-amyloid can arise from neuron-specific expression of high levels of APP. Since the source of vascular amyloid has been argued to be from circulating peripheral Aβ [20] or from smooth muscle cells [21,22], the observation of CAA with a predominant neuronal source in three different plaque-forming transgenic lines strengthens the argument that neuron-derived Aβ deposits in plaques or vessels in AD. However, while neurons are almost certainly the major source of Aβ in the APP transgenic mouse amyloid, this does not imply the lack of involvement of vascular smooth muscle cells (SMC), which may be a significant sink for neuronal Aβ. SMC readily take up exogenous Aβ riding on lipoproteins using the α_2-macroglobulin / low-density lipoprotein receptor-related protein receptor

(LRP) [23]. Alternatively, amyloid binding extracellular matrix proteins associated with the basal lamina may be the initial trap for soluble, secreted neuronal Aβ.

$A\beta_{1-40}$. As in AD, $A\beta_{40}$ is the predominant vascular species although $A\beta_{42}$ is present in the early vascular deposits. Unlike the APP717 and presenilin mutations, the APPsw mutations raise $A\beta_{40}$ and $A\beta_{42}$ levels 6-9 fold [24,25] The significant vascular amyloid in the APPsw, but not the APP717F may reflect the elevated $A\beta_{40}$ in the APPsw mice- consistent with the large amounts of $A\beta_{40}$ in vascular amyloid. The few APPsw cases that have come to autopsy have also had significant vascular amyloid which, in contrast to the parenchymal deposits, was predominantly $A\beta_{40}$ [26].

3.2 THERE ARE FOCAL SOURCES AND SINKS FOR Aβ.

Regional patterns of plaque deposition are similar in the APP transgenics and in AD and consistent with deposition in the immediate vicinity of neuronal sources [17,27]. The observation that neurons can be the major source for vascular Aβ might therefore be explained as due to local Aβ derived from neurons innervating blood vessels. Alternatively, the observations are also compatible with the hypothesis that $A\beta_{40}$ produced by neurons may be able to travel some distance without depositing and eventually become trapped and deposited in the vessel wall in or near the basal lamina, which serves as a focal sink. How can we distinguish between these possibilities? In contrast to $A\beta_{40}$, $A\beta_{42}$ is widely established as the earliest and major Aβ species in plaque deposits. For example, in most models, $A\beta_{42}$ is deposited in the hippocampus in the outer molecular layer (OML) of the dentate, forming a diffuse cloud- or band-like series of deposits; this is very clear in the PDAPP mouse. Similar diffuse hippocampal deposition of $A\beta_{42}$ also occurs, but at later ages, in the Tg2576 APPsw line. Hippocampal OML Aβ is largely derived from entorhinal cortical projection neurons terminating in the OML as demonstrated by entorhinal cortex lesioning studies [28]. Similarly, the earliest lesions in Down's syndrome are almost exlusively $A\beta_{42}$ [29,30] and very often centered directly on neurons [31]. We have also observed this type of neuron-centered $A\beta_{42}$ deposit in Tg2576 mice (unpublished observations). In addition to other observations strongly arguing for a local neuronal source of plaque Aβ [32], the data suggest a simple hypothesis: $A\beta_{42}$ self-aggregates sufficiently rapidly at relatively low concentrations that substantial deposits develop locally near the site of $A\beta_{42}$ release from neurons giving rise to diffuse plaques; these typically do not contain detectable $A\beta_{40}$. Because $A\beta_{40}$ aggregates more slowly at higher concentrations, it more readily diffuses over considerable distances and deposits in association with non-Aβ_{40} cofactors or cells such as those present in the vessel wall.

3.3 Aβ_{42} SEEDING IS INSUFFICIENT FOR Aβ_{40} DEPOSITION.

Many intense Aβ deposits in AD brains and in PDAPP717 transgenics are primarily A$\beta_{X\text{-}42}$ and appear to lack Aβ_{40}. Aβ_{42} deposit seeding [33,34] of Aβ_{40} deposition appears to be inefficient by itself and not sufficient to cause robust Aβ_{40} deposition at the site of production and release of both Aβ_{40} and Aβ_{42}. If Aβ_{42} seeding of Aβ_{40} was efficient, one would expect Aβ_{40} to typically co-distribute with the Aβ_{42} and show at least weak labeling of most plaques. This is not the case in either AD or APP transgenics. Instead, by 16 months in older APPsw mice, one typically sees intensely Aβ_{42} positive plaques which don't show detectable Aβ_{40} labeling and other plaques, isolated or in clusters, which are heavily Aβ_{40} labeled. The presence of diffuse Aβ_{42} deposits is evidently not sufficient to ensure Aβ_{40} deposition and this is dramatically illustrated in the PDAPP mice [15]. Conversely, Aβ_{40} deposits are invariably present in many, but not all of the major vessels of the leptomeninges and often present in cortex and hippocampus of the aged APPsw transgenics. This is most consistent with local factors other than Aβ_{42} or the mere presence of a vessel wall, controlling Aβ_{40} deposition. Therefore, it seems probable that Aβ_{40} is often not trapped and deposited locally at the site of neuronal production, e.g. at the site of Aβ_{42} deposits, but is able to diffuse over some distance to be either degraded, or reach vessel walls. This is consistent with observations in human CSF by many authors which show that Aβ_{42}, but not Aβ_{40}, is typically reduced in CSF from AD cases by roughly 50% compared to controls. This finding is consistent with the selective deposition of a significant percentage of the steady state nascent Aβ_{42}, but not Aβ_{40}. Because Aβ_{42} is also present from very early stages in the vascular amyloid of mice with neuronal APP expression, one can assume that a portion of the Aβ_{42} derived from neurons innervating blood vessels or in CSF also deposits in the vessel walls or escapes into the perivascular spaces.

3.4 Aβ MAY BE TRANSPORTED FROM BRAIN TO PLASMA.

One implication of endogenous Aβ diffusion to the vessels over significant distances is that a significant flux of soluble neuronal Aβ should make it not only to the vessel walls, but also to the perivascular spaces connecting into the lymphatics and thence to the plasma [35]. Because substantial pools of oligomeric, but soluble $A\beta_{42}$ have been purified from AD brain [36], both vascular and plasma Aβ would thus be expected to contain neuronally derived $A\beta_{40}$ and $A\beta_{42}$ species.

3.5 INJURY-INDUCED TGFβ1 UPREGULATION OF HSPG AND APOE MAY PROMOTE CAA.

If $A\beta_{42}$ is not sufficient to induce efficient seeding of $A\beta_{40}$ deposition, what non-Aβ local factors may play a role in determining the deposition of diffusible Aβ of neuronal origin? Focal CNS injury induces increases in TGFβ1 which upregulate extracellular matrix (ECM) proteins [37,38] including those produced by astrocytes [39]. Based on this work, Frautschy and collaborators reported that TGFβ1 promoted plaque-like Aβ deposition *in vivo* [40,41] and in hippocampal slices [42], suggesting common mechanisms may be involved in promoting vascular and plaque amyloid. As discussed above, data from transgenics show that the ability of the vessel wall to trap Aβ is markedly increased by local TGFβ1 expression, which may be related to TGFβ1 induction of apolipoprotein E or ECM components [43]. Perlecan is a well-studied, vessel wall heparan sulfate proteoglycan (HSPG) which can bind Aβ and promote its deposition and fibril formation [44].

In addition to focal capture of Aβ by ECM, TGFβ1 may influence Aβ uptake by cells. HSPGs also play a prominent role in ApoE isoform-dependent cell binding and internalization by LRP and other lipoprotein receptors [45]. TGFβs markedly increase human ApoE levels assayed by ELISA in human ApoE hippocampal slice culture media (Harris-White, Galaskc, Cole and Frautschy, manuscript in preparation). This may impact Aβ uptake by SMC which is LRP family receptor-dependent [23] and consistent with this, ApoE addition increases Aβ accumulation by cultured SMC [46,47]. The relationship of TGFβ1 to ApoE is particularly interesting because ApoE4 selectively increases $A\beta_{40}$ in AD brain [48,49] and is a risk factor for both vascular and plaque amyloid [50-52]. Since TGFβ1 in the CNS is primarily injury-induced, this may explain the frequent occurrence of β-amyloid deposits near or at the site of vascular injury or malformations [53]. These data also suggest that other causes of vascular injury may be able to selectively promote vascular β-amyloid deposition in individuals with an ApoE4 genotype. In particular, Aβ-induced vascular injury may result in increased TGFβ1 expression which in turn promotes amyloid deposit formation. If the injured cellular component is vascular smooth muscle (SMC), this may contribute to the Aβ-induced increase in SMC Aβ production previously demonstrated [22] and the SMC cell surface assembly of Aβ [54]. Finally, in the FVB strain background, APP overexpression results in endothelial cell injury in the absence of amyloid aggregates; this injury can be protected against by crossing to superoxide dismutase transgenics [55]. These results suggest that elevated Aβ near the vessels may be damaging even in the absence of amyloid. Toxic species of Aβ could induce injury-related TGFβ1, leading to a vascular amyloid cascade.

4. $A\beta_{40}$ ACCUMULATION IS CELL-DEPENDENT: A HYPOTHESIS.

The majority of $A\beta_{40}$ immunostaining in AD brain [56] and APP transgenics is not found evenly distributed over $A\beta_{42}$ deposits, but in much more developed compact plaques with activated microglia or in vessels with CAA, where SMC appear to play an important role in its deposition. Analogous to SMC accumulation of Aβ, TGFβ1-induced, cell-dependent Aβ accumulation also occurs in microglia in rat brain and hippocampal slices [41,42]. In the huAPPsw Tg2576 mice, microglia are immunoreactive for $A\beta_{40}$, but not $A\beta_{42}$, suggesting preferential accumulation of $A\beta_{40}$ [57]. Unlike $A\beta_{42}$ which can readily self-aggregate and deposit, $A\beta_{40}$ deposition may require cell-mediated uptake, concentration and association with cofactors such as HSPGs before aggregation occurs. If the majority of Aβ secreted is $A\beta_{40}$, the majority of soluble carrier-associated Aβ in the brain should be $A\beta_{40}$. $A\beta_{40}$ may then accumulate through a lipoprotein-dependent, cell-mediated process. $A\beta_{42}$ could, of course, accumulate via the same process, but may not predominate when carrier-mediated cell accumulation becomes the rate-limiting step, rather than self-aggregation and deposition in the extracellular matix in diffuse plaques. Both microglia and pericytes possess scavenger receptors for Aβ aggregates [58-61] which can capture and sequester oligomers and small aggregates, preferentially enriched for the more readily aggregating $A\beta_{42}$. The lack of SMC lipoprotein-mediated $A\beta_{40}$ uptake and concentration can explain why capillary amyloid is often enriched for $A\beta_{42}$ relative to arterial amyloid [62]. Lipoprotein-dependent Aβ interactions can also occur with microglia [63], suggesting that a cell-dependent Aβ sequestration mechanism may be involved in the ApoE4-dependent selective accumulation of $A\beta_{40}$ [48,49,64] in plaques with microglia [56]. Consistent with cell-dependent $A\beta_{40}$ accumulation, antibodies for several epitopes from $A\beta_{1\text{-}5}$ to $A\beta_{34\text{-}40}$ labeled plaque-associated microglia in Tg2576 mice while $A\beta_{42}$ specific antibodies did not [57].

5. CONCLUSIONS.

APPsw transgenic mice with high level neuron-specific expression develop significant leptomeningeal, cortical and hippocampal CAA as they age. As in AD, $A\beta_{40}$ labeling predominates in vessels, although $A\beta_{42}$ is present, even at early stages. The amount of vascular amyloid increases dramatically as the mice age. Some mice have predominantly vascular amyloid for unknown reasons. Whether there are functional or degenerative consequences of CAA in the Tg2576 mice remains unknown. These mice are a useful tool for testing mechanistic hypotheses. For example, one might hypothesize that $A\beta_{40}$ from all sources (but primarily neurons) accumulates through ApoE-dependent, cell-mediated events which are microglia-dependent in plaques, but pericyte- and SMC-dependent in capillary and vascular amyloid, respectively. Alternatively, one might propose a critical role for injury-induced TGFβ1-mediated increases in extracellular matrix protein trapping of diffusible Aβ. Because the Tg2576 mice are available to academic investigators and widely disseminated, they represent a convenient model for testing these and other hypotheses and exploring mechanisms underlying vascular amyloidosis. Ongoing investigations of these mice should reveal whether CAA in the mice can lead to SMC or endothelial damage and alterations in the blood brain barrier or even cerebral hemorrhage. Crosses with human ApoE isotype mice will create models for investigating the mechanism of ApoE genotype on selective $A\beta_{40}$ deposition in vessels. Finally, the APPsw mice have the potential to aid in the development of preventive or therapeutic strategies for vascular amyloidosis.

REFERENCES.

1 Glenner, G.G. and Wong, C.W., (1984) Alzheimer's disease: initial report of the purification and characterization of a novel cerebrovascular amyloid protein, *Biochem. Biophys. Res. Commun.* **120**, 885-890.

2 Masters, C.L., SImms, G., Weinman, N.A., Multhaup, G., McDonald, B.L. and Beyreuther, K., (1985) Amyloid plaque core protein in Alzheimer disease and Down syndrome, *Proc. Natl. Acad. Sci. USA* , **82**, 4245-4249.

3 Gravina, S.A., Ho, L., Eckman, C.B., Long, K.E., Otvos, L.,Jr, Younkin, L.H., Suzuki, N. and Younkin, S.G., (1995) Amyloid β protein (Aβ) in Alzheimer's Disease brain, *J. Biol. Chem.* **270**, 7013-7016.

4 Roher, A.E., Lowenson, J.D., Clarke, S., Wolkow, C., Wang, R., Cotter, R.J., Reardon, I.M., Zurcher-Neely, H.A., Heinrikson, R.L., Ball, M.J. and Greenburg, B.D., (1993) Structural alterations in the peptide backbone of b-amyloid core protein may account for its deposition and stability in Alzheimer's disease, *J. Biol. Chem.* **268**, 3072-3083.

5 Suzuki, N., Iwatsubo, T., Odaka, A., Ishibashi, Y., Kitada, C. and Ihara, Y., (1994) High tissue content of soluble $\beta_{1\text{-}40}$ is linked to cerebral amyloid angiopathy, *Am. J. Pathol.* **145**, 452-460.

6 Akiyama, H., Mori, H., Sahara, N., Kondo, H., Ikeda, K., Nishimura, T., Oda, T. and McGeer, P.L., (1997) Variable deposition of amyloid β-protein (Aβ) with the carboxy-terminus that ends at residue valine$_{40}$ (Aβ40) in the cerebral cortex of patients with Alzheimer's disease: A double-labeling immunohistochemical study with antibodies specific for Aβ40 and the Aβ that ends at residues alanine$_{42}$/threonine$_{43}$ (Aβ42), *Neurochem. Res.*, **22**,1499-1506.

7 Suzuki, N., Cheung, T.T., Cai, X.D., Odaka, A., Otvos, L., Eckman, C., Golde, T.E. and Younkin, S.G., (1994) An Increased percentage of long amyloid β protein precursor (bAPP717) mutants, *Science* **264**, 1336-1340.

8 Selkoe, D.J., (1996) Amyloid beta-protein and the genetics of Alzheimer's disease, *J. Biol. Chem.* **271**, 18295-18298.

9 Castano, E.M., Prelli, F., Soto, C., Beavis, R., Matsubara, E., Shoji, M. and Frangione, B., (1996) The length of amyloid-beta in hereditary cerebral hemorrhage with amyloidosis, Dutch type. Implications for the role of amyloid-beta 1-42, *J. Biol. Chem.* **271**, 32185-32191.

10 Walker, L.C., (1997) Animal models of cerebral β-amyloid angiopathy, *Brain Res. Reviews* **25**, 70-84.

11 Higgins, L.S., Catalano, R., Quon, D. and Cordell, B., Transgenic mice expressing human β-APP751, but not mice expressing β-APP695, display early Alzheimer's disease-like histopathology. In R.M. Nitsch, J.H. Growdon, S. Corkin and R.J. Wurtman (Eds.) *Alzheimer's disease: Amyloid precursor proteins, signal transduction, and neuronal transplantation*, Ann. N.Y. Acad. Sci., New York, 1993, pp. 321-324.

12 Games, D., Adams, D., Alessandrini, R., Barbour, R., Berthelette, P., Blackwell, C., Carr, T., Clemens, J., Donaldson, T., Gillespie, F., Guido, T., Hagoplan, S., Johnson-Wood, K., Khan, K., Lee, M., Leibowitz, P., Lieberburg, I., Little, S., Masliah, E., McConlogue, L., Montoya-Zavala, M., Mucke, L., Paganini, L., Penniman, E., Power, M., Schenk, D., Seubert, P., Snyder, B., Soriano, F., Tan, H., Vitale, J., Wadsworth, S., Wolozin, B. and Zhao, J., (1995) Alzheimer-type neuropathology in transgenic mice overexpressing V717F β-amyloid precursor protein, *Nature* **373**, 523-527.

13 Hsiao, K., Chapman, P., Nilsen, S., Eckman, C., Harigaya, Y., Younkin, S., Yang, F. and Cole, G., (1996) Correlative memory deficits, Aβ elevation and amyloid plaques in transgenic mice, *Science* **274**, 99-102.

14 Sturchler-Pierrat, C., Abramowski, D., Duke, M., Wiederhold, K-H., Mistl, C., Rothacher, S., Ledermann, B., Bürki, K., Frey, P., Paganetti, P.A., Waridel, C., Calhoun, M.E., Jucker, M., Staufenbiel, M. and Sommer, B., (1997) Two amyloid precursor protein transgenic mouse models with Alzheimer disease-like pathology, *Proc. Natl. Acad. Sci. USA* **94**, 13287-13292.

15 Johnson-Wood, K., Lee, M., Motter, R., Hu, K., Gordon, G., Barbour, R., Khan, K., Gordon, M., Tan, H., Games, D., Lieberburg, I., Schenk, D., Seubert, P. and McConlogue, L., (1997) Amyloid precursor protein processing and Aβ42 deposition in a transgenic mouse model of Alzheimer disease, *Proc. Natl. Acad. Sci. USA* **94**, 1550-1555.

16 Masliah, E., Sisk, A., Mallory, M., Mucke, L., Schenk, D. and Games, D., (1996) Comparison of neurodegenerative pathology in transgenic mice overexpressing V717F beta-amyloid precursor protein and Alzheimer's disease, *J. Neurosci.* **16**, 5795-5811.

17 Irizarry, M.C., McNamara, M., Fedorchak, K., Hsiao, K. and Hyman, B.T., (1997) APP_{sw} Transgenic Mice develop age-related Aβ deposits and neuropil abnormalities, but no neuronal loss in CA1, *J. Neuropathol. Exp. Neurol.* **56**, 965-973.
18 Calhoun, M.E., Burgermeister, P., Phinney, A.L., Stalder, M., Tolnay, M., Wiederhold, K.H., Abramowski, D., Sturcheler-Pierrat, C. Sommer, B., Staufenbiel, M. and Jucker, M. (1999) Neuronal expression of mutant APP results in prominent deposition of cerebrovascular amyloid, *PNAS (USA)* in press
19 Hartmann, H., Busciglio, J., Baumann, K,H., Staufenbiel, M. and Yankner, B.A., (1997) Developmental regulation of presenilin-1 processing in the brain suggests a role in neuronal differentiation, *J. Biol. Chem.* **272**, 14505-14508.
20 Glenner, G.G. and Murphy, M.A., (1989) Amyloidosis of the nervous system, *J. Neurol. Sci.* **94**, 1-28.
21 Frackowiak, J., Mazur-Kolecka, B., Wisniewski, H.M., Potempska, A., Carroll, R.T., Emmerling, M.R. and Kim, K.S., (1995) Secretion and accumulation of Alzheimer's β-protein by cultured vascular smooth muscle cells from old and young dogs, *Brain Res.* **676**, 225-230.
22 Davis-Salinas, J., Saporito-Irwin, S.M., Cotman, C.W. and Van Nostrand, W.E., (1995) Amyloid β-Protein induces its own production in cultured degenerating cerebrovascular smooth muscle cells, *J. Neurochem.* **65**, 931-934.
23 Urmoneit, B., Prikulis, I., Wihl, G., D'Urso, D., Frank, R., Heeren, J., Beisiegel, U. and Prior, R., (1997) Cerebrovascular smooth muscle cells internalize Alzheimer amyloid beta protein via a lipoprotein pathway: implications for cerebral amyloid angiopathy, *Lab. Invest.* **77**, 157-166.
24 Haass, C., Lemere, C.A., Capell, A., Citron, M., Seubert, P., Schenk, D. and Selkoe, D.J., (1995) The Swedish mutation causes early-onset Alzheimer's disease by β-secretase cleavage within the secretory pathway, *Nature Med* **1**, 1291-1296.
25 Cai, X.D., Golde, T.E. and Younkin, S.G., (1993) Release of excess amyloid β protein from a mutant amyloid β protein precursor, *Science* **259**, 514-516.
26 Mann, D.M.A., Iwatsubo, T., Ihara, Y., Cairns, N.J., Lantos, P.L., Bogdanovic, N. and Winblad, B., (1996) Predominant deposition of amyloid-β42(43) in plaques in cases of Alzheimer's disease and hereditary cerebral hemorrhage associated with mutations in the amyloid precursor protein gene, *Am. J. Pathol.* **148**, 1257-1266.
27 Irizarry, M.C., Soriano, F., McNamara, M., Page, K.J., Schenk, D., Games, D. and Hyman, B.T., (1997) Aβ deposition is associated with neuropil changes, but not with overt neuronal loss in the human amyloid precursor protein V717F (PDAPP) transgenic mouse, *J. Neurosci.* **17**, 7053-7059.
28 Chen, K., Soriano, F., Lyn, W., Grajeda, H., Masliah, E. and Games, D., (1998) Effects of entorhinal cortex lesions on hippocampal β-amyloid deposition in PDAPP transgenic mice, *Soc. Neurosci. Abstr.* **24** (#592.6), 1502.
29 Iwatsubo, T., Mann, D.M.A., Odaka, A., Suzuki, N. and Ihara, Y., (1995) Amyloid β protein (Aβ) deposition: Aβ42(43) Precedes Aβ40 in Down's Syndrome, *Ann. Neurol.* **37**, 294-299.
30 Lemere, C.A., Blusztajn, J.K., Yamaguchi, H., Wisniewski, T., Saido, T. and Selkoe, D., (1996) Sequence of deposition of heterogeneous amyloid beta-peptides and Apo E in Down syndrome: Implications for inital events in amyloid plaque formation, *Neurobiol. Dis.* **3**, 26-32.
31 Allsop, D., Haga, S., Haga, C., Ikeda, S., Mann, D.M. and Ishii, T., (1989) Early senile plaques in Down's syndrome brains show a close relationship with cell bodies of neurons, *Neuropathol. Appl. Neurobiol.* **15**, 531-542.

32 Mann, D.M.A., Jones, D., South, P.W., Snowden, J.S. and Neary, D., (1992) Deposition of amyloid beta protein in non-Alzheimer dementias: evidence for a neuronal origin of parenchymal deposits of beta protein in neurodegenerative disease, *Acta Neuropathol.* **83**, 415-419.

33 Jarrett, J.T., Berger, E.P. and Lansbury, P.T., The C-terminus of the β protein is critical in amyloidogenesis. In R.M. Nitsch, J.H. Growdon, S. Corkin and R.J. Wurtman (Eds.) *Alzheimer's disease: Amyloid precursor proteins, signal transduction, and neuronal transplantation*, Annals New York Academy Sciences, New York, 1993, pp. 285-289.

34 Giulian, D. and Baker, T.J., (1986) Characterization of ameboid microglia isolated from developing mammalian brain, *J. Neurosci* **6(8)**, 2163-2178.

35 Weller, R.O., Massey, A., Newman, T.A., Hutchings, M., Kuo, Y-M. and Roher, A.E., (1998) Cerebral amyloid angiopathy: amyloid β accumulates in putative interstitial fluid drainage pathways in Alzheimer's disease, *Am. J. Pathol.* **153**, 725-733.

36 Kuo, Y-M., Emmerling, M.R., Vigo-Pelfrey, C., Kasunic, T.C., Kirkpatrick, J.B., Murdoch, G.H., Ball, M.J. and Roher, A.E., (1996) Water-soluble Aβ (N-40,N-42) oligomers in normal and Alzheimer Disease brains, *J. Biol. Chem.* **271**, 4077-4081.

37 Logan, A., Frautschy, S.A., Gonzalez, A.M., Sporn, M.B. and Baird, A., (1992) Enhanced expression of transforming growth factor β1 in the rat brain after a localized cerebral injury, *Brain Res.* **587**, 216-225.

38 Logan, A., Berry, M., Gonzalez, A.M., Frautschy, S.A., Sporn, M.B. and Baird, A., (1994) Effects of transforming growth factor β1 on scar formation in the injured central nervous system, *Eur. J. Neuroscience* **6**, 355-363.

39 Baghdassarian, D., Toru-Delbauffe, D., Gavaret, J.M. and Pierre, M., (1993) Effects of transforming growth factor-β1 on the extracellular matrix and cytoskeleton of cultured astrocytes, *GLIA*, **7**, 193-202.

40 Frautschy, S.A., Albright, T., Dvorak, C., Wolfe, D.S. and Baird, A., (1994) Transforming growth factor β (TGFβ) modification of β protein immunoreactivity in the hippocampus with and without β-protein infusion and the resulting ultrastructural neuropathology, *Neurobiol. Aging* **15**(S1), S55-S56.

41 Frautschy, S.A., Yang, F., Calderón, L. and Cole, G.M., (1996) Rodent models of Alzheimer's disease: rat Aβ infusion approaches to amyloid deposits, *Neurobiol. Aging* **17**, 311-321.

42 Harris-White, M.E., Chu, T., Balverde, Z., Sigel, J.J., Flanders, K.C. and Frautschy, S.A., (1998) Effects of TGFβs(1-3) on Aβ deposition and inflammation and cell-targeting in organotypic hippocampal slice cultures, *J. Neurosci.* **18**, 10366-10374.

43 Wyss-Coray, T., Masliah, E., Mallory, M., McConlogue, L., Johnson-Wood, K., Lin, C. and Mucke, L., (1997) Amyloidogenic role of cytokine TGF-beta-1 in transgenic mice and in Alzheimer's Disease, *Nature* **389**, 603-605.

44 Castillo, G.M., Ngo, C., Cummings, J., Wight, T.N., Snow, A.D. (1997) Perlecan binds to the β-amyloid proteins (Aβ) of Alzheimer's disease, accelerates Aβ fibril formation, and maintains Aβ fibril stability. J. Neurochem. **69**, 24552-2465

45 Ji, Z.S., Pitas, R.E. and Mahley, R.W., (1998) Differential cellular accumulation/retention of apolipoprotein E mediated by cell surface heparan sulfate proteoglycans, *J. Biol. Chem.* **273**, 13452-13460.

46 Mazur-Kolecka, B., Frackowiak, J., Krzeslowska, J., Ramakrishna, N., Haske, T., Emmerling, M.R., Zhang, W., Kim, K.S. and Wisniewski, H.M., (1999) Apolipoprotein E alters metabolism of AbetaPP in cells engaged in beta-amyloidosis, *J. Neuropathol. Exp. Neurol.* **58**, 288-295.

47 Mazur-Kolecka, B., Frackowiak, J. and Wisniewski, H.M., (1995) Apolipoproteins E3 and E4 induce, and transthyretin prevents accumulation of the Alzheimer's beta-amyloid peptide in cultured vascular smooth muscle cells, *Brain Res.* **698**, 217-222.

48 Mann, D.M.A., Iwatsubo, T., Pickering-Brown, S.M., Owen, F., Saido, T.C. and Perry, R.H., (1997) Preferential deposition of amyloid β protein (Aβ) in the form of Aβ40 in Alzheimer's disease is associated with a gene dosage effect of the apolipoprotein E ε4 allele, *Neurosci. Lett.* **221**, 81-84.

49 Ishii, K., Tamaoka, A., Mizusawa, H., Shoji, S., Ohtake, T., Fraser, P.E., Takahashi, H., Tsuji, S., Gearing, M., Mizutani, T., Yamada, S., Kato, M., St.George-Hyslop, P.H., Mirra, S.S. and Mori, H., (1997) $A\beta_{1-40}$ but not Aβ1-42 levels in cortex correlate with apolipoprotein E E4 allele dosage in sporadic Alzheimer's disease, *Brain Research* **748**, 250-252.

50 Greenberg, S.M., Rebeck, G.W., Vonsattel, J.P.G., Gomez-Isla, T. and Hyman, B.T., (1995) Apolipoprotein E E4 and cerebral hemorrhage associated with amyloid angiopathy, *Ann. Neurol.* **38**, 254-259.

51 Greenberg, S.M., (1998) Cerebral amyloid angiopathy: prospects for clinical diagnosis and treatment, *Neurology* **51**, 690-694.

52 Zarow, C., Zaias, B., Lyness, S.A. and Chui, H., (1999) Cerebral amyloid angiopathy in Alzheimer disease is associated with apolipoprotein E4 and cortical neuron loss, *Alzheimer Disease and Associated Disorders* **13**, 1-8.

53 Hart, M.N., Merz, P., Bennett-Gray, J., Menezes, A.H., Goeken, J.A., Schelper, R.L. and Wisniewski, H.M., (1988) β-amyloid protein of Alzheimer's disease is found in cerebral and spinal cord vascular malformations, *Am. J. Pathol.* **132**, 167-172.

54 Van Nostrand, W.E., Melchor, J.P. and Ruffini, L., (1998) Pathologic amyloid β-protein cell surface fibril assembly on cultured human cerebrovascular smooth muscle cells, *J. Neurochem.* **70**, 216-223.

55 Iadecola, C., Zhang, F., Niwa, K., Eckman, C., Turner, S.K., Fischer, E., Younkin, S., Borchelt, D.R., Hsiao, K.K. and Carlson, G.A., (1999) SOD1 rescues cerebral endothelial dysfunction in mice overexpressing amyloid precursor protein, *Nat. Neurosci.* **2**, 157-161.

56 Mann, D.M., Iwatsubo, T., Fukumoto, H., Ihara, Y., Odaka, A. and Suzuki, N., (1995) Microglial cells and amyloid beta protein (A beta) deposition; association with A beta 40-containing plaques, *Acta Neuropathol.* **90**, 472-477.

57 Frautschy, S.A., Yang, F., Irrizarry, M., Hyman, B., Saido, T.C., Hsiao, K. and Cole, G.M., (1998) Microglial response to amyloid plaques in APPsw transgenic mice, *Am. J. Pathol.* **152**, 307-317.

58 Khoury, J.E., Hickman, S.E., Thomas, C.A., Cao, L., Silverstein, S.C. and Loike, J.D., (1996) Scavenger receptor-mediated adhesion of microglia to β-amyloid fibrils, *Nature* **382**, 716-719.

59 Paresce, D.M., Ghosh, R.N. and Maxfield, F.R., (1996) Microglial cells internalize aggregates of the Alzheimer's disease amyloid beta-protein via a scavenger receptor, *Neuron* **17**, 553-565.

60 Mato, M., Ookawara, S., Sakamoto, A., Aikawa, E., Ogawa, T., Mitsuhasi, U., Masuzawa, T., Suzuki, H., Honda, M., Yazaki, Y., Watanabe, E., Luoma, J., Yla-Herttuala, S., Fraser, I., Gordon, S. and Kodama, T., (1996) Involvement of specific macrophage-lineage cells surrounding arterioles in barrier and scavenger function in brain cortex, *Proc. Natl. Acad. Sci. USA* **93**, 3269-3274.

61 Honda, M., Akiyama, H., Yamda, Y., Kondo, H., Kawabe, Y., Takeya, J., Takahashi, K., Suzuki, H., Doi, T., Sakamoto, A. and *et al.*, , (1998) Immunohistochemical evidence for a

macrophage scavenger receptor in Mato cells and reactive microglia of ischemia and Alzheimer's disease, *Biochem. Biophys. Res. Commun.* **245**, 734-740.

62 Nakamura, S., Tamaoka, A., Sawamura, N., Shoji, S., Nakayama, H., Ono, F., Sakakibara, I., Yoshikawa, Y., Mori, H., Goto, N. and Doi, K., (1995) Carboxyl end-specific monoclonal antibodies to amyloid β protein (Aβ) subtypes (Aβ40 and Aβ42(43) differentiate Aβ in senile plaques and amyloid angiopathy in brains of aged cynomolgus monkeys, *Neurosci. Lett.* **201**, 151-154.

63 Cole, G.M., Beech, W., Frautschy, S.A., Sigel, J.J., Glasgow, C. and Ard, M.D., (1999) Lipoprotein effects on Aβ accumulation and degradation by microglia *in vitro*, *J. Neurosci. Res.* **57**, 504-520.

64 Gearing, M., Mori, H. and Mirra, S.S., (1996) Aβ-peptide length and apolipoprotein E genotype in Alzheimer's disease *Ann. Neurol.*, **39**, 395-399.

Chapter 19

CEREBRAL AMYLOID ANGIOPATHY IN AGED DOGS AND NONHUMAN PRIMATES

LARY C. WALKER
Neuropathology Laboratory, Neuroscience Therapeutics, Parke-Davis Pharmaceutical Research Division, Warner-Lambert, Ann Arbor, MI, USA

Cerebral amyloid angiopathy (CAA) is a common finding in aged dogs and nonhuman primates. As in humans, cerebrovascular amyloid in these animals is composed fundamentally of the Aβ peptide, along with various associated substances. The amount and distribution of CAA vary among brain regions and among animals of equivalent age. All vessel types can be involved, although amyloidotic venules are relatively rare. In nonhuman primates, capillaries are frequently affected; these small vessels accumulate almost exclusively the 42-amino acid peptide ($A\beta_{42}$), whereas larger vessels contain a mix of $A\beta_{42}$ and $A\beta_{40}$. Compared to aged rhesus monkeys, which usually develop a preponderance of senile plaques, squirrel monkeys manifest mostly CAA. This species-difference is not due to differences in apolipoprotein E type or to known disease-causing polymorphisms in the β-amyloid precursor protein gene; however, squirrel monkeys have an Icelandic-like mutation in the cystatin C gene that could influence the tendency of these monkeys to accrue Aβ in the cerebral vasculature. Aged nonhuman primates and dogs are being used to test cerebral amyloid-targeting strategies and, along with emerging transgenic mice, are beneficial models for validating new diagnostic and therapeutic approaches to CAA.

1. NATURALLY OCCURRING CEREBRAL AMYLOID IN NONHUMAN SPECIES

Amyloid was first described in the brain of a nonhuman species by von Braunmühl [1], who reported in the 1950's the occurrence of senile plaques and congophilic angiopathy (cerebral amyloid angiopathy, CAA) in aged dogs. Sporadic descriptions of canine cerebral amyloidosis followed von Braunmühl's paper [2], but there was little development in the study of animal models of the disorder until the 1970's. Since then, it has been

M.M. Verbeek et al. (eds.),
Cerebral Amyloid Angiopathy in Alzheimer's Disease and Related Disorders, 313–324.

established that most amyloid arising in the senescent brain is composed of a peptide designated Aβ [3], the β-amyloid precursor protein has been characterized and found to be highly conserved in mammals [4-7], and a variety of species have been confirmed to manifest cerebral β-amyloid deposits as they age [6-23]. In addition, studies have demonstrated that prion-based amyloid can be induced in the brain by the intracerebral injection of afflicted brain tissue [24-27], and there is now evidence that β-amyloid deposition in the parenchyma and vasculature can be seeded by injection of Alzheimeric brain tissue into monkeys [28] and APP-transgenic mice [29]. Finally, in addition to developing senile plaques, older mice that are transgenic for human APP also manifest CAA (see Cole *et al*, this volume) [30]. Transgenic mouse models promise to accelerate research on the pathogenesis of CAA; however, natural models also provide significant advantages for research. The two most studied models of naturally occurring CAA are aged dogs and nonhuman primates. CAA in these animals closely resembles that found in aged humans, and their relatively large brains are advantageous for the development of diagnostic imaging strategies for cerebral amyloidosis.

2. CAA IN DOGS

Von Braunmühl's original description of CAA in aged dogs has been confirmed in numerous laboratories [2,13,19,23,31-35] (Figure 1). While occasional classical senile plaques also can be found, most canine parenchymal amyloid is diffuse in nature. In a study of 30 mongrel dogs between 6.5 and 26.5 years of age, Wegiel *et al.* [35] report that all dogs over 13.2 years exhibit amyloid angiopathy, but the density of CAA varies widely among animals. Canine CAA afflicts particularly the leptomeningeal arteries [35], but also involves parenchymal vessels, including capillaries [13,23,31,36]. In arteries, atrophy of smooth muscle cells is evident in advanced CAA [37]. Canine amyloid angiopathy appears to begin in larger, brain-supplying arteries, whence it progresses into smaller vessels [35]. As is the case in humans, the presence of CAA in dogs is associated with an increased incidence of intracerebral hemorrhage [36,38], and possibly with blood-brain barrier dysfunction as well [39].

Canine vascular amyloid deposits are found most frequently in the intercellular spaces of the tunica media (similar to the deposits in humans [37,40] and nonhuman primates [12,20]), underscoring the hypothesized role of vascular smooth muscle cells in the generation of CAA [41-44]. The amino acid sequence of canine $A\beta_{1-42}$ is identical to that in humans and several other mammalian species [4], including nonhuman primates [5,6].

Although diffuse parenchymal Aβ deposits in dogs contain mostly $A\beta_{42}$ rather than $A\beta_{40}$, vascular amyloid contains mainly $A\beta_{40}$ [19,45]. Amyloid in the blood vessels and parenchyma is intensely stained by specific lectins, suggesting that sugar residues recognized by the lectins may play a role in Aβ deposition [46]. Some amyloidotic vessels also are immunoreactive with antibodies to apolipoprotein E (ApoE), cystatin C, cathepsin D, and α_1-antichymotrypsin [47]. The consistent presence of CAA in a variety of dogs after the age of 13 years, the correlation of CAA with intracerebral hemorrhage, and the biochemical similarity of canine and human Aβ all bolster the utility of aged dogs in the investigation of CAA (see also Prior and Urmoneit, this volume). The relatively large canine brain facilitates *in vivo* imaging studies as well [39]. As in other animal models of CAA, however, the variability in the amount and distribution of vascular β-amyloid in animals of similar age requires the analysis of relatively large groups of subjects to achieve adequate statistical power.

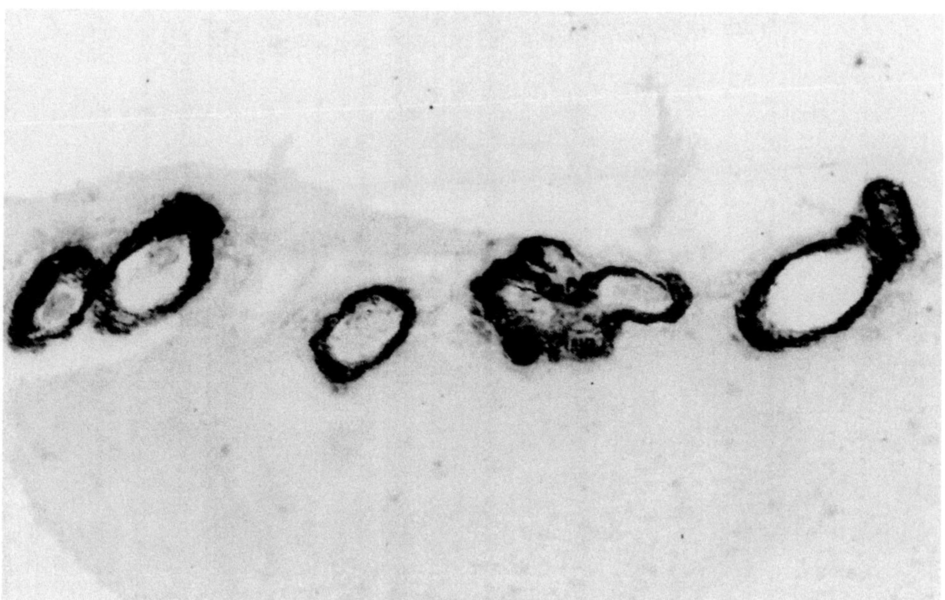

Figure 1. β-amyloid in superficial blood vessels of an aged dog, immunostained with monoclonal antibody 6E10 to Aβ (photo courtesy of Jerzy Wegiel, Staten Island, New York).

3. CAA IN NONHUMAN PRIMATES

Cerebrovascular β-amyloidosis is a common finding with age in many species of nonhuman primates, from prosimians to great apes [48] (Figure 2). Some primate species are particularly prone to CAA, for example squirrel monkeys [22]. In monkeys, CAA most frequently afflicts the neocortex. The lower brainstem, cerebellum, diencephalon and globus pallidus are unaffected. A number of other brain areas can manifest moderate amounts of CAA in aged monkeys, including the amygdala, hippocampus, and the nucleus accumbens. Both the amount of CAA and its anatomical distribution vary among animals. Across neocortical regions, CAA often is severe in some areas and absent in others. For example, in aged rhesus and squirrel monkeys, both CAA and Aβ-plaques are usually relatively sparse in the occipital lobe compared to the anterior frontal and temporal lobes. The scarcity of CAA in the occipital lobe distinguishes monkeys from humans with AD or hereditary cerebral hemorrhage with amyloidosis-Dutch type (HCHWA-D), in which the vasculature of the occipital lobe can be amply involved [49-52].

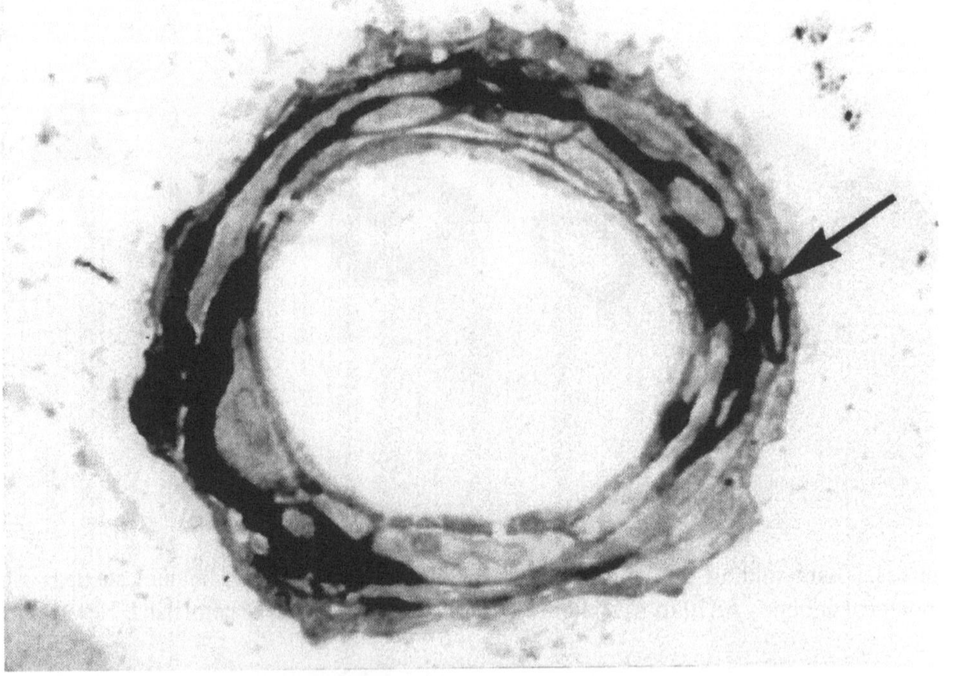

Figure 2. Cerebrovascular β-amyloid (arrow) in the tunica media of a cortical arteriole from an aged chimpanzee, immunostained with a polyclonal antibody to Aβ (antibody courtesy of Colin Masters, Melbourne, Australia).

Human vascular β-amyloid has relatively less of the longer, 42/43-amino acid Aβ-peptide than of the 40 amino acid form [53-55]. In rhesus monkeys, β-amyloid in arterioles was shown to be immunoreactive comparably for both $A\beta_{40}$ and $A\beta_{42}$, but many small cortical vessels were positive only for $A\beta_{42}$ [56]. Similar findings have been reported in cynomolgus monkeys (*Macaca fascicularis*) [15] and squirrel monkeys [48], suggesting that capillaries and arterioles may contain different ratios of the C-terminal $A\beta_{40}$ and $A\beta_{42}$ peptides (see below). However, enzyme-linked immunosorbent assays suggest that the quantitation of Aβ by immunohistochemistry may underestimate the levels of $A\beta_{42}$ in the primate brain [57].

3.1 The squirrel monkey as a model of CAA

Squirrel monkeys (*Saimiri sciureus*) are small, well-characterized New World monkeys with large, lissencephalic brains [58,59]. Squirrel monkeys generally develop some degree of CAA by around 15 years of age [22,60]. Cerebral capillaries are more frequently amyloidotic in *Saimiri* than in macaques (Figure 3) and humans. Interestingly, in Congo Red-stained tissue from squirrel monkeys, capillaries are seldom birefringent when viewed with cross-polarized light, and definite amyloid fibrils are rarely evident in capillaries by electron microscopy (personal observations). As noted above, amyloid in large vessels (mostly arterioles) is immunoreactive with C-terminal antibodies to both $A\beta_{40}$ and $A\beta_{42}$ in *Saimiri*, but capillaries contain mainly $A\beta_{42}$ [48], the type of Aβ that predominates in diffuse senile plaques. Capillary Aβ in squirrel monkeys thus may be soluble, or prefibrillar, in nature. Because the vessels also stain robustly with monoclonal antibody 10D5 to amino acids 1-16 of Aβ (Figure 3B), it is likely that a significant fraction of the capillary Aβ is full-length $A\beta_{1\text{-}42}$.

In a series of collaborative studies with Efrat Levy and her colleagues at New York University, we have sought a genetic basis for the proclivity of aged squirrel monkeys to develop CAA. These investigations have shown that mutations of APP that are linked to CAA in humans, either at codon 692 [61] or codon 693 [62] do *not* occur in squirrel monkeys [5]. Squirrel monkeys, like other nonhuman primates, are homozygous for apolipoprotein E ε4 according to the human numbering and nomenclature [14,63-67]. However, whereas humans have arginine at position 61 of ApoE, squirrel monkeys and rhesus monkeys have threonine [64], which causes ApoE4 to resemble ApoE3 in its interactions with lipoproteins. Hence, simian ApoE4 functionally imitates human ApoE3, possibly explaining why nonhuman primates are not predisposed to Alzheimer-like neurodegeneration despite being homozygous for *APOE* ε4. A missense mutation in cystatin C causes the Icelandic form of hereditary CAA in humans [68,69] (see Olafsson and

Thorsteinsson, this volume). While Icelandic cerebrovascular amyloid contains cystatin C and not Aβ, an Icelandic-like mutation has been implicated in a case of sporadic CAA in which *both* Aβ and cystatin C were deposited in the vascular walls [70]. Surprisingly, squirrel monkeys (but not rhesus monkeys) also have a missense mutation in cystatin C at the Icelandic locus [71]. Thus, there may be a pathogenic link between cystatin C and Aβ in some cases of human CAA, a possibility that could be explored experimentally in mice transgenic for human cystatin C and APP.

4. UTILITY OF ANIMAL MODELS OF CAA

There is currently a paucity of available animal models for the experimental analysis of CAA. However, canine, primate and rodent models are emerging that will be of considerable benefit in deciphering the pathogenesis of cerebral amyloidoses. β-amyloid deposits in monkeys have been labeled *in vivo* by intracisternally infused monoclonal antibodies to Aβ [72], asserting the feasibility of delivering diagnostic and therapeutic agents to amyloidotic cerebral vessels. Significantly, studies in monkeys have also shown that radiolabeled Aβ infused into the bloodstream can cross the blood-brain barrier and bind selectively to brain amyloid deposits [73]. In addition, blood-brain barrier permeability and vascular sequestration of Aβ are augmented in aged animals [74].

In another study, β-amyloid-containing human AD brain homogenates were injected into the brains of young marmosets (~2 years old), and the brains were analyzed 6-7 years later [28]. CAA and senile plaques were found in several of the experimental animals, but not in age-matched controls. The long time frame needed to complete such studies in primates attenuates the utility of the model, however. In a recent series of experiments, we have found that CAA and senile plaques can be induced by the intracerebral injection of dilute AD-brain homogenates into APP-transgenic mice within five months [29]. These experiments in monkeys and mice support the contention that β-amyloid deposition in plaques and blood vessels can be triggered by exogenously administrated material, and also identify *in vivo* models in which the pathogenesis of CAA can be profitably analyzed. Animal models are increasingly important for bridging the gap between *in vitro* studies of amyloidogenesis and the ontogeny of cerebral amyloidoses in humans. They also will be essential to testing new methods for the diagnosis and treatment of CAA.

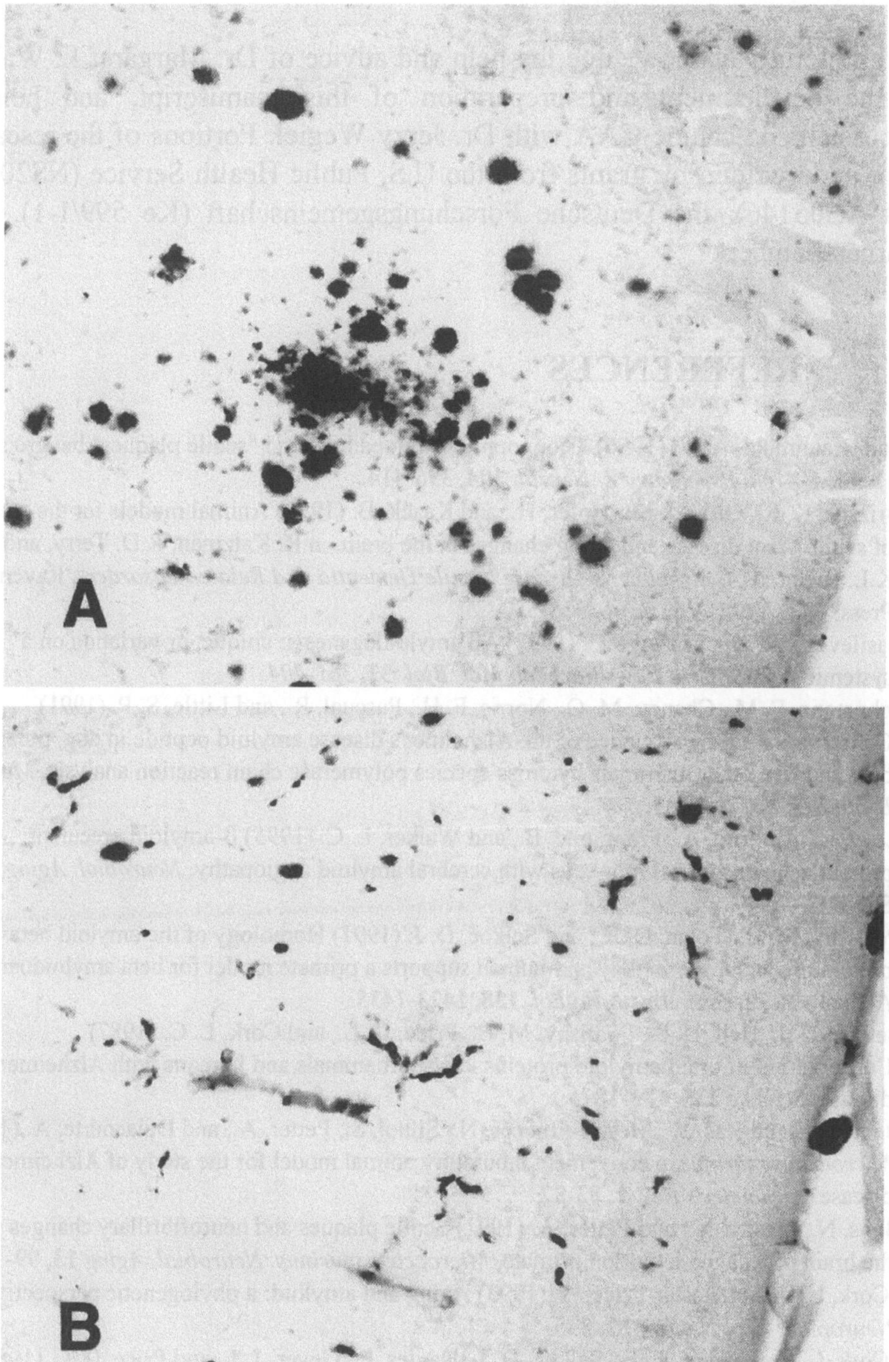

Figure 3. β-amyloid immunoreactivity in the neocortices of an aged rhesus monkey (A) and squirrel monkey (B), labeled with monoclonal antibody 10D5 to Aβ (antibody courtesy of Dale Schenk, South San Francisco, California). Note the predominance of vascular Aβ in the squirrel monkey.

ACKNOWLEDGEMENTS

I gratefully acknowledge the help and advice of Dr. Margaret L. Walker in the development and preparation of this manuscript, and helpful discussions on canine CAA with Dr. Jerzy Wegiel. Portions of the research were underwritten by grants from the U.S. Public Health Service (NS20471 and AG05146), the Deutsche Forschungsgemeinschaft (Ke 599/1-1), and Warner-Lambert.

REFERENCES

1. von Braunmühl, A. V. (1956) "Kongophile Angiopathie" und "senile plaques" bei greisen Hunden. *Arch. Psychiatry Z. Neurol.* **194,** 396-414.
2. Brizzee, K. R., Ordy, J. M., Hofer, H., and Kaack, B. (1978) Animal models for the study of senile brain disease and aging changes in the brain. in R. Katzman, R.D. Terry, and K.L. Bick (eds.), *Alzheimer's Disease: Senile Dementia and Related Disorders*, Raven Press, New York, Vol. 7, pp. 515-553.
3. Kisilevsky, R. and Fraser, P.E. (1997) Aβ amyloidogenesis: unique, or variation on a systemic theme? *Crit. Rev. Biochem. Mol. Biol.* 32, *361-404.*
4. Johnstone, E. M., Chaney, M. O., Norris, F. H., Pascual, R., and Little, S. P. (1991) Conservation of the sequence of the Alzheimer's disease amyloid peptide in dog, polar bear and five other mammals by cross-species polymerase chain reaction analysis. *Mol. Brain Res.* **10,** 299-305.
5. Levy, E., Amorim, A., Frangione, B., and Walker, L. C. (1995) β-amyloid precursor protein gene in squirrel monkeys with cerebral amyloid angiopathy. *Neurobiol. Aging* **16,** 805-808.
6. Podlisny, M. B., Tolan, D. R., and Selkoe, D. J. (1991) Homology of the amyloid beta protein precursor in monkey and human supports a primate model for beta amyloidosis in Alzheimer's disease. *Am. J. Pathol.* **138,** 1423-1435.
7. Selkoe, D. J., Bell, D. S., Podlisny, M. B., Price, D. L., and Cork, L. C. (1987) Conservation of brain amyloid proteins in aged mammals and humans with Alzheimer's disease. *Science* **235,** 183-187.
8. Bons, N., Jallageas, V., Mestre-Francés, N., Silhol, S., Petter, A., and Delacourte, A. (1995) *Microcebus murinus*, a convenient laboratory animal model for the study of Alzheimer's disease. *Alzheimer's Res.* **1,** 83-87.
9. Bons, N., Mestre, N., and Petter, A. (1991) Senile plaques and neurofibrillary changes in the brain of an aged lemurian primate, *Microcebus murinus. Neurobiol. Aging* **13,** 99-105.
10. Cork, L. C. and Hester-Price, A. (1993) Aging and amyloid: a phylogenetic perspective. *J. Neuropath. Exp. Neurol.* **52,** 335.
11. Cork, L. C., Powers, R. E., Selkoe, D. J., Davies, P., Geyer, J. J., and Price, D. L. (1988) Neurofibrillary tangles and senile plaques in aged bears. *J. Neuropath. Exp. Neurol.* **47,** 629-641.
12. Cork, L. C. and Walker, L. C. (1993) Age-related lesions, nervous system. in T.C. Jones, U. Mohr, and R.D. Hunt (eds.) *Nonhuman Primates II*, Springer-Verlag, New York, pp. 173-183.

13. Cummings, B. J., Su, J. H., Cotman, C. W., White, R., and Russell, M. J. (1993) β-amyloid accumulation in aged canine brain: A model of early plaque formation in Alzheimer's disease. *Neurobiol. Aging* **14,** 547-560.
14. Gearing, M., Rebeck, G. W., Hyman, B. T., Tigges, J., and Mirra, S. S. (1994) Neuropathology and apolipoprotein E profile of aged chimpanzees: implications for Alzheimer disease. *Proc. Natl. Acad. Sci., USA* **91,** 9382-9386.
15. Nakamura, S., Tamaoka, A., Sawamura, N., Shoji, S., Nakayama, H., Ono, F., Sakakibara, I., Yoshikawa, Y., Mori, H., Goto, N., and Doi, K. (1995) Carboxyl end-specific monoclonal antibodies to amyloid β protein (Aβ) subtypes (Aβ42-43) differentiate Aβ in senile plaques and amyloid angiopathy in brains of aged cynomolgus monkeys. *Neurosci. Lett.* **201,** 151-154.
16. Nakamura, S., Nakayama, H., Uetsuka, K., Sasaki, N., Uchida, K., and Goto, N. (1995) Senile plaques in an aged two-humped (Bactrian) camel (Camelus bactrianus). *Acta Neuropath.* **90,** 415-418.
17. Roertgen, K. E., Parisi, J. E., Clark, H. B., Barnes, D. L., O'Brien, T. D., and Johnson, K. H. (1996) Aβ-Associated cerebral angiopathy and senile plaques with neurofibrillary tangles and cerebral hemorrhage in an aged wolverine (*Gulo gulo*). *Neurobiol. Aging* **17,** 243-248.
18. Strittmatter, W. J., Saunders, A. M., Schmechel, D., Pericak-Vance, M., Enghild, J., Salvesen, G. S., and Roses, A. D. (1993) Apolipoprotein E: high-avidity binding to β-amyloid and increased frequency of type 4 allele in late-onset familial Alzheimer disease. *Proc. Natl. Acad. Sci., USA* **90,** 1977-1981.
19. Tekirian, T. L., Cole, G. M., Russell, M. J., Yang, F., Wekstein, D. R., Patel, E., Snowdon, D. A., Markesbery, W. R., and Geddes, J. W. (1996) Carboxy terminal of β-amyloid deposits in aged human, canine, and polar bear brains. *Neurobiol. Aging* **17,** 243-248.
20. Uno, H., Alsum, P. B., Dong, S., Richardson, R., Zimbric, M. L., Thieme, C. S., and Houser, W. D. (1996) Cerebral amyloid angiopathy and plaques, and visceral amyloidosis in aged macaques. *Neurobiol. Aging* **17,** 275-282.
21. Walker, L. C., Kitt, C. A., Struble, R. G., Wagster, M. V., Price, D. L., and Cork, L. C. (1988) The neural basis of memory decline in aged monkeys. *Neurobiol. Aging* **9,** 657-666.
22. Walker, L. C., Masters, C., Beyreuther, K., and Price, D. L. (1990) Amyloid in the brains of aged squirrel monkeys. *Acta Neuropathol.* **80,** 381-387.
23. Wisniewski, H., Johnson, A. B., Raine, C. S., Kay, W. J., and Terry, R. D. (1970) Senile plaques and cerebral amyloidosis in aged dogs. A histochemical and ultrastructural study. *Lab. Invest.* **23,** 287-296.
24. Bruce, M. E., Dickinson, A. G., and Fraser, H. (1976) Cerebral amyloidosis in scrapie in the mouse: effect of agent strain and mouse genotype. *Neuropathol. Appl. Neurobiol.* **2,** 471-478.
25. Bruce, M. E. and Fraser, H. (1975) Amyloid plaques in the brains of mice infected with scrapie: morphological variation and staining properties. *Neuropathol. Appl. Neurobiol.* **1,** 189-202.
26. Wisniewski, H. M., Bruce, M. E., and Fraser, H. (1975) Infectious etiology of neuritic (senile) plaques in mice. *Science* **190,** 1108-1110.
27. Wisniewski, H. M., Moretz, R. C., and Lossinsky, A. S. (1981) Evidence for induction of localized amyloid deposits and neuritic plaques by an infectious agent. *Ann. Neurol.* **10,** 517-522.

28. Baker, H. F., Ridley, R. M., Duchen, L. W., Crow, T. J., and Bruton, C. J. (1993) Evidence for the experimental transmission of cerebral β-amyloidosis to primates. *Int. J. Exp. Path.* **74,** 441-454.
29. Kane, M. D., Lipinski, W. J., Callahan, M. J., Bian, F., Durham, R. A., Schwarz, R. D., Roher, A. E., and Walker, L. C. (submitted) β-Amyloid induction by intracerebral injection of Alzheimer brain homogenate in βAPP-transgenic mice.
30. Jucker, M., Stalder, M., Tolnay, M., Wiederhold, K. H., Abramowski, D., Sommer, B., Staufenbiel, M., and Calhoun, M. E. (1998) Cerebral amyloid angiopathy occurs in conjunction with amyloid plaque formation in APP transgenic mice. *Neurobiol. Aging* **19(4S),** S276-S277.
31. Giaccone, G., Verga, L., Finazzi, M., Pollo, B., Tagliavini, F., Frangione, B., and Bugiani, O. (1990) Cerebral preamyloid deposits and congophilic angiopathy in aged dogs. *Neurosci. Lett.* **114,** 178-183.
32. Hirai, T., Kojima, S., Shimada, A., Umemura, T., Sakai, M., and Itakura, C. (1996) Age-related changes in the olfactory system of dogs. *Neuropathol. Appl. Neurobiol.* **22,** 531-539.
33. Tekirian, T. L., Saido, T. C., Markesbery, W. R., Russell, M. J., Wekstein, D. R., Patel, E., and Geddes, J. W. (1998) N-terminal heterogeneity of parenchymal and cerebrovascular Aβ deposits. *J. Neuropath. Exp. Neurol.* **57,** 76-94.
34. Uchida, K., Miyauchi, Y., Nakayama, H., and Goto, N. (1990) Amyloid angiopathy with cerebral hemorrhage and senile plaque in aged dogs. *Jpn. J. Vet. Sci.* **52,** 605-611.
35. Wegiel, J., Wisniewski, H. M., Dziewiatkowski, J., Tarnawski, M., Nowakowski, J., Dziewiatkowska, A., and Soltysiak, Z. (1995) The origin of amyloid in cerebral vessels of aged dogs. *Brain Res.* **705,** 225-234.
36. Uchida, K., Nakayama, H., and Goto, N. (1991) Pathological studies on cerebral amyloid angiopathy, senile plaques and amyloid deposition in visceral organs in aged dogs. *J. Vet. Med. Sci.* **53,** 1037-1042.
37. Pauli, B. and Luginbühl, H. (1971) Fluorescenzmikroskopische Untersuchungen der cerebralen Amyloidose bei alten Hunden und senilen Menschen. *Acta Neuropath.* **19,** 121-128.
38. Dahme, E. and Schröder, B. (1979) Kongophile Angiopathie, cerebrovasculäre Mikroaneurysmen und cerebrale Blutungen beim alten Hund. *Zbl. Vet. Med. A.* **26,** 601-613.
39. Su, M.-Y., Head, E., Brooks, W. M., Wang, Z., Muggenburg, B. A., Adam, G. E., Sutherland, R., Cotman, C. W., and Nalcioglu, O. (1998) Magnetic resonance imaging of anatomic and vascular characteristics in a canine model of human aging. *Neurobiol. Aging* **19,** 479-485.
40. Walker, L.C. and Durham, R.A. (1999) Cerebrovascular amyloidosis: Experimental analysis *in vitro* and *in vivo*. *Histol. Histopathol.* (in press).
41. Davis, J. and Van Nostrand, W. E. (1996) Enhanced pathologic properties of Dutch-type mutant amyloid β-protein. *Cell Biol.* **93,** 2996-3000.
42. Frackowiak, J., Mazur-Kolecka, B., Wisniewski, H. M., Potempska, A., Carroll, R. T., Emmerling, M. R., and Kim, K. S. (1995) Secretion and accumulation of Alzheimer's β-protein by cultured vascular smooth muscle cells from old and young dogs. *Brain Res.* **676,** 225-230.
43. Kawai, M., Kalaria, R. N., Cras, P., Siedlak, S. L., Velasco, M. E., Shelton, E. R., Chan, H. W., Greenberg, B. D., and Perry, G. (1993) Degeneration of vascular muscle cells in cerebral amyloid angiopathy of Alzheimer disease. *Brain Res.* **623,** 142-146.

44. Wisniewski, H. M. and Wiegel, J. (1994) β-Amyloid formation by myocytes of leptomeningeal vessels. *Acta Neuropath.* **87,** 233-241.
45. Yamada, M., Itoh, Y., Shintaku, M., Kawamura, J., Jensson, Ó., Thornsteinsson, L., Suematsu, N., Matsushita, M., and Otomo, E. (1996) Immune reactions associated with cerebral amyloid angiopathy. *Stroke* **27,** 1155-1162.
46. Kiatipattanasakul, W., Nakayama, H., Nakamura, S., and Koi, K. (1998) Lectin histochemistry in the aged dog brain. *Acta Neuropath.* **95,** 261-268.
47. Uchida, K., Kuroki, K., Yoshino, T., Yamaguchi, R., and Tateyama, S. (1997) Immunohistochemical study of constituents other than β-protein in canine senile plaques and cerebral amyloid angiopathy. *Acta Neuropath.* **93,** 277-284.
48. Walker, L. C. (1997) Animal models of cerebral β-amyloid angiopathy. *Brain Res. Rev.* **25,** 70-84.
49. Tomonaga, M. (1981) Cerebral amyloid angiopathy in the elderly. *J. Am. Geriatr. Soc.* **29,** 151-157.
50. Vinters, H. V. (1987) Cerebral amyloid angiopathy: a critical review. *Stroke* **18,** 311-324.
51. Wattendorff, A. R., Frangione, B., Luyendijk, W., and Bots, G. T. A. M. (1995) Hereditary cerebral hemorrhage with amyloidosis, Dutch type (HCHWA-D): clinicopathological studies. *J. Neurol. Neurosurg. Psychiatr.* **58,** 699-705.
52. Wong, C. W., Quaranta, V., and Glenner, G. G. (1985) Neuritic plaques and cerebrovascular amyloid in Alzheimer disease are antigenically related. *Proc. Natl. Acad. Sci., USA* **82,** 8729-8732.
53. Prelli, F., Castaño, E. M., van Duinen, S. G., Bots, G. T. A. M., Luyendijk, W., and Frangione, B. (1988a) Different processing of Alzheimer's β-protein precursor in the vessel wall of patients with hereditary cerebral hemorrhage with amyloidosis-Dutch type. *Biochem. Biophys. Res. Comm.* **151,** 1150-1155.
54. Prelli, F., Castaño, E., Glenner, G. G., and Frangione, B. (1988b) Differences between vascular and plaque core amyloid in Alzheimer's disease. *J. Neurochem.* **51,** 648-651.
55. Roher, A. E., Lowenson, J.D., Clarke, S., Woods, A. S., Cotter, R. J., Gowing, E., and Ball, M. J. (1993) β-Amyloid (1-42) is a major component of cerebrovascular amyloid deposits: Implications for the pathology of Alzheimer disease. *Proc. Natl. Acad. Sci., USA* **90,** 10836-10840.
56. Gearing, M., Tigges, J., Mori, H., and Mirra, S. S. (1996) Aβ40 is a major form of β-amyloid in nonhuman primates. *Neurobiol. Aging* **17,** 903-308.
57. Sawamura, N., Tamaoka, A., Shoji, S., Koo, E. H., Walker, L. C., and Mori, H. (1997) Characterization of amyloid beta protein species in cerebral amyloid angiopathy of a squirrel monkey by immunocytochemistry and enzyme-linked immunosorbent assay. *Brain Res.* **764,** 225-229.
58. Emmers, R. and Akert, K. (1963) *A Stereotaxic Atlas of the Brain of the Squirrel Monkey (Saimiri sciureus),* University of Wisconsin Press, Madison, Wisconsin.
59. Rosenblum, L. A. and Coe, C. L. (1985) *Handbook of Squirrel Monkey Research,* Plenum Press, New York.
60. Walker, L. C. (1993) Comparative neuropathology of aged nonhuman primates. *Neurobiol. Aging* **14,** 667.
61. Hendriks, L., van Duijn, C. M., Cras, P., Cruts, M., Van Hul, W., van Harskamp, F., Warren, A., McInnis, M. G., Antonarakis, S. E., Martin, J-J., Hofman, A., and Van Broeckhoven, C. (1992) Presenile dementia and cerebral hemorrhage linked to a mutation at codon 692 of the β-amyloid precursor protein gene. *Nature Genet.* **1,** 218-221.
62. Levy, E., Carman, M. D., Fernandez-Madrid, I. J., Power, M. D., Lieberburg, I., van Duinen, S. G., Bots, G. T. A. M., Luyendijk, W., and Frangione, B. (1990) Mutation of the

Alzheimer's disease amyloid gene in hereditary cerebral hemorrhage, Dutch type. *Science* **248,** 1124-1126.
63. Calenda, A., Jallegeas, V., Silhol, S., Bellis, M., and Bons, N. (1995) Identification of a unique apolipoprotein E allele in Microcebus murinus; apoe brain distribution and co-localization with β-amyloid and tau proteins. *Neurobiol. Dis.* **2,** 169-176.
64. Morelli, L., Wei, L., Amorim, A., McDermid, J., Abee, C. R., Frangione, B., Walker, L. C., and Levy, E. (1996) Cerebrovascular amyloidosis in squirrel monkeys and rhesus monkeys: apolipoprotein E genotype. *FEBS Letters* **379,** 132-134.
65. Mufson, E. J., Benzing, W. C., Cole, G. M., Wang, H., Emerich, D. F., Sladek, J. R., Morrison, J. H., and Kordower, J. H. (1994) Apolipoprotein E-immunoreactivity in aged rhesus monkey cortex: Colocalization with amyloid plaques. *Neurobiol. Aging* **15,** 621-627.
66. Poduri, A., Gearing, M., Rebeck, G. W., Mirra, S. S.,Tigges, J., and Hyman, B. T. (1994) Apolipoprotein E4 and beta amyloid in senile plaques and cerebral blood vessels of aged rhesus monkeys. *Am. J. Pathol.* **144,** 1183-1187.
67. Weisgraber, K. H., Pitas, R. E., and Mahley, R. W. (1994) Lipoproteins, neurobiology, and Alzheimer's disease: structure and function of apolipoprotein E. *Curr. Opin. Struct. Biol.* **4,** 507-515.
68. Levy, E., Lopez-Otin, C., Ghiso, J., Geltner, D., and Frangione, B. (1989) Stroke in Icelandic patients with hereditary amyloid angiopathy is related to a mutation in the cystatin C gene, an inhibitor of cysteine proteases. *J. Exp. Med.* **169,** 1771-1778.
69. Olafsson, Í., Thorsteinsson, L., and Jensson, Ó. (1996) The molecular pathology of hereditary cystatin C amyloid angiopathy causing brain hemorrhage. *Brain Path.* **6,** 121-126.
70. Graffagnino, C., Herbstreith, M. H., Schmechel, D. E., Levy, E., Roses, A. D., and Alberts, M. J. (1995) Cystatin C mutation in an elderly man with sporadic amyloid angiopathy and intracerebral hemorrhage. *Stroke* **26,** 2190-2193.
71. Wei, L., Walker, L. C., and Levy, E. (1996) Cystatin C: Icelandic-like mutation in an animal model of cerebrovascular β-amyloidosis. *Stroke* **27,** 2080-2085.
72. Walker, L. C., Price, D. L., Voytko, M. L., and Schenk, D. B. (1994) Labeling of cerebral amyloid *in vivo* with a monoclonal antibody. *J. Neuropath. Exp. Neurol.* **53,** 377-383.
73. Ghilardi, J. R., Catton, M., Stimson, E. R., Rogers, S., Walker, L. C., Maggio, J. E., and Mantyh, P. W. (1996) Intra-arterial infusion of ^{125I}-Aβ_{1-40} labels amyloid deposits in the aged primate brain *in vivo*. *Neuroreport* **7,** 2607-2611.
74. Mackic, J. B., Weiss, M. H., Miao, W., Kirkman, E., Ghiso, J., Calero, M., Bading, J., Frangione, B., and Zlokovic, B. V. (1998) Cerebrovascular accumulation and increased blood-brain barrier permeability to circulating Alzheimer's amyloid β peptide in aged squirrel monkey with cerebral amyloid angiopathy. *J. Neurochem.* **70** 210-215.

Chapter 20

VASCULAR TRANSPORT OF ALZHEIMER'S AMYLOID β PEPTIDES AND APOLIPOPROTEINS

BERISLAV V. ZLOKOVIC*, JORGE GHISO[†] and BLAS FRANGIONE[†]
*Department of Neurological Surgery, USC School of Medicine, Los Angeles, CA, USA and
[†]Department of Pathology, New York University Medical Center, New York, NY 10016

Deposition of amyloid β (Aβ) in the central nervous system (CNS) occurs during normal aging and is accelerated by Alzheimer's disease. Aβ accumulation in the CNS is considered a central part in the pathogenesis of Alzheimer's disease. Recent studies from our laboratory and other laboratories suggest a major role of the blood-brain barrier (BBB) in determining the concentrations of Aβ in the CNS. The BBB has a dual role: i) to control the entry of plasma-derived Aβ and it's binding transport proteins into the CNS, and ii) to regulate the levels of brain-derived Aβ via clearance mechanisms. Using *in vivo* rodent and non-human primate models, and an *in vitro* model of the human BBB, we have demonstrated transcytosis of plasma-derived $A\beta_{1-40}$ and $A\beta_{1-42}$ across the BBB suggesting that putative Aβ receptors in brain endothelial cells, e.g., the 'receptor for advanced glycation end products' (RAGE), scavenger receptor (SR), and lipoprotein receptors, e.g., gp330/megalin and possibly other members of the low density lipoprotein (LDL) receptor family, may regulate the BBB transport of Aβ free and/or complexed to apolipoproteins J and E (ApoJ, ApoE). Since ApoJ and ApoE receptors, and Aβ putative receptors/transporters are potential drug targets, understanding their function *in vivo* and *in vitro* in different animal and human models may help developing strategies to prevent and/or decelerate brain accumulation of Aβ, amyloid formation and associated cytotoxic effects. In this chapter we will discuss the role of vascular CNS transport of Aβ and its binding apolipoproteins in relation to its importance in the development of brain amyloidosis and AD pathology.

1. INTRODUCTION

Deposition of Aβ in the CNS occurs during normal aging and is accelerated by AD [1,2]. AD is the most common form of human

M.M. Verbeek et al. (eds.),
Cerebral Amyloid Angiopathy in Alzheimer's Disease and Related Disorders, 325–346.

amyloidosis and the major cause of dementia affecting 5-10% of the population > 65 yrs old, and ~ 50% of the population > 85 yrs old. Neuropathological characteristics of AD include: i. Intraneuronal deposits of neurofibrillary tangles (NFT); ii. Parenchymal amyloid deposits - neuritic plaques; iii. Vascular amyloidosis; and iv. Synaptic loss. Aβ is implicated in neuropathology of AD and related disorders [1-8]. Recent studies from our and other laboratories suggest a major role of the BBB in determining the concentrations of Aβ in the CNS [9-20]. The BBB has a dual role: i) to control the entry of plasma-derived Aβ and its binding transport proteins into the CNS, and ii) to regulate the levels of brain-derived Aβ via clearance mechanisms. The process of Aβ amyloid formation is highly tissue specific, and normally with aging occurs in brain extracellular space and in the walls of cortical and leptomeningeal vessels in humans, non-human primates and several mammalian species [1,21]. Specific predisposing genetic factors, e.g., ApoE4 genotype, mutations in amyloid-β-precursor protein (APP), presenilin 1 and 2 genes, α_2-macroglobulin, α_1-antichymotrypsin and age-dependent mechanisms, e.g., expression of specific receptors, transporters, binding proteins, may affect the CNS transport, increase the aggregation, and/or enhance the cytotoxicity of Aβ [1-8]. Among those, ApoJ and ApoE critically influence fibrillogenesis, cytotoxicity, transport across biological membranes and cell-specific uptake of Aβ within the CNS.

2. AMYLOID β PROTEIN

Aβ is produced by most somatic cells as a soluble peptide that circulates in blood, urine and cerebrospinal fluid (CSF) [22-24] and is present in brain [25,26]. It is not known exactly how much of the Aβ in the circulation, CSF and other extracellular body fluids is free, and how much is transported on ApoJ [27,28], and ApoE [29], transthyretin [30], albumin [31] and/or lipoproteins [32]. Aβ is generated through still unclear proteolytic mechanisms as an internal degradation product (39-43 residue) of APP encoded by a gene on chromosome 21 [33-35]. The Aβ extracted from AD senile plaques contains mainly 42/43 amino acids [36]. The vascular amyloid is 3 residues shorter [37], but $A\beta_{1-42}$ is also found [38-40]. A major soluble secreted form of Aβ is $A\beta_{1-40}$, whereas $A\beta_{1-42}$ accounts for about 10% [41]. An increase in soluble Aβ in AD and Down's syndrome brains [25,42,43] precedes amyloid plaque formation [44] and correlates with the development of vascular pathology [43].

2.1 Origin

The *"neuronal"* theory argues that soluble Aβ produced and secreted locally in the CNS is a precursor of Aβ deposits. Neuronal cells in culture secrete Aβ [45] which supports this view. It has been speculated that defective CNS clearance of brain-derived Aβ via ApoE and ApoJ, and possibly via some other transport proteins may either precede and/or exacerbate amyloid CNS deposition [1,19,20,46], but this hypothesis has never been tested in animal models or confirmed in humans.

The discovery of circulating Aβ has raised the possibility that the amyloid in AD may originate from blood, and that brain amyloidosis and systemic amyloidoses have unifying features [52]. The *"vascular"* theory is supported by pathological changes in patients with hereditary cerebral hemorrhage with amyloidosis-Dutch type (HCHWA-D), a genetic form of cerebral amyloid angiopathy (CAA) [53,54]. Recent studies from our and other laboratories support the concept that transport across the BBB may critically regulate the levels of plasma-derived Aβ/apolipoproteins in the CNS [9-20].

2.2 Amyloidogenesis

The exact mechanisms of amyloid fibril formation and deposition, as well as their tissue-specific localization, are poorly understood [1]. Spontaneous fibril formation has been demonstrated *in vitro* with synthetic Aβ peptides [52]. The early onset familial forms of AD in certain Dutch and Swedish pedigrees indicated that the primary structure and concentration of Aβ may be critical in influencing fibril formation [55,56]. A mutation at codon 693 (APP770) of Gln substituting Glu at residue 22 in HCHWA-D [57] converts $A\beta_{1-40}$ into a highly amyloidogenic and pathologic form, Aβ Q22 [58,59]. Post-translational modification of amino acids (e.g., oxidation, phosphorylation, glycosylation, methylation) may be important [52]. According to the 'amyloid hypothesis' Aβ is first deposited as preamyloid lesions [60] associated with few or no dystrophic neurites. In Down's syndrome with three copies of the APP genes, preamyloid lesions can appear as early as the age of 12 [61-63]. These lesions become compacted over a period of many years, and by the end of third decade acquire the characteristics of amyloid plaques associated with neuronal damage and NFT [60-63]. Clinically non-demented aged individuals can develop extensive preamyloid deposits which may herald the later development of AD [64]. The Aβ-associated proteins, ApoE and ApoJ (see below), amyloid P-component, heparan sulfate proteoglycans, α_1-antichymotrypsin, vitronectin, transthyretin, may act to either inhibit amyloid formation, or to stabilize the β-pleated sheet structure. The role of complement, C1q, cytokines and inflammation has been shown by several groups. High degree of insolubility may preclude complete proteolytic degradation of fibrils *in vivo*.

2.3 Cytotoxicity of Aβ

Aβ accumulation in the CNS is considered a central part in the pathogenesis of AD [1-8]. The pathogenic role of Aβ is supported by demonstration that Aβ peptides have neurotoxic properties *in vitro* [65-67] and *in vivo* [68-70]. An endoplasmic reticulum-associated binding protein can mediate Aβ cellular toxicity [71]. Aβ induces neuronal oxidative stress directly and indirectly by activating microglia [72,73]. Activated microglia may play a central role in the inflammatory process associated with amyloid plaques [74]. The cells of macrophage/monocyte lineage are associated with CAA [75] and found along vessel walls in HCHWA-D [76]. Vascular damage, leukocyte activation and migration across the vessel wall in response to Aβ were shown *in vivo* [77]. Aβ may produce endothelial injury by generating superoxide radicals [78]. Some forms of Aβ may alter the BBB permeability and produce apoptosis in endothelial cells [79]. Cerebrovascular deposition of Aβ is associated with degeneration of vascular smooth muscle cells [80-82]. The length, form (wild or mutant) and conformational state (e.g. aggregated or soluble) importantly influence the cytotoxic effects of Aβ. Thus, the accumulation of Aβ seems to be the starting point for the development of several cellular pathologies.

2.4 Aβ-receptors

RAGE [73] and SR [83,84] have been recently identified as receptors for free, unbound Aβ. RAGE, a member of the immunoglobulin superfamily of cell-surface molecules, is expressed in the CNS vascular endothelium, smooth muscle cells and phagocytes [85]. RAGE can mediate pathophysiological responses in the vasculature (e.g. oxidant stress) when occupied with glycated ligands or Aβ [17,18,73,85]. RAGE expression is up-regulated in AD brain [73,88]. SR, a homotrimeric cell surface molecule [86] is expressed on macrophages, microglia and vascular endothelium. The SR type I and II are both expressed in cerebral capillaries [87]. SR promotes endocytosis and degradation of modified oxidized LDL and glycated ligands, and its expression is up-regulated in AD brain [73,88]. In AD, SR is involved in signaling microglia to accumulate at sites of Aβ deposition [73,83,88]. Both RAGE and SR mediate endocytosis and transcytosis of macromolecules [85,86] and can bind and internalize AGE and Aβ at the BBB [17,18,89]. In addition, RAGE is involved in mediating transcytosis of unbound Aβ across the BBB *in vitro* and *in vivo* [17,18].

2.5 BBB-transport of Aβ

We and others have suggested that specific transport systems and/or receptors at the luminal membrane of the BBB, mediate microvascular sequestration and blood-to-brain transfer of plasma-derived unbound Aβ peptides in several mammalian species [9-14,17,18,90]. Briefly, the N-terminal portion of Aβ is important for its recognition at the luminal BBB transport site (Figure 1A), while RAGE has a major role in *in vivo* transcytosis of free circulating Aβ across the BBB (Figure 1B). The SR is involved in cell surface binding and internalization of Aβ at the luminal side of the BBB (not shown), but does not take a part in luminal-to-abluminal transcytosis (Figure 1B) . Neither integrin receptors nor APP interact with plasma-derived Aβ, while the lipoprotein receptors mediate Aβ transport via ApoJ and ApoE4 (see below). The role of RAGE and SR in binding and internalization of free Aβ has been confirmed in an *in vitro* human model of the BBB, and in RAGE- and SR- transfected cells [18]. The BBB permeability to circulating Aβ and vascular uptake were significantly increased in aged squirrel monkeys with CAA, while the barrier remains intact to reference tracers such as cerebrovascular space marker inulin [14]. Transport of circulating $A\beta_{1-40}$ into brain of aged squirrel monkey resulted in significant co-localization of radiolabeled peptide onto vascular amyloid lesions [13]. Degradation during BBB transport significantly reduces the amount of intact peptide that enters the brain from periphery [11,90,91] (Figure 1C). On the other hand, the role of clearance mechanisms in keeping low levels of brain-derived Aβ cannot be ruled out at this point.

3. APOLIPOPROTEIN J

Recent findings indicated several possible relationships between ApoJ (or clusterin) and AD. ApoJ is a component of the senile plaques in AD and Down's syndrome [92,93]. In a rodent model of neurodegeneration ApoJ expression is increased in astrocytes [94,95]. ApoJ has been shown to be the major carrier protein of Aβ in the CSF and plasma [27]. *In vitro* studies have demonstrated that the $A\beta_{1-40}$ -ApoJ complex cannot be dissociated by ApoE2, ApoE3, ApoE4, α_1-antichymotrypsin, transthyretin and vitronectin at their physiological plasma concentrations [28]. Plasma Aβ-ApoJ complexes are normally incorporated into HDL3 particles [96]. In its native

HDL form, ApoJ is fully active to interact with Aβ and protects Aβ from proteolytic degradation [97]. In addition to forming soluble heterodimers, ApoJ may also form highly aggregated complexes with Aβ [98], whereas lipidated ApoJ prevents aggregation and polymerization of synthetic Aβ *in vitro* [108]. The ApoE ε4 allele dose correlates positively with ApoJ load and inversely with ApoE load in AD brains [99].

3.1 Glycoprotein 330/megalin

A receptor for ApoJ has been identified, the gp330/megalin or LDL receptor (LDLR) related protein 2, LRP-2, that is capable of mediating its endocytosis with subsequent lysosomal degradation [100], or transcytosis in certain tissues [15] similar to LDLR [101]. Gp330/megalin is expressed in the choroid plexus, the BBB endothelium and ventricular ependyma [15,102]. A recent study indicated that gp330/megalin mRNA and protein are expressed in brain capillaries and choroid plexus in smaller amounts than in kidney, yet sufficient to carry important functions of the receptor in CNS ApoJ/Aβ homeostasis [103]. Astrocytes secrete a novel ApoE-ApoJ lipoprotein particle [104]. In AD brains, the cells containing significant nuclear DNA fragmentation express the highest level of cell surface gp330-like immunoepitopes [99].

3.2 BBB-transport of ApoJ

Figure 2A illustrates the cerebrovascular permeability surface (PS) products of circulating 125I-labeled ApoJ, ApoE and of their complexes. The PS products were based on the levels of intact apolipoproteins or intact complexes and/or peptides in brain. Thus, the PS products were corrected for test-tracers distribution into the vascular space by subtracting the value of brain vascular space marker 99Tc-albumin, and for the degradation during BBB transport by correcting for the amount of Aβ dissociated from the complexes and/or the amount of metabolized peptide determined in brain parenchyma (i.e., capillary-depleted brain) by the HPLC or SDS-PAGE analysis.

The PS value of human ApoJ or its complex with $A\beta_{1-40}$ were higher than for many other peptides and proteins that cross the BBB by specialized mechanisms [15]. For example, the cerebrovascular permeability to $A\beta_{1-40}$-ApoJ was 9- to 14-fold higher than for insulin, a peptide transported across the BBB via insulin receptor [107,108]. In comparison to transferrin and nerve growth factor, transport rate of the complex was 45- to 99-fold higher [107], and 17-fold higher than for OX26 antibody to transferrin receptor which mediates BBB transport of Aβ-OX26 complex [109].

A. ^{125}IAβ 1-40 BRAIN INFLUX KINETIC INHIBITION

Inhibitor (40 nM)	K_i (nM)	K_i/K_m	K_i/K_i (Aβ 1-40)
Aβ 1-40*	44 ± 14	1.05	1.00
Aβ 1-42*	83 ± 13	1.99	1.90
Aβ 1-28*	106 ± 18	2.52	2.40
Aβ 5-8^{ns}	∞		
Aβ 17-40^{ns}	∞		
Aβ 15-28^{ns}	∞		
Aβ 40-1^{ns}	∞		

$K_m = 49 \pm 10$ nM Vmax = 111 ± 19 fmol/min/g

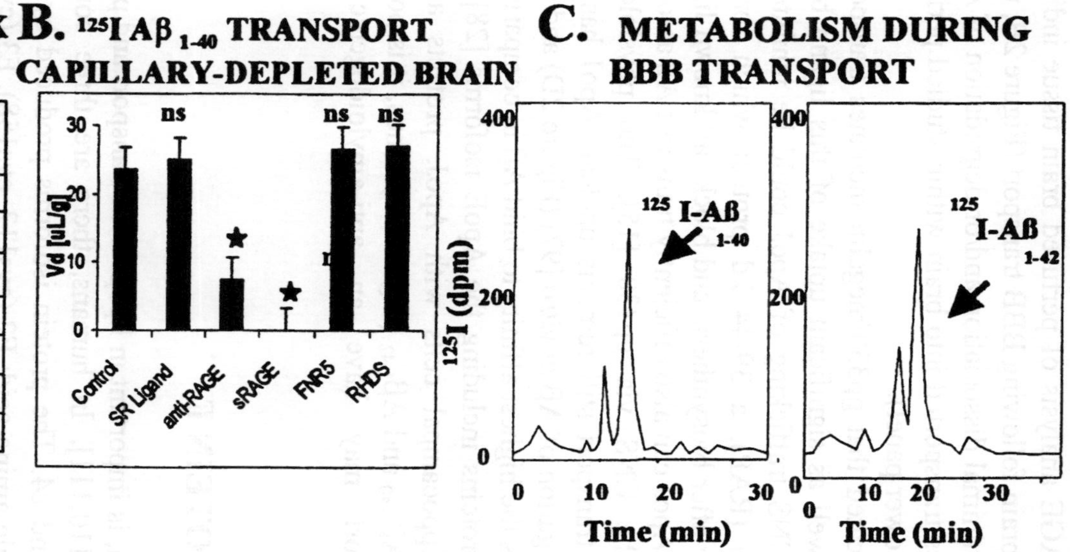

Fig. 1. **A.** The inhibitory constant, K_I, for different forms of Aß peptides vs. $Aß_{1\text{-}40}$ determined at 40 nM and computed using published K_M and V_{MAX} values for $Aß_{1\text{-}40}$ transport at the BBB (10). Mean ± SE, n = 4-7. ***B.*** Brain parenchymal uptake of 125I-$Aß_{1\text{-}40}$ (1 nM) in the presence of SR ligand fucoidan (200 mg/ml), α-RAGE (10 mg/ml), sRAGE (10 molar excess), anti-integrin b antibody FnR5 (15 mg/ml) and Aß 5-8 sequence, RHDS (40 nM). Mean ± SE, n = 3 - 7. ***C.*** HPLC elution profile of brain tissue perfused for 10 min with either ^{125}I-$Aß_{1\text{-}40}$ or ^{125}I-$Aß_{1\text{-}40}$. $^{*}p < 0.005$ for $Aß_{1\text{-}40}$ in the absence vs. presence of different molecular reagents. ^{ns}Non-significant vs. control uptake.

The HPLC and SDS-PAGE analysis of perfused brain tissue indicated that ApoJ remains intact in brain following BBB transport (Figure 2B, upper panel), and there was also minimal dissociation and/or degradation of $A\beta_{1-40}$ from $A\beta_{1-40}$-ApoJ complexes transported into brain within studied period of time of 10 min (Figure 2B, lower panel).

Subsequent studies indicated that gp330/megalin mediates transport of ApoJ across the BBB, as well as significant uptake of this ligand by the choroid plexus [15]. The CNS trafficking of ApoJ could be inhibited by receptor associated protein (RAP), a 39-44 kd protein which forms a complex with gp330 shortly after biosynthesis and blocks all known ligands to gp330 and LRP-1 [106]. ApoJ can also efficiently transport Aβ across the vascular membranes of the CNS via gp330 [15]. The physiological importance of ApoJ BBB transport still remains unclear. ApoJ has been reported to inhibit the aggregation of Aβ *in vitro* [97] (Figure 2D) and has at physiological concentrations the highest affinity to bind Aβ in comparison to other amyloid-associated proteins including the ApoE isoforms [28]. Pre-incubation of neuronal hippocampal cells with ApoJ, protects against neurotoxic effects of both $A\beta_{1-40}$ and $A\beta_{1-42}$ (Figure 2E). Thus, it is possible that ApoJ CNS transport may have an anti-amyloidogenic and neuroprotective role.

4. APOLIPOPROTEIN E

ApoE, a 34.2 kd protein, is important in lipoprotein transport and plasma cholesterol homeostasis [110,111]. In humans, there are three common alleles of ApoE: ε2, ε3, and ε4. The protein isoforms produced by these alleles differ by only a single amino acid: E2 (cys112, cys158), E3 (cys112, arg158), which is the most common, and E4 (arg112, arg158) [110,111]. ApoE4 is considered a risk for atherosclerosis, cardiovascular disease, stroke and AD [112,113]. In addition to prominent expression in the liver, ApoE is also highly expressed in the brain [114], where it is synthesized predominantly by astrocytes and some microglia [115-117], and secreted mainly within HDL-like particles [104]. ApoE containing lipoproteins in the CSF and CNS bind efficiently to the lipoprotein receptors in brain. In plasma, ApoE resides in VLDL, IDL and HDL particles and chylomicron remnants.

4.1 Lipoprotein receptors

ApoE promotes the recognition and catabolism of ApoE-containing lipoproteins such as chylomicron remnants, bVLDL, hypertriglyceridemic

VLDL, IDL and HDL by the LDLR [118], the ApoE/α_2-macroglobulin receptor LRP1 [119,120], gp330/megalin [100,105], and possibly the VLDL receptor [125]. Brain cells contain all four types of receptors [101-103,105,119,121-125]. Lipoprotein receptors are also present in brain endothelium [15,101,103,123], the choroid plexus epithelium [15,102,103], and the ependymal cells lining the CNS ventricles [102]. The role of these receptors in lipid transport and cholesterol homeostasis and ApoE-mediated Aβ transport in the CNS is poorly understood [125].

4.2 ApoE in AD

ApoE colocalizes with the senile plaques and congophilic angiopathy of AD [126,127]. Genetic epidemiologic studies have shown that the ε4 allele is a risk factor for late onset familial and sporadic AD [126,128-130], its frequency is approximately two to three times higher in patients with AD than in non-AD subjects, and there is a dosage effect of the ε4 allele on lifetime risk for AD and the age at onset of the disease [128,130]. The amount of deposited $A\beta_{1\text{-}40}$ is increased in AD brains according to the number of copies of ε4 alleles [131], as well as their vascular Aβ load [2,132,133]. Major ethnic differences with no effect of ApoE ε4 allele on the development of AD among African Americans and Hispanics have also been reported [134].

Isoform-specific properties of ApoE on binding to Aβ [29,46,135-140], binding to tau [141-144], cholinergic deficit [145], neuronal morphology, cytoskeletal structure, neuronal degeneration and dendritic remodeling [146-150], binding to soluble APP and excitotoxicity [151], activation of microglial cells [152] etc. have been reported. Several theories have been formulated to explain the role of ApoE isoforms in the pathogenesis of AD.

The first theory is focused on interactions of ApoE with Aβ (see below). The second theory implies that the ε4 allele is less efficient in neuronal repair mechanisms. The third implies the lack of ApoE4 interaction with the microtubule associated protein tau [141,143,144]. The fourth theory suggests that neuronal degeneration in AD results from impairments in lipid transport caused by ApoE4 [145,147,153]. The fifth theory implies that different ApoE isoforms differentially regulate the CNS and BBB transport mechanisms for Aβ [16,154].

A **BBB Permeability**

^{125}I-Test-Tracer	PS (mL/g /s x 10^6)
apo E2	0.42 ± 0.13^{ns}
apo E3	1.81 ± 0.41^{ns}
apo E4	1.40 ± 0.31^{ns}
Aβ1-40-apo E2	$2.21 \pm 0.81^{ns,a}$
Aβ1-40-apo E3	$0.00 \pm 0.56^{ns,a}$
Aβ1-40-apo E4	8.10 ± 1.46^{a}
apo J	67.7 ± 15.5^{b}
Aβ1-40-apo J	87.6 ± 27.9^{b}
Aβ 1-40	20.1 ± 2.42^{c}
Aβ 40-1	1.07 ± 0.69

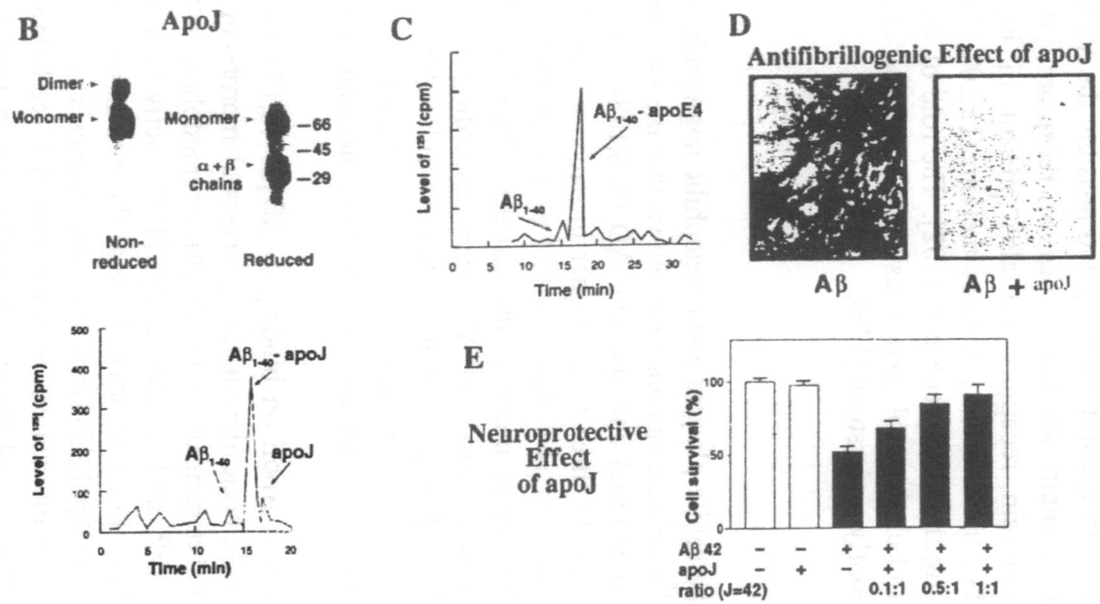

Fig. 2. ***A.*** The PS products of intact apolipoproteins and of their complexes with Aß (mean ± SE; n = 6 - 18 guinea pigs) determined within 10 min of arterial infusion and analysis of brain radioactivity by the HPLC and/or SDS-PAGE. ns, denotes - non-significant difference between apoE2, apoE3, apoE4, $Aß_{1-40}$-apoE2 and $Aß_{1-40}$-apoE3 vs. Aß40-1. $^a p < 0.05$ to 0.001 for $Aß_{1-40}$-apoE4 vs. $Aß_{40-1}$; $Aß_{1-40}$-apoE2 vs. $Aß_{1-40}$; $Aß_{1-40}$-apoE3 vs. $Aß_{1-40}$; $Aß_{1-40}$-apoE4 vs. $Aß_{1-40}$. $^b p < 0.002$ for $Aß_{1-40}$-apoJ or apoJ vs. $Aß_{1-40}$; $^c p < 0.01$ for $Aß_{1-40}$ vs. $Aß_{40-1}$ by ANOVA. ***B.*** ***Upper panel:*** SDS/PAGE analysis of brain tissue perfused for 10 min with ^{125}I-apo J. The radioactivity in the brain was eluted as a single apoJ peak on HPLC, and reduced and non-reduced aliquots subjected to 10% Tris Tricine SDS/PAGE. ***Lower panel:*** HPLC elution profile of brain tissue perfused for 10 min with ^{125}I-$Aß_{1-40}$-apoJ (both $Aß_{1-40}$ and apoJ were labeled). ***C.*** HPLC elution profile of brain tissue perfused for 10 min with 125I-$Aß_{1-40}$-apoE4 (the label was on $Aß_{1-40}$). ***D.*** Aß fibril formation (10 uM) abolished by apoJ. ***E.*** Neuroprotective effect of apoJ against Aß42 toxicity (2.5 mM) on E10 rat primary hippocampal neurons.

4.3 ApoE-Aβ interactions

There have been conflicting reports as to whether there is greater or similar binding of ApoE3 and ApoE4 to Aβ [29,155], and whether or not ApoE enhances or inhibits Aβ fibril formation and seeding *in vitro* in isoform-dependent fashion [156-158]. The isoform-specific binding of ApoE varies depending on whether ApoE is lipidated or not [136,137], with lipidation enhancing the binding of Aβ to ApoE3 relative to ApoE4 [138]. Our recent studies indicated that delipidation decreases the affinity of Aβ peptides by 5- to 10-fold, and also abolishes the isoform specificity. ApoE2 and ApoE3 bind Aβ with higher affinity than ApoE4 [159], and this may prevent its polymerization and inhibit fibrillogenesis if Aβ-ApoE complexes are efficiently cleared from the CNS [154]. Nascent ApoE4 binds more weakly to Aβ, which may favor polymerization and aggregation of free $A\beta_{1-40}$ and $A\beta_{1-42}$ in ε4 homozygous patients [159]. The initial observations indicated that ApoE may be a pathological "chaperone" stabilizing Aβ in a β-pleated conformation [127,139].

4.4 ApoE in transgenic models of brain amyloidosis

Transgenic mice with an APPV717F mutation develop age-dependent thioflavine positive Aβ plaques, loss of synaptophysin immunoreactivity and astrogliosis [161]. Transgenic mice that express the APP695 double Swedish mutation (APPsw+/-) develop Aβ-immunoreactive CNS deposits and plaques progressively beginning at 8-9 months [162], a correlative memory deficit [162], vascular amyloid and endothelial dysfunction [163], neuropil abnormalities [164], impaired synaptic plasticity [165], and may exhibit plaque-associated regional neuronal loss [166], thus resembling Alzheimer's type amyloidosis. Doubly transgenic APP/PS1 mice develop accelerated Alzheimer's type amyloidosis [167]. Recent studies indicated that transgenic APPV717 mice bred onto a mouse apoE –/– background, have a significant decrease in Aβ deposition and thioflavine S–positive (grossly fibrillar) deposits compared with animals expressing endogenous mouse ApoE [168]. Thus, the expression of mouse ApoE in some way influences Aβ degradation to promote Aβ deposition. Recently, it has been reported that human ApoE can influence deposition of human Aβ *in vivo* in transgenic APP mice [169]. In contrast to the effects of mouse ApoE, the expression of human ApoE suppresses early deposition of Aβ both in brains of transgenic APPV717+/- [169] and transgenic APPsw+/- mice. Thus, human ApoE may clear Aβ more efficiently from the mouse CNS suppressing its initial deposition, but at later stages significant deposition of Aβ may still occur in the presence of ApoE4, but not ApoE3, suggesting differential effects on Aβ clearance *in vivo*.

4.5 BBB-transport of ApoE

We have recently demonstrated negligible brain uptake of circulating ApoE isoforms (Figure 2A), and low but detectable capillary sequestration and significant uptake by the choroid plexus [16,154]. Complexing of Aβ to ApoE2 and ApoE3 abolished the uptake of the peptide into brain parenchyma (Figure 2A), by cerebral microvessels and choroid plexus [16]. In contrast, ApoE4 mediates moderate transport of circulating Aβ across the BBB, and protects the peptide from degradation (Figure 2C). The difference between ApoE4 and its complex with $A\beta_{1\text{-}40}$ may be explained by a conformational transformation induced by ApoE4 binding to Aβ that allows uptake at the BBB. Although gp330 has been ruled out as a likely candidate for transport of these complexes, the role of other lipoprotein receptors has been suggested [160]. By preventing Aβ degradation, favoring transport of plasma-derived peptide, and reducing the clearance of brain-derived Aβ, ApoE4 relative to E2 or E3 isoforms may predispose to Aβ accumulation and cytotoxicity in the aging brain. If ApoE4 increases the rate of fibrillogenesis as suggested [156], the alterations in ApoE-mediated BBB and CNS transport of Aβ could also contribute directly to amyloid formation.

ACKNOWLEDGMENT

This work was supported by grants NS 34467 and AG 16233 from the National Institute of Health to B.V. Z. and grants AG 05891 and AG 08721 to B.F.

REFERENCES

1. Wisniewski T, Ghiso J, Frangione B. (1997) Biology of Aβ amyloid in Alzheimer's disease, *Neurobiol. Dis.* **4**, 311-328.
2. Selkoe DJ. (1997) Alzheimer's disease: genotype, phenotype, and treatments, *Science* **275**, 630-631.
3. Selkoe DJ. (1998) The cell biology of beta-amyloid precursor protein and presenilin in Alzheimer's disease, *Trends Cell. Biol.* **8**, 447-453.
4. Younkin SG. (1998) The role of A beta 42 in Alzheimer's disease, *J. Physiol. (Paris)* **92**, 289-292.

5. Roses AD. (1998) Alzheimer disease: a model of gene mutations and susceptibility polymorphisms for complex psychiatric diseases, *Am. J. Med. Gen.* **81**, 49-57.
6. Hardy J, Duff K, Hardy KG, Perez-Tur J, Hutton M. (1998) Genetic dissection of Alzheimer's disease and related dementias: amyloid and its relationship to tau, *Nat. Neurosci.* **1**, 355-358.
7. Dickson DW. (1997) The pathogenesis of senile plaques, *J. Neuropathol. Exp. Neurol.* **56**, 321-339.
8. Blacker D, Wilcox MA, Laird NM, Rodes L, Horvath SM, Go RC, Perry R, Watson Jr B, Bassett SS, McInnis MG, Albert MS, Hyman BT, Tanzi RE. (1998) Alpha-2 macroglobulin is genetically associated with Azlheimer's disease, *Nature Gen.* **19**, 357-360.
9. Zlokovic BV. (1997) Can blood-brain barrier play a role in the development of cerebral amyloidosis and Alzheimer's disease pathology, Neurobiol. Dis. **4(1)**, 23-26.
10. Zlokovic BV, Ghiso J, Mackic JB, McComb JG, Weiss MH, Frangione B. (1993) Blood-brain barrier transport of circulating Alzheimer's amyloid β, *Biochem. Biophys. Res. Commun.* **197**, 1034-1040.
11. Maness LM, Banks WA, Podlisny MB, Selkoe DJ, Kastin AJ. (1994) Passage of human amyloid-β protein 1-40 across the murine blood-brain barrier, *Life Sci.* **55**, 1643-1650.
12. Poduslo JF, Curran GL, Haggard JJ, Biere AL, Selkoe DJ. (1997) Permeability and residual plasma volume of human, Dutch variant, and rat amyloid β-protein 1-40 at the blood-brain barrier, *Neurobiol. Dis.* **4(1)**, 27-34.
13. Ghilardi JR, Catton M, Stimson ER, Rogers S, Walker LC, Maggio JE, Mantyh PW. (1996) Intra-arterial infusion of $[^{125}I]A\beta_{1-40}$ labels amyloid deposits in the aged primate brain *in vivo*, *Neuroreport* **7**, 2607-2611.
14. Mackic JB, Weiss MH, Miao W, Ghiso J, Calero M, Bading J, Frangione B, Zlokovic BV. (1998) Cerebrovascular accumulation and increased blood-brain barrier permeability to circulationg Alzheimer's amyloid-β peptide in aged squirrel monkey with cerebral amyloid angiopathy, *J. Neurochem.* **70**, 210-215.
15. Zlokovic BV, Martel CL, Matsubara E, McComb JG, Zheng G, McCluskey RT, Frangione B, Ghiso J. (1996) Glycoprotein 330/megalin: Probable role in receptor-mediated transport of apolipoprotein J alone and in a complex with Alzheimer's disease amyloid β at the blood-brain and blood-cerebrospinal fluid barriers, *Proc. Natl. Acad. Sci. USA* **93**, 4229-4236.
16. Martel CL, Mackic JB, Matsubara E, Governale S, Miguel C, Miao W, McComb JG, Frangione B, Ghiso J, Zlokovic BV. (1997) Isoform-specific effects of apolipoproteins E2, E3, E4 on cerebral capillary sequestration and blood-brain barrier transport of circulating Alzheimer's amyloid β, *J. Neurochem.* **69**, 1995-2004.
17. Miao W, Mackic JB, Ghiso J, McComb JG, Frangione B, Van Nostrand W, Yan SD, Stern D, Zlokovic BV. (1998) Possible role of RAGE in transport and sequestration of amyloid-β 1-40 and 1-42 at the blood-brain barrier in rodents, Cereb. Vasc. Biol. Conf. March 26-28, Portland.
18. Mackic JB, Stins M, McComb JG, Calero M, Ghiso J, Kim KS, Stern D, Frangione B, Zlokovic BV. (1998) Human blood-brain barrier receptors for Alzheimer's amyloid-β 1-40: asymmetrical binding, endocytosis and transcytosis at the apical side of brain microvascular endothelial cell monolayer, *J. Clin. Invest.* **102**, 734-743.
19. Zlokovic BV. (1996) Cerebrovascular transport of Alzheimer's amyloid-β and apolipoproteins J and E: possible anti-amyloidogenic role of the blood-brain barrier, *Life Sci.* **59(18)**, 1483-1497.

20. Ghersi-Egea JF, Gorevic PD, Ghiso J, Frangione BF, Patlak CS, Fenstermacher JD. (1996) Fate of cerebrospinal fluid-borne amyloid β-peptide: rapid clearance into blood and appreciable accumulation by cerebral arteries, *J. Neurochem.* **67**, 880-883.
21. Walker LC. (1997) Animal models of cerebral β-amyloid agiopathy, *Brain. Res. Rev.* **25**, 70-84.
22. Seubert P, Vigo-Pelfrey C, Esch F, Lee M, Dovey H, Davis D, Sinha S, Schlossmacher M, Whaley J, Swindlehurst C, McCormack R, Wolfert R, Selkoe DJ, Lieberburg I, Schenk D. (1992) Isolation and quantification of soluble Alzheimer's β-peptide from biological fluids, *Nature* **359**, 325-327.
23. Shoji M, Golde T, Ghiso J, Cheung T, Estus S, Shaffer L, Cai XD, McKay D, Tintner R, Frangione B, Younkin S. (1996) Production of the Alzheimer amyloid β protein by normal proteolytic processing, *Science* **258**, 126-129.
24. Vigo-Pelfrey C, Lee D, Keim P, Lieberburg I, Schenk DB. (1993) Characterization of β-amyloid peptide from human cerebrospinal fluid, *J. Neurochem.* **61**, 965-968.
25. Tabaton M, Nunzi MG, Xue R, Usiak M, Autilio-Gambetti L, Gambetti P. (1994) Soluable amyloid beta-protein is a marker of Alzheimer amyloid in brain but not in cerebrospinal fluid, *Biochem. Biophys. Res. Commun.* **200**, 1598-1603.
26. Kuo YM, Emmerling MR, Vigo-Pelfrey C, Kasunic TC, Kirkpatrick JB, Murdoch GH, Ball MJ, Roher AE. (1996) Water-soluble Aβ (N-40, N-42) oligomers in normal and Alzheimer disease brains. *J. Biol. Chem.* **271**, 4077-4081.
27. Ghiso J, Matsubara E, Koudinov A, Choi-Miura NH, Tomita M, Wisniewski T, Frangione B. (1993) The cerebrospinal-fluid soluble form of Alzheimer's amyloid beta is complexed to SP-40,40 (apolipoprotein J), and inhibitor of the complement membrane-attack complex, *Biochem. J.* **293**, 27-30.
28. Matsubara E, Frangione B, Ghiso J. (1995) Characterization of apolipoprotein J-Alzheimer's Aϑ interaction, *J. Biol. Chem.* **270**, 7563-7567.
29. Yang DS, Smith JD, Zhou Z, Gandy, SE, Martins RN. (1997) Characterization of the binding of amyloid-β peptide to cell culture-derived native apolipoprotein E2, E3, and E4 isoforms and to isoforms from human plasma., *J. Neurochem.* **68**, 721-725.
30. Schwarzman AL, Gregori L, Vitek MP, Lyubski S, Strittmatter WJ, Enghilde JJ, Bhasin R, Silverman J, Weisgraber KH, Coyle PK, Goldgaber D. (1994) Transthyretin sequesters amyloid β protein and prevents amyloid formation, *Proc. Natl. Acad. Sci. USA* **91**, 8368-8372.
31. Biere AL, Ostazewski B, Stimson ER, Hyman BT, Maggio JE, Selkoe DJ. (1996) Amyloid β-peptide is transported on lipoproteins and albumin in human plasma, *J. Biol. Chem.* **271**, 32916-32922.
32. Matsubara E, Ghiso J, Frangione B, Amari M, Tomidokoro Y, Ikeda Y, Harigaya Y, Okamoto K, Shoji M. (1999) Lipoprotein-free amyloidogenic peptides in plasma are elevated in patients with sporadic Alzheimer's disease and Down's syndrome, *Ann. Neurol.* **45**, 537-541.
33. Kang J, Lemaire HG, Unterbeck A, Salbaum JM, Masters CL, Grzeschik KH, Malthaup G, Beyreuther K, Muller-Hill B. (1987) The precursor of Alzheimer's disease A4 protein resembles a cell-surface receptor, *Nature* **325**, 733-736.
34. Tanzi RE, Gusella JF, Watkins PC, Bruns GAP, St. George-Hyslop P, Van Keuren ML, Patterson D, Pagan S, Kurnit DM, Neve RL. (1987) Amyloid β-protein gene: cDNA, mRNA distribution and genetic linkage near the Alzheimer locus, *Science* **23**, 880-884.
35. Goldgaber D, Lerman MI, McBride OW, Saffioti U, Gajdusek DC. (1987) Characterization and chromosomal localization of a cDNA encoding brain amyloid of Alzheimer's disease, *Science* **235**, 877-80.

36. Masters CL, Simms G, Weinman NA, Multhaup G, McDonald BL, Beyreuther K. (1985) Amyloid plaque core protein in Alzheimer disease and Down syndrome, *Proc. Natl. Acad. Sci. USA* **82**, 4245-4249.
37. Prelli F, Castano EM, Glenner GG, Frangione B. (1988) Differences between vascular and plaque core amyloid in Alzheimer's disease, *J. Neurochem.* **51**, 648-651.
38. Roher AE, Lowenson JD, Clarke S, Woods AS, Cotter Rj, Gowing E, Balland MJ. (1193) β-amyloid (1-42) is a major component of cerebrovascular amyloid deposits: implications for the pathology of Alzheimer disease, *Proc. Natl. Acad. Sci. USA* **90**, 10836-10840.
39. Shinkai Y, Yoshimura M, Ito Y, Odaka A, Suzuki N, Yanagisawa K, Ihara Y. (1995) Amyloid β-proteins 1-40 and 1-42(43) in the soluable fraction of extra- and intracranial blood vessels, *Ann. Neurol.* **38**, 421-428.
40. Castaño EM, Prelli F, Soto C, Beavis R, Matsubara E, Shoji M, Frangione B. (1996) The length of amyloid-beta in hereditary cerebral hemorrhage with amyloidosis, Dutch type. Implications for the role of amyloid-β 1-42 in Alzheimer's disease, *J. Biol. Chem.* **271**, 32185-32191.
41. Citron M, Diehl TS, Gordon G, Biere Al, Seubert P, Selkoe DJ. (1996) Evidence that the 42- and 40-amino acid forms of amyloid β protein are generated from the β-amyloid precursor protein by different protease activities, *Proc. Natl. Acad. Sci. USA* **93**, 13170-13175.
42. Naslund J, Schierhorn A, Hellman U, Lannfelt L, Roses AD, Tjernberg LO, Silberring J, Gandy SE, Winblad B, Greengard P, Nordstedt C, Terenius L. (1994) Relative abundance of Alzheimer Aβ amyloid peptide variants in Alzheimer disease and normal aging, *Proc. Natl. Acad. Sci. USA* **91**, 8378-8382.
43. Suzuki N, Iwatsubo T, Odaka A, Ishibashi Y, Kitada C, Ihara Y. (1994) High tissue content of soluble beta 1-40 is linked to cerebral amyloid angiopathy, *Am. J. Pathol.* **145**, 452-460.
44. Teller JK, Russo C, DeBusk LM, Angelini G, Zaccheo D, Dagna-Bricarelli F, Scartezzini P, Bertolini S, Mann DM, Tabaton M, and Gambetti P. (1996) Presence of soluble amyloid β-peptide precedes amyloid plaque formation in Down's syndrome, *Nat. Med.* **2**, 93-95.
45. Busciglio J, Gabuzda DH, Matsudaira P, Yanker BA. (1993) Generation of β-amyloid in the secretory pathway in neuronal and nonneuronal cells, *Proc. Natl. Acad. Sci. USA* **90**, 2092-2096.
46. Aleshkov S, Abraham CR, Zannis VI. (1997) Interaction of nascent apoE2, apoE3, and apoE4 isoforms expressed in mammalian cells with amyloid peptide β (1-40). Relevance to Alzheimer's disease, *Biochemistry* **36**, 10571-10580.
52. Ghiso J, Wisniewski T, Frangione B. (1994) Unifying features of systemic and cerebral amyloidosis, *Mol. Neurobiol.* **8**, 49-64.
53. Van Duinen SG, Castano EM, Prelli F, Bots GTAM, Luyendijk W, Frangione B. (1987) Hereditary cerebral hemorrhage with amyloidosis in patients of Dutch origin is related to Alzheimer's disease, *Proc. Natl. Acad. Sci. USA* **84**, 5991-5994.
54. Maat-Schieman MLC, Duinen SG, van Bornebroek M, Haan J, Roos RAC. (1996) Hereditary cerebral hemorrhage with amyloidosis-Dutch type (HCHWA-D): II - A review of histopathological aspects, *Brain Pathol.* **6**, 115-120.
55. Citron M, Oltersdorf T, Haass C, McConlogue L, Hung AY, Seubert P, Vigo-Pelfrey C, Lieberburg I, Selkoe DJ. (1992) Mutation of the beta-amyloid precursor protein in familial Alzheimer's disease increases beta-protein production, *Nature* **360**, 672-674.
56. Citron M, Vigo-Pelfrey C, Teplow DB, Miller C, Schenk D, Johnston J, Winblad B, Venizelos N, Lannfelt L, Selkoe DJ. (1994) Excessive production of amyloid beta-protein

by peripheral cells of symptomatic and presymptomatic patients carrying the Swedish familial Alzheimer disease mutation, *Proc. Natl. Acad. Sci. USA* **91**, 11993-11997.

57. Levy E, Carman MD, Fernandez-Madrid IJ, Power MD, Lieberburg I, van Duinen SG, Bots GTAM, Luyendijk W, Frangione B. (1990) Mutation of the Alzheimer's disease amyloid gene in hereditary cerebral hemorrhage, Dutch type, *Science* **248**, 1124-1126.
58. Castano EM, Prelli F, Wisniewski T, Golabek A, Kumar RA, Soto C, Frangione B. (1995) Fibrillogenesis in Alzheimer's disease of amyloid beta peptides and apolipoprotein E, *J. Biochem.* **306** (2), 599-604.
59. Davis J, Van Nostrand WE. (1996) Enhanced pathologic properties of Dutch-type mutant amyloid β-protein, *Proc. Natl. Acad. Sci. USA* **93**, 2996-3000.
60. Hardy J. (1997) Amyloid, the presenilins and Alzheimer's Disease, *Trends Neurosci.* **20**, 154-159.
61. Wisniewski HM, Wiegel J and Popovitch ER. (1994) Age-associated development of diffuse and thioflavine-S-positive plaques in Down syndrome, *Dev. Brain Dysfunct.* **7**, 330-339.
62. Kida E, Choi-Miura NH, Wisniewski KE. (1995) Deposition of apolipoproteins E and J in senile plaques is topographically determined in both Alzheimer's disease and Down's syndrome brain, *Brain Res.* **685**, 211-216.
63. Lemere CA, Blusztajn JK, Yamaguchi H, Wisniewski T, Saido TC, Selkoe DJ. (1996) Sequence of deposition of heterogeneous amyloid β-peptides and APO E in Down syndrome: implications for initial events in amyloid plaque formation, *Neurobiol. Dis.* **3**, 16-32.
64. Gallo G, Wisniewski T, Choi-Miura N-H, Ghiso J, Frangione B. (1994) Potential role of ApoE in fibrillogenesis, *Am. J. Pathol.* **145**, 526-530.
65. Pike CJ, Burdick D, Walencewicz AJ, Glabe CJ, Cotman CW. (1993) Neurodegeneration induced by β-amyloid peptides *in vitro*: the role of peptide assembly state, *J. Neurosci. 13*, 1676-1687.
66. Ueda K, Fukui , Kageyama H. (1994) Amyloid beta protein-induced neuronal cell death: neurotoxic properties of aggregated amyloid beta protein, *Brain Res.* **639**, 240-244.
67. Lorenzo A, Yakner BA. (1994) Beta-amyloid neurotoxicity requires fibril formation and is inhibited by congo red, *Proc. Natl. Acad. Sci. USA* **91**, 12243-12247.
68. Kowall NW, Beal MF, Busciglio J, Duffy LK, Yankner BA. (1991) An *in vivo* model for the neurodegenerative effects of β amyloid and protection by substance P, *Proc. Natl. Acad. Sci. USA* **88**, 7247-7251.
69. Frautschy SA, Baird A, Cole GM. (1991) Effects of injected Alzheimers beta-amyloid cores in rat brain, *Proc. Natl. Acad. Sci. USA* **88**, 8362-8366.
70. Kowall NW, McKee AC, Yankner BA, Beal MF. (1992) *In vivo* neurotoxicity of beta-amyloid [β(1-40)] and the β(25-35) fragment, *Neurobiol. Aging* **13**, 537-542.
71. Yan SD, Fu J, Soto C, Chen X, Zhu H, Al-Mohanna F, Collison K, Zhu A Stern E, Saido T, Tohyama M, Ogawa S, Roher A, Stern D. (1997) ERAB: A novel intracellular amyloid-beta peptide binding protein which mediates neurotoxicity in Alzheimer's disease, *Nature* **389**, 689-695.
72. Smith MA, Sayre LM, Monnier VM, Perry G. (1995) Radical AGEing in Alzheimer's disease, *Trends Neurosci.* **18**, 172-176.
73. Yan SD, Chen X, Fu J, Chen M, Zhu H, Roher A, Slattery T, Zhao L, Nagashima M, Morser J, Migheli A, Nawroth P, Stern D, Schmidt AM. (1996) RAGE and amyloid-β peptide neurotoxicity in Alzheimer's disease, *Nature* **382**, 685-691.

74. McGeer PL, McGeer EG. (1995) The inflammatory response system of brain: implications for therapy of Alzheimer and other neurodegenerative diseases, *Brain Res. Rev.* **21**, 195-218.
75. Yamada M, Itoh Y, Shintaku M, Kawamura J, Jensson O, Thornsteinsson L, Suematsu N, Matsushita M, Otomo E. (1996) Immune reactions associated with cerebral amyloid angiopathy, *Stroke* **27**, 1155-1162.
76. Maat-Schieman MLC, van Duinen SG, Rozemuller AJM, Haan J, Roos RAC. (1997) Association of vascular amyloid β and cells of the mononuclear phagocyte system in hereditary cerebral hemorrhage with amyloidosis (Dutch) and Alzheimer disease, *J. Neuropathol. Exp. Neurol.* **56**, 273-284.
77. Thomas T, Sutton ET, Bryant MW, Rhodin JAG. (1997) *In vivo* vascular damage, leukocyte activation and inflammatory response induced by β-amyloid, *J. Submicrosc. Cytol. Pathol.* **29**, 293-304.
78. Thomas T, Thomas G, McLendo C, Sutton T, Mullan M. (1996) β-Amyloid-mediated vasoactivity and vascular endothelial damage, *Nature* **380**, 115-118.
79. Blanc EM, Toboreck M, Mark RJ, Hennig B, Mattson MP. (1997) Amyloid β-peptide induces cell monolayer albumin permeability, impairs glucose transport, and induces apoptosis in vascular endothelial cells, *J. Neurochem.* **68**, 1870-1881.
80. Coria F, Larrondo-Lillo M, Frangione B. (1989) Degeneration of smooth muscle cells in β-amyloid angiopathies, *J. Neuropathol. Exp. Neurol.* **48**, 368-375.
81. Kawai M, Kalaria RN, Cras P, Siedlak SL, Velasco ME, Shelton ER, Chan HW, Greenberg BD, Perry G. (1993) Degeneration of vascular muscle cells in cerebral amyloid angiopathy of Alzheimer disease, *Brain Res.* **623**, 142-146.
82. Vinters HV, Secor DL, Read SL, Frazee JG, Tomiyasy U, Stanley TM, Ferreiro JA, Akers MA. (1994) Microvasculature in brain biopsy specimens from patients with Alzheimer's disease: an immunohistochemical and ultrastructural study, *Ultrastruct. Path.* **18**, 333-348.
83. El Khoury J, Hickman SE, Thomas CA, Cao L, Silverstein SC, Loike JD. (1996) Scavenger receptor-mediated adhesion of microglia to β-amyloid fibrils, *Nature* **382**, 716-719.
84. Christie RH, Freeman M, Hyman BT. (1996) Expression of the macrophage scavenger receptor, a multifunctional lipoprotein receptor, in microglia associated with senile plaques in Alzheimer's disease, *Am. J. Pathol.* **148**, 399-403.
85. Brett J, Schmidt AM, Yan SD, Zou YS, Weidman E, Pinsky D, Nowygrod R, Neeper M, Przysiecki C, Shaw A, Migheli A, Stern D. (1993) Survey of the distribution of a newly characterized receptor for advanced glycation end products in tissues, *Am. J. Pathol.* **143**, 1699-1712
86. Krieger M, Herz J. (1994) Structures and functions of multiligand lipoprotein receptors: macrophage scavenger receptors and LDL receptor-related protein (LRP*), Annu. Rev. Biochem.* **63**, 601-637.
87. Lucarelli M, Gennarelli M, Cardeli R, Cardeli R, Novelli G, Scarpa S, Dallapiccola B, Strom R. (1997) Expression of receptors for native and chemically modified low-density lipoproteins in brain microvessels, *FEBS Lett.* **401**, 53-58.
88. Mattson MP, Rydel RE. (1996) Amyloid ox-tox transducers, Nature 382, 674-5.
89. Schmidt AM, Hasu M, Popov D, Zhang JH, Chen J, Yan SD, Brett J, Cao R, Kuwabara K, Gostache G, Simionescu N, Simionescu M, Stern D. (1994) Receptor for advanced glycation end products (AGE) has a central role in vessel wall interactions and gene activation in response to circulating AGE proteins, *Proc. Natl. Acad. Sci. USA* **91**, 8807-8811.

90. Martel CL, Mackic JB, McComb JG, Ghiso J, Zlokovic BV. (1996) Blood-brain barrier uptake of the 40 and 42 amino acid sequences of circulating Alzheimer's amyloid β in guinea-pigs, *Neurosci. Lett.* **206**, 157-160.
91. Saito Y, Buciak J, Yang J, Pardridge WM. (1995) Vector-mediated delivery of ^{125}I-labeled β-amyloid peptide $A\beta_{1-40}$ through the blood-brain barrier and binding to Alzheimer's disease amyloid to the $A\beta_{1-40}$ /vector complex, *Proc. Natl. Acad. Sci. USA* **92**, 10227-10231.
92. McGeer PL, Kawamata T, Walker DG. (1992) Distribution of clusterin in Alzheimer brain tissue, *Brain Res.* **579**, 337-341.
93. Choi-Miura NH, Ihara Y, Kukuchi K, Takeda M, Nakano Y, Tobe T, Tomita T. (1992) SP-40,40 is a constituent of Alzheimer's amyloid, *Acta Neuropathol.* **83**, 260-264.
94. May PC, Finch CE. (1992) Sulfated glycoprotein-2: new relationships of this multifunctional protein to neurodegeneration, *TINS* **15**, 391-396.
95. Pasinetti GM, Cheng HW, Morgan DG, Olampert-Etchells M, McNeill TH, Finch CE. (1993) Astrocytic responses to striatal deafferentation in male rat, *Neuroscience* **53**, 199-211.
96. Koudinov A, Matsubara E, Frangione B, Ghiso J. (1994) The soluble form of Alzheimer's amyloid β protein is complexed to high density lipoprotein 3 and very high density lipoprotein in normal human plasma, *Biochem. Biophys. Res. Commun.* **205**, 1164-1171.
97. Matsubara E, Soto C, Governale S, Frangione B, Ghiso J. (1996) Apolipoprotein J and Alzheimer's amyloid β solubility, *Biochem. J.* **316**, 671-679.
98. Oda T, Wals P, Osterburg HH, Johnson SA, Pasinetti GM, Morgan TE, Rozovsky I, Stine WB, Snyder SW, Holzman TF. (1995) Clusterin (apoJ) alters the aggregation of amyloid beta-peptide ($A\beta_{1-42}$) and forms slowly sedimenting Aβ complexes that cause oxidative stress, *Exp. Neurol.* **136**, 22-31.
99. Bertrand P, Poirier J, Oda T. (1995) Association of apolipoprotein E genotype with brain levels of apolipoprotein E and apolipoprotein J (clusterin) in Alzheimer disease, *Mol. Brain Res.* **33**, 174-178.
100. Kounnas MZ, Loukinova EB, Stefansson S. (1995) Identification of glycoprotein 330 as an endocytic receptor for apolipoprotein J/clusterin, *J. Biol. Chem.* **270**, 13070-13075.
101. Dehouck B, Fenart L, Dehouck MP, Pierce A, Torpier g, Cecchelli R. (1997) A new function for the LDL receptor: transcytosis of LDL across the blood-brain barrier, *J. Cell. Biol.* **138**, 877-889.
102. Zheng G, Bachinsky DR, Stamenkovic I. (1994) Organ distribution in rats of two members of the low-density lipoprotein receptor gene family, gp330 and LRP/alpa 2MR, and the receptor-associated protein (RAP), *J. Histochem. Cytochem.* **42**, 531-542.
103. Chun JT, Wang L. (1999) Glycoprotein 330/megalin (LRP-2) has low prevalence as mRNA and protein in brain microvessels and choroid plexus, *Exp. Neurol.* **157**, 194-201
104. LaDu MJ, Gilligan SM, Lukens JR, Cabana VG, Reardon CA, Van Eldik LJ, Holzman DM. (1998) Nascent astrocyte particles differ from lipoproteins in CSF, *J. Neurochem.* **70**, 2070-2081.
105. LaFerla FM, Troncoso JC, Strickland DK, Kawas CH, Jay G. (1997) Neuronal cell death in Alzheimer's disease correlates with apoE uptake and intracellular Aβ stabilization, *J. Clin. Invest.* **100**, 310-320.
106. Farquhar MG, Saito A, Kerjaschki D. (1995) The Heymann nephritis antigenic complex: megalin (gp330) and RAP, *J. Am. Soc. Nephrol.* **6**, 35-47.
107. Poduslo JF, Curran GL, Berg C. (1994) Macromolecular permeability across the blood-nerve and blood-brain barriers, *Proc. Natl. Acad. Sci. USA* **91**,5705-5709.

108. Pardridge WM, Kang YS, Buciak JL, Yang J. (1995) Human insulin receptor monoclonal antibody undergoes high affinity binding to human rain capillaries *in vitro* and rapid transcytosis through the blood-brain barrier *in vivo* in the primate, *Pharm. Res.* **12**, 807-816.
109. Zlokovic BV. (1995) Cerebrovascular permeability to peptides: manipulations of transport systems at the blood-brain barrier, *Pharm. Res.* **12**, 1395-1406.
110. Mahley, RW. (1988) Apolipoprotein E: Cholesterol transport protein with expanding role in cell biology. *Science* **240**, 622-630.
111. Plump AS, Breslow JL. (1995) Apolipoprotein E and the Apolipoprotein E-Deficient mouse, *Annu. Rev. Nutr.* **15**, 495-518.
112. Roses AD. (1997) Apolipoprotein E, a gene with complex biological interactions in the aging brain, *Neurobiol. Dis.* **4**, 170-186.
113. Laskowitz DT, Horsburgh K, Roses AD. (1998) Apolipoprotein E and the CNS response to injury, *J. Cereb. Blood F. Metab.* **18**, 465-471.
114. Newman TC, Dawson PA, Rudel LL, William DL. (1985) Quantitation of Apolipoprotein E mRNA in the liver and peripheral tissues of nonhuman primates, *J. Biol. Chem.* **260**, 2452-2457.
115. Boyles JK, Pitas RE, Wilson E, Mahley RW, Taylor JM. (1985) Apolipoprotein E associated with astrocytic glia of the central nervous system and with nonmyelinating glia of the peripheral nervous system, *J. Clin. Invest.* **76**, 1501-1513.
116. Nakai M, Kawamata T, Maeda K, Tanaka, C. (1996) Expression of apoE mRNA in rat microglia. *Neurosci. Lett.* **143**, 313-318.
117. Strittmatter WJ, Roses AD. (1996) Apolipoprotein E and Alzheimer's Disease. *Annu. Rev. Neurosci.* **19**, 53-77.
118. Handelmann GE, Boyles JK, Weisgraber KH, Mahley RW, Pitas RE. (1992) Effects of apolipoprotein E, β-very low density lipoproteins, and cholesterol on the extension of neurites by rabbit dorsal root ganglion neurons *in vitro*, *J. Lipid Res.* **33**, 1677-1688.
119. Wolf BB, Lopes MB, VandenBerg SR, Gonias SL. (1992) Characterization and immunohistochemical localization of α-2-microglobulin receptor (low density lipoprotein receptor-related protein) in human brain, *Am. J. Pathol.* **141**, 37-42.
120. Kounnas MZ, Loukinova EB, Stefansson S. (1995) Identification of glycoprotein 330 as an endocytic receptor for apolipoprotein J/clusterin, *J. Biol. Chem.* **270**, 13070-13075.
121. Takahashi S, Kawarabayasi Y, Nakai T, Sakai J, Yamamoto T. (1992) Rabbit very low density lipoprotein receptor: A low density lipoprotein receptor-like protein with distinct ligand specificity, *Proc. Natl. Acad. Sci. USA* **89**, 9252-9256.
122. Okuizumi K, Onodera O, Namba Y, Ikeda K, Yamamoto T, Seki K, Ueki A, Nanko S, tanaka H, Takahashi H. (1995) Genetic association of the very low density lipoprotein (VLDL) receptor gene with sporadic Alzheimer's disease, *Nat. Genet.* **11**, 207-209.
123. Wyne KI, Pathak K, Seabra MC. (1996) Expression of the VLDL receptor in endothelial cells, *Arterioscler. Thromb. Vasc. Biol.* **16**, 407-415.
124. Christie RH, Chung H, Rebeck GW, Strickland D, Hyman BT. (1996) Expression of the very low-density lippoprotein receptor in the central nervous system and in Alzheimer's disease. *J. Neuropathol. Exp. Neurol.* **55**, 491-498.
125. Rebeck GW, Bradley TH. (1999) Lipoprotein receptors in brain, in C. Finch (ed.),. *Clusters in Normal Brain Function & During Neurodegeneration*, W. Saunders, pp. 49-59.
126. Strittmatter W, Saunders A, Schmecher D, Pericak-Vance M, Enghild J, Salversen G, Roses A. (1993) Apolipoprotein E: High-avidity binding to β- amyloid and increased

frequency of type 4 allele in late-onset familial Alzheimer's disease, *Proc. Natl. Acad. Sci. USA* **90**, 1977-1981.

127. Wisniewski, Frangione B. (1992) Apolipoprotein E: a pathologic chaperone protein in patients with cerebral and systemic amyloid, *Neurosci. Lett.* **135**, 235-238.

128. Corder EH, Saunders AM, Risch NJ, Strittmatter WJ Schmechel DE, Gaskell Jr PC, Rimmler JB, Locke PA. Conneally PM, Schmader KE, Small GW, Roses AD, Haines JL, Pericak-Vance MA. (1994) Protective effect of apolipoprotein E type 2 allele for late-onset Alzheimer disease, *Nat. Genet.* **7**, 180-184.

129. Saunders AM, Strittmatter WJ, Schmechel D, St. George-Hyslop PH, Pericak-Vance MA, Joo SH, Rosi B, Gusella JF, Crapper-MacLachlan DR, Alberts MJ, Hulette C, Crain B, Goldgaber D, Roses AD. (1994) Association of apolipoprotein E allele E4 with late-onset familial and sporadic Alzheimer's disease, *Neurology* **43**, 1467-1472.

130. Corder EH, Saunders AM, Strittmatter WJ, Schmechel DE, Gaskell PC, Small GW, Rases AD, Haines JL, Pericak-Vance MA. (1993) Gene dose of apolipoprotein E type 4 allele and the risk of Alzheimer's disease in late onset families, *Science* **261**, 921-923.

131. Mann DM, Iwatsubo T, Pickering-Brown SM, Owen F, Saido TC, Perry RH. (1997) Preferential deposition of amyloid β protein (Aβ) in the form Aβ40 in Alzheimer's disease is associated with a gene dosage effect of the apolipoprotein E,E4 allele, *Neurosci. Lett.* **221**, 81-84.

132. Premkumar DR, Cohen DL, Hedera P, Friedland RP, Kalaria RN. (1996) Apolipoprotein β-E4 alleles in cerebral amyloid angiopathy and cerebrovascular pathology associated with Alzheimer's disease, *Am. J. Pathol.* **148**, 2083-2095.

133. Greenberg SM, Rebeck GW, Vonsattel JPG, Gomez-Isla T, Hyman BT. (1995) Apolipoprotein E(epsilon)4 and cerebral hemorrhage associated with amyloid angiopathy, *Ann. Neurol.* **38**, 254-259.

134. Tang MX, Stern Y, Marder K, Bell K, Gurland B, Lantigua R, Andrews H, Feng L, Tycko B, Mayeux R. (1998) The ApoE-epsilon4 allele and the risk of Alzheimer disease among African Americans, whites, and Hispanics, *JAMA* **279**, 751-755.

135. Naslund J, Thyberg J, Tjemberg LO, Wernstedt C, Kadstrom AR, Bodganovic N, Gandy SE, Lannfelt L, Terenius L, and Nordstedt C. (1995) Characterization of stable complexes involving apoplipoprotein E and the amyloid β peptide in Alzheimer's disease brain, *Neuron* **15**, 219-228.

136. Zhou Z, Smith JD, Greengard P, Gandy S. (1996) Alzheimer amyloid-β peptide forms denaturant-resistant complex with type E3 but not type E4 isoform of native apolipoprotein E, *Molecul. Med.* **2**, 175-180.

137. LaDu MJ, Falduto MT, Manelli AM, Reardon CA, Getz GS, Frail DE. (1994) Isoform-specific binding of apolipoprotein E to β-amyloid, *J. Biol. Chem.* **269**, 23403-23406.

138. LaDu MJ, Pederson TM, Frail DE, Reardon CA, Getz GS, Falduto MT. (1995) Purification of apolipoprotein E attenuates isoform-specific binding to β-amyloid, *J. Biol. Chem.* 9039-9042

139. Wisniewski T, Golabek A, Matsubara E, Shiso J, Frangione B. (1993) Apolipoprotein E: Binding to soluble Alzheimer's β-amyloid, *Biochem. Biophys. Res. Commun.* **192**, 359-365.

140. Sanan DA, Weisgraber KH, Russel SJ, Mahley RW, Huang D, Saunders A, Schmochol D, Wianlowski T, Frangione B, Roses AD, Strimatter WJ. (1994) Apolipoprotein E associates more efficiently than apoE3, *J. Clin. Invest.* **94**, 860-869.

141. Strittmatter WJ, Saunders AM, Goedert M, Weisgraber KH, Dong LM, Jakes R, Huang D, Pericak-Vance M, Schmechel D, Roses AD. (1994) Isoform-specific interactions of

apolipoprotein E with microtubule-associated protein tau: Implications for Alzheimer's disease, *Proc. Natl. Acad. Sci. USA* **94**, 11183-11186.

142. Strittmatter W, Roses AD. (1995) Apolipoprotein E and Alzheimer disease, *Proc. Natl. Acad. Sci. USA* **92**, 4725-4727.
143. Huang DY, Goedert M, Jakes R, Weisgraber KH, Garner CC, Saunders AM, Pericak-Vance MA, Schmechel DE, Roses AD. (1994) Isoform-specific interactions of apolipoprotein E with the microtubule-associated protein MAP2C: Implications for Alzheimer's disease, *Neurosci. Lett.* **182**, 55-58.
144. Strittmatter WJ, Weisgraber KH, Godert M, Saunders AM, Huang D, Corder EH, Dong LM, Jakes R, Alberts MJ, Gilbert JR, Han SH, Hulette C, Einstein G, Schmechel DE, Pericak-Vance MA, Roses AD. (1994) Hypothesis: Microtubule instability and paired helical filament formation in the Alzheimer disease brain are related to apolipoprotein E genotype, *Exp. Neurol.* **125**, 163-171.
145. Poirier J, Delisle MC, Quirion R, Aubert I, Farlow M, Lahiri D, Hui S, Bertrand P, Nalbantoglu J, Gilfix BM, Gauthier S. (1995) Apolipoprotein E4 allele as a predictor of cholinergic deficits and treatment outcome in Alzheimer disease, *Proc. Natl. Acad. Sci. USA* **92**, 12260-12264.
146. Weisgraber KH, Mahley RE. (1996) Human apolipoprotein E: The Alzheimer's disease connection, *FASEB J* **10**, 1485-1494.
147. Nathan BP, Bellosta S, Sanan DA, Weisgraber KH, Mahley RW, Pitas RE. (1994) Differential effects of apolipoproteins E3 and E4 on neuronal growth *in vitro*, *Science* **264**, 850-852.
148. Nathan BP, Chang KC, Bellosta S, Brisch E, Ge N, Nahley RW, Pitas RE. (1994) The inhibitory effect of apolipoproteins E3 and E4 on neuronal growth *in vitro*, *Science* **264**, 850-852.
149. Bellosta S, Nathan BP, Orth M, Dong LN, Mahley RW, Pitas RE. (1995) Stable expression and secretion of apolipoproteins E3 and E4 in mouse neuroblastoma cells produces differential effects on neurite outgrowth, *J. Biol. Chem.* **270**, 27063-27071.
150. Arendt T, Schindler C, Bruckner MK, Eschrich K, Bigl V, Zedlick D, Marcova I. (1997) Plastic neuronal remodeling is impaired in patients with Alzheimer's disease carrying apolipoprotein E4 allele, *J. Neurosci.* **17**, 516-529.
151. Barger SW, Mattson MP. (1997) Isoform-specific modulation by apolipoprotein E of the activities of secreted β-amyloid precursor protein, *J. Neurochem.* **69**, 60-67.
152. Barger SW, Harmon AD. (1997) Microglial activation by Alzheimer amyloid precursor protein and modulation by apolipoprotein E, *Neurosci. Lett.* **388**, 878-881.
153. Poirier J. (1994) Apolipoprotein E in animal models of CNS injury and in Alzheimer's disease, *Trends Neurosci.* **17**, 525-530.
154. Martel Cl, Ghiso J, Frangione B. (1997) Transport of apolipoproteins across the blood-brain barrier: relevance to Alzheimer's Diease, *STP Pharma Sci.* **7**, 28-36.
155. Chan W, Fornwald J, Brawner M, Wetzel R. (1996) Native complex formation between apolipoprotein E isoforms and the Alzheimer's disease peptide Aβ, *Biochemistry* **35**, 7123-7130.
156. Castano EM, Prelli F, Golabek R, Kuma A, Soto C, Frangione B. (1995) Fibrillogenesis in Alzheimer's disease of amyloid β peptides and apolipoprotein E, *Biochem. J.* **306**, 599-604.
157. Evans KC, Berger EP, Cho CG, Weisgraber KH, Lansbury Jr PT. (1995) Apolipoprotein E is a kinetic but not a thermodynamic inhibitor of amyloid formation: implications for the pathogenesis and treatment of Alzheimer's disease, *Proc. Natl. Acad. Sci. USA* **92**, 763-767.

158. Wood SJ, Chan W, Wetzel R. (1996) Seeding of Aβ fibril formation is inhibited by all three isotypes of apolipoprotein E, *Biochemistry* **35**, 12623-12628.
159. Harper JD, Wong SS, Lieber CM, Lansbury Jr PT. (1997) Observation of metastable Aβ amyloid produced by atomic force microscopy, *Chemistry & Biology* **4**, 119-125.
160. Urmoneit B, Prikulis I, Wihl G, D'Urso D, Frank R, Heeren J, Beisiegel U, Prior R. (1997) Cerebrovascular smooth muscle cells internalize Alzheimer amyloid beta protein via a lipoprotein pathway: implications for cerebral amyloid angiopathy, *Lab. Invest.* **77**, 157-166.
161. Games D, Adams D, Tan H, Zhao J. (1995) Alzheimer-type neuropathology in transgenic mice overexpressing V717F β-amyloid precursor protein, *Nature* **373**, 523-527.
162. Hsiao K, Chapman P, Nilsen S, Eckman C, Harigaya Y, Youkin S, Yang F, Cole G. (1996) Correlative memory deficits, Aβ elevation, and amyloid plaques in transgenic mice, *Science* **274**, 99-102.
163. Iadecola C, Zhang F, Hsiao K, Carlson G. (1999) SOD1 rescues cerebral endothelial dysfunction in mice overexpressing amyloid precursor protein, *Nat. Neurosci.* **2**, 157-161.
164. Irizarry MC, McNamara M, Fedorchak K, Hsiao KK, Hyman BT. (1997) APP_{SW} transgenic mice develop age-related Aβ deposits and neuropil abnormalities, but no neuronal loss in CA1, *J. Neuropathol. Exp. Neurol.* **56**, 965-973.
165. Chapman P, Whie G, Jones M, Hsiao K. (1999) Impaired synaptic plasticity and learning in aged amyloid precursors protein transgenic mice, *Nat. Neurosci.* **2**, 271-276.
166. Staufenbiel M, Sommer B, Jucker M. (1998) Neuron loss in APP transgenic mice, *Nature* **395**, 755-756.
167. Holcomb L, Gordon M, McGowan E, Yu X, Younkin S, Hsiao K, Duff K. (1998) Acelerated Alzheimer-type phenotype in transgenic mice carrying both mutant *Amyloid Precursor Protein* and *Presenilin 1* transgenes, *Nature Med* **4**, 97-100.
168. Bales KR, Verina T, Dodel RC, Du Y, Altstiel L, Bender M, Hyslop P, Johnstone EM, Little SP, Cummins DJ, Piccardo P, Ghetti B, Paul SM. (1997) Lack of apolipoprotein E dramatically reduces amyloid β-peptide deposition, *Nat Gen* **1**, 263-264.
169. Holtzman DM, Bales KR, Wu S, Bhat P, Parsadanian M, Fagan Am, Chang LK, Sun Y, Paul SM. (1999) *In Vivo* expression of apolipoprotein E reduces amyloid-β -deposition in a mouse model of Alzheimer's Disease, *J. Clin. Invest.* **103**, R15-21.

Index

MIX
Papier aus verantwortungsvollen Quellen
Paper from responsible sources
FSC® C105338

If you have any concerns about our products,
you can contact us on
ProductSafety@springernature.com

In case Publisher is established outside the EU,
the EU authorized representative is:
Springer Nature Customer Service Center GmbH
Europaplatz 3, 69115 Heidelberg, Germany

Printed by Libri Plureos GmbH
in Hamburg, Germany